Frequency Synthesizers

Frequency Synthesizers
Theory and Design

Third Edition

Vadim Manassewitsch
Consultant

A Wiley-Interscience Publication
JOHN WILEY & SONS
New York · Chichester · Brisbane · Toronto · Singapore

Library of Congress Cataloging in Publication Data:

Manassewitsch, Vadim, 1927-
  Frequency synthesizers.

  "A Wiley-Interscience publication."
  Bibliography: p.
  Includes index.
  1. Frequency synthesizers. I. Title.

TK7872.F73M37 1987   621.3815'363   86-18968
ISBN 0-471-01116-9

Printed in the United States of America

10 9 8 7 6 5 4 3

Preface

The purpose of this book is to provide training and reference material for designers and users of frequency synthesizers. It is suitable as a text for graduate-level courses in frequency generation and control.

The objective in the selection and presentation of the material has been to formulate basic design principles and to demonstrate design procedures for frequency synthesizers meeting a number of stringent requirements simultaneously. The emphasis is on high-speed synthesis. The design of such synthesizers requires a thorough knowledge of several disciplines.

The order of presentation has been made as logical as possible. Chapter 1 describes numerous approaches to frequency synthesis and shows how various building blocks such as mixers, oscillators, and frequency multipliers and dividers are used in frequency synthesis. In light of the increasing importance of high-speed synthesizers, such approaches as direct digital synthesis are included in the text. This introduction makes the following chapters more meaningful to the student. The description of frequency synthesis techniques is followed by system analysis in Chapter 2, since a knowledge of the system design problems such as phase noise control is essential to choosing the best approach for a given set of requirements. Chapter 2 also describes various techiques of measuring spurious outputs, phase noise, and tuning time. Chapter 3 deals with the control of the propagation of spurious signals generated in frequency synthesis. It demonstrates the use of important design aids applicable to rf shielding. Chapters 4, 5, and 6 present circuit design in detail. Phase-locked loops are given special attention because of their widespread use. The design and testing of these circuits are developed in Chapters 4 and 5 and in the concluding chapters of the book. The material of the preceding chapters is fully used in Chapter 7 to describe state-of-the-art synthesizers. An extensive discussion of reference sources is given in Chapter 8 to familiarize the student with the system problems to be considered when using this equipment. Both rubidium and cesium atomic frequency standards are included in the discussion. A theory of troubleshooting for direct and indirect synthesizers is developed in Chapter 9. It contains many practical suggestions to aid in identifying the design and manufacturing problems associated with these systems. This material is presented in a way useful to technicians as well as to engineers.

Chapter 10, which covers the design of high-speed synthesizers, deals extensively with fast-switching-time phase-locked loops.

To meet the current needs of engineers in this rapidly developing field, the book has been revised twice since it was originally published in 1976. Several novel frequency synthesis techniques are added to the text. The principles of high-speed synthesis are developed. New synthesizers utilizing important design approaches are described. Achievements in the performance of reference sources are indicated, and outdated test equipment is replaced with newly developed instruments best fitted for the task. The book presents a summary of the most recent developments in frequency generation and control. It is firmly based on the realities of current design practices in the United States as well as abroad.

I would like to thank former and present colleagues who contributed to this work by discussing with me various aspects of synthesizer design and by familiarizing me with circuit design techniques described in the book. I am indebted to Professor J. A. Connelly of the Georgia Institute of Technology and other users for providing helpful suggestions that facilitate the reading of the book. I am especially grateful to Mrs. Virginia Spriggs for her careful reading of the initial and revised editions and for her helpful comments on writing style.

VADIM MANASSEWITSCH

New York, New York
January 1987

Contents

1 Frequency Synthesis

A frequency synthesis is a combination of system elements that results in the generation of one or many frequencies from one or a few reference sources. In its early form it was a crystal-controlled oscillator with a bank of crystals switched in manually. The frequency accuracy and stability of this device were determined by the accuracy and stability of the crystal and, to a lesser extent, the circuit.

The crystal-controlled oscillator was superseded, but not replaced, by an approach presently known as incoherent synthesis. This utilized a number of crystal-controlled oscillators combined in such a fashion as to generate many frequencies with relatively few crystals.

While these improvements took place, the rapidly growing field of communications developed requirements for more sophisticated frequency generation systems displaying accuracies and stabilities higher by orders of magnitude than incoherent synthesis could provide. To meet the requirements, a new family of approaches, grouped under the name of coherent synthesis, emerged. As the name implies, these approaches provide means for producing many frequencies from one reference source of the required accuracy and stability.

These techniques resulted in the generation of spurious outputs, which had to be eliminated by a proper choice of frequencies used in synthesis and suppressed by filtering. Phase noise no longer could be neglected.

As more was learned about the techniques, the possibility of utilizing frequency synthesis in systems other than ground and space communications, such as Doppler radar, was recognized. The capability of programmed high-speed frequency switching extended the use of synthesizers to laboratory and production testing of various electronic circuits and systems, for example, to the testing of narrowband crystal filters and the checking out of multichannel telemetry systems.

In meeting the demands stemming from various applications and changing technology, frequency-generating systems became difficult to implement, complex, and costly. Chapter 7 discusses the requirements for such systems in some detail and describes synthesizers presently in use. This chapter familiarizes the student with basic concepts of frequency-increment generation.

1-1 Incoherent Synthesis

The manner in which output frequencies are generated from input frequencies in incoherent synthesis varies, depending on the application. Output frequency range, value of the smallest frequency increment, frequency stability and accuracy, levels of spurious outputs, size, cost, and power consumption—these are the factors governing the choice of an approach. The main goal of this technique, though, remains the same in all cases; it is to minimize the number of crystals and basic building blocks such as oscillators, mixers, and filters used in synthesis in order to reduce cost.

A typical example of incoherent synthesis is discussed below to point out the limitations of the approach and the design problems associated with it. The interested student is referred to the works of Lindholm, Ross, and Baltas (Refs. 1 to 3) for more extensive presentations of the subject.

Figure 1-1 is a block diagram of an incoherent synthesis approach that is similar to the system proposed by H. Granger (Ref. 3). The synthesis utilizes the method of successive heterodyning. The output of the left-hand-side mixer is the sum of the two input frequencies,

$$f_{n-1} + f_n + \frac{(\Delta f_{0-9})_{n-1}}{10^{n-2}} + \frac{(\Delta f_{0-9})_n}{10^{n-1}}.$$

Throughout the text Δf_{0-9} means 10 increases from a base frequency with $\Delta f_0 = 0$. For example, $f_n + (\Delta f_{0-9})_n / 10^{n-1}$ means that the nth oscillator can be set to generate 10 frequencies:

$$f_n, f_n + \frac{(\Delta f_1)_n}{10^{n-1}}, f_n + \frac{(\Delta f_2)_n}{10^{n-1}}, \dots, f_n + \frac{(\Delta f_9)_n}{10^{n-1}}.$$

This notation was chosen to describe the relative weights of individual digits of the final output frequency at the places of their generation and to trace the propagation of the digits along the path of frequency synthesis. Similarly, every other oscillator in the scheme can be set independently because it has its own control. Hence one can have

$$f_n + \frac{(\Delta f_0)_n}{10^{n-1}} \quad \text{and} \quad f_{n-1} + \frac{(\Delta f_9)_{n-1}}{10^{n-2}}$$

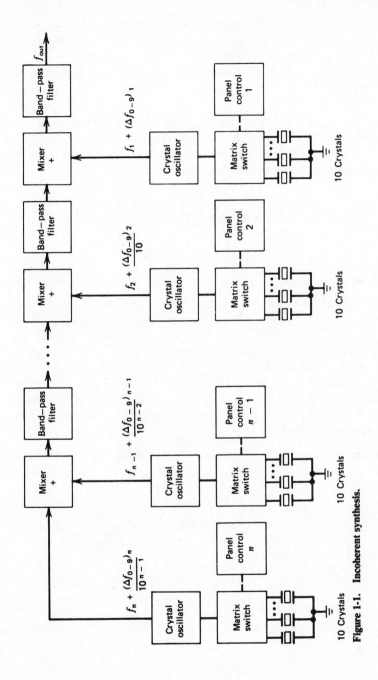

Figure 1-1.   Incoherent synthesis.

3

generated simultaneously. Parentheses with a subscript indicate the position of Δf_{0-9} in the expression of the output frequency. Thus $(\Delta f_{0-9})_n$ always refers to the least significant figure in this expression.

The output of the following mixer, omitted in Fig. 1-1, is again a sum of its two inputs:

$$f_{n-2}+f_{n-1}+f_n+\frac{(\Delta f_{0-9})_{n-2}}{10^{n-3}}+\frac{(\Delta f_{0-9})_{n-1}}{10^{n-2}}+\frac{(\Delta f_{0-9})_n}{10^{n-1}}.$$

The final output is the sum of all frequencies:

$$f_{\text{out}}=f_1+f_2+\cdots+f_{n-1}+f_n+(\Delta f_{0-9})_1+\frac{(\Delta f_{0-9})_2}{10}$$

$$+\cdots+\frac{(\Delta f_{0-9})_{n-1}}{10^{n-2}}+\frac{(\Delta f_{0-9})_n}{10^{n-1}}. \tag{1-1}$$

For illustrative purposes 10 crystals are assigned to every oscillator; however, any number of crystals can be used with equal success (Refs. 1 and 3).

A numerical example in Fig. 1-2 demonstrates how this approach works in practice. It is assumed that the only requirements are the output frequency range, 58 to 59 MHz, and the smallest frequency increment, 1 kHz. To generate 10^3 frequencies, $n=3$. Usually, the choice of f_1 through f_n is governed by the specified levels of spurious outputs, but for simplicity we omit this requirement.

Let

$$f_1=47.0 \text{ MHz,}$$

$$f_2=6.0 \text{ MHz,}$$

$$f_3=5.0 \text{ MHz.}$$

To generate 1 kHz increments,

$$\frac{(\Delta f_1)_n}{10^{n-1}}=\frac{(\Delta f_1)_3}{10^2}=1 \text{ kHz}$$

or

$$\Delta f_1=0.1 \text{ MHz.}$$

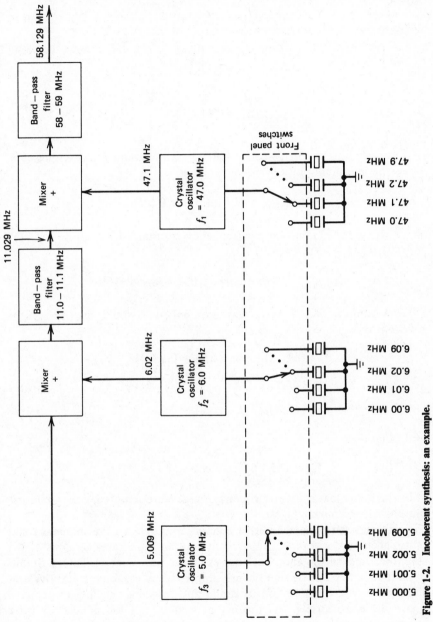

Figure 1-2. Incoherent synthesis: an example.

5

Hence

$$
\left.\begin{array}{l}
\Delta f_0 = 0.0 \\
\Delta f_1 = 0.1 \\
\Delta f_2 = 0.2 \\
\vdots \\
\Delta f_9 = 0.9
\end{array}\right\} \text{MHz.}
$$

Setting (Δf_{0-9})'s to $(\Delta f_0)_1 = (\Delta f_0)_2 = (\Delta f_0)_3 = 0$ and using Eq. 1-1, one computes the minimum output frequency:

$$(f_{out})_{min} = 47 + 6 + 5 = 58.0 \text{ MHz.}$$

Similarly, the maximum output frequency is obtained by setting (Δf_{0-9})'s to $(\Delta f_9)_1 = (\Delta f_9)_2 = (\Delta f_9)_3 = 0.9$ MHz:

$$(f_{out})_{max} = 47.9 + 6.09 + 5.009 = 58.999 \text{ MHz.}$$

If, for example, 58.129 MHz was required, one would set (Δf_{0-9})'s as follows:

$$(\Delta f_{0-9})_1 \quad \text{to} \quad (\Delta f_1)_1 = 0.1 \text{ MHz,}$$

$$(\Delta f_{0-9})_2 \quad \text{to} \quad (\Delta f_2)_2 = 0.2 \text{ MHz,}$$

$$(\Delta f_{0-9})_3 \quad \text{to} \quad (\Delta f_9)_3 = 0.9 \text{ MHz,}$$

and obtain

$$f_{out} = 47.1 + 6.02 + 5.009 = 58.129 \text{ MHz.}$$

In the event that a wider frequency range were required, one would use more oscillator-mixer stages.

This approach utilizes additive heterodyning. Hence the frequency stability, accuracy, and phase noise of the output signal are the sums of the stabilities, accuracies, and phase noises of the individual oscillators. Stability and accuracy can be significantly improved by combining additive and subtractive heterodyning and by aging and frequency pulling of crystals. Phase noise, because of its incoherent nature, can be reduced only by improving the noise performance of individual circuits.

Special attention should be given to spurious outputs. Mixing processes generate an infinite number of undesirable intermodulation products, some of which are of low order and may fall in band. Proper choice of frequencies will move the products out of band, where they can be suppressed by filtering. Spurious-output analysis, which is discussed in great detail in Chapter 2, is therefore of prime importance and is the first step in design. The choice of f_1 through f_n in Fig. 1-1 is governed mainly by this requirement.

The important advantage of incoherent synthesis is low cost.

1-2 Coherent Direct Synthesis

The main difference between incoherent and coherent synthesis is the number of frequency sources utilized in the process of frequency generation. In the first approach there are numerous crystal-controlled oscillators; in the second only one reference source is used. Hence the stability and accuracy of the output frequency in a coherent direct synthesis system are the same as the stability and accuracy of the reference source.

To develop n crystal-controlled oscillators with 10 times n switchable crystals having stabilities of 1 part in 10^8/day and accuracies of ± 5 parts in 10^7 (the requirement of 5 years ago) would be very challenging, and the excessive cost of such a venture would be difficult to justify. To develop one such oscillator with only one crystal and to use it to drive the rest of the circuitry is a much easier task. At the present time reference sources with stabilities of 1 part in 10^9 are available on the market at relatively low cost.

The feature that all frequencies are generated from one reference source makes coherent synthesis indispensable.

Brute-Force Approach

Whenever the generation of a small number of frequencies is required, it is customary to use a brute-force or a harmonic approach. A brute-force approach is preferred whenever these frequencies have to be generated simultaneously.

The basic building blocks utilized in a brute-force approach are frequency multipliers, dividers, and mixers and a reference source. Figure 1-3 is an example of such an approach. For clarity amplifiers and filters are not shown. The six frequencies, 20.0, 21.5, 22.0, 23.5, 24.0, and 25.5 MHz, can be generated in many other ways, and it will be a good exercise for the interested student to show that the way in which these frequencies are generated in Fig. 1-3 is not optimum, that is, there is another way that requires fewer building blocks. Unfortunately for the designer, the definition of an optimum system is not so clear cut. It may happen that the system with fewer building blocks will be the more expensive one. This is

8

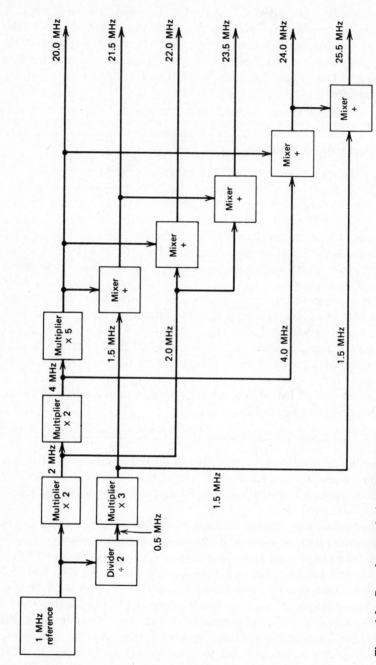

Figure 1-3. Brute-force synthesis: an example.

usually discovered when the filters, amplifiers, and isolators needed to make the system operational are defined and included in cost estimates. It is recommended, therefore, that one develop various approaches, perform detailed system and circuit analysis, define all circuits, determine the cost of each approach, and only then select the least expensive one.

Brute-force techniques can be implemented with both analog and digital circuitry (Refs. 4 to 6).

Two basic problems are associated with the brute-force approach: (*a*) spurious outputs generated in mixing, multiplication, and division, and (*b*) phase noise. These are discussed in Chapter 2.

The technique is extensively used in synthesizers to generate auxiliary frequencies.

Harmonic Approach

A harmonic approach is preferred to a brute-force approach whenever spacing between any two adjacent output frequencies is the same number throughout; these frequencies are some multiple of the spacing, and only one frequency has to be provided at a particular time. For example, a harmonic approach would be used if five frequencies—20, 21, 22, 23, and 24 MHz—were to be generated one at a time. This approach consists of two basic steps: (*a*) generation of a signal with a high harmonic content (usually a train of pulses) whose fundamental frequency is equal to the spacing, and (b) selection of desired harmonics. There are many ways of converting the sine wave of a reference source into a train of pulses. Two of these are described in Chapter 6. In this section three practical ways of selecting desired harmonics are discussed. They are presented in Fig. 1-4.

In the figure the expression $\sum_{m=x_1}^{x_2} mf_r + R$ describes an input signal with a high harmonic content. The harmonics of the signal that are passed by the filter one at a time are $x_1 f_r$ through $x_2 f_r$. The rest of the harmonics suppressed by the filter is denoted as R.

The passive filter, Fig. 1-4*a*, is used to select harmonics whenever the spacing between the adjacent harmonics is large in relation to the output frequency, mf_r, so that required attenuation of unwanted harmonics is achieved with a small number of poles. Narrowband, tunable filters with a large number of poles are costly and difficult to implement.

When the spacing is small, the double-mix approach is used, Fig. 1-4*b*. In this approach a train of pulses is applied to the signal port of a difference mixer. Local oscillator input is provided by a tuned oscillator whose frequency, f_{TO}, is smaller than the output frequency, mf_r, always by a fixed amount, f_{IF}. A band-pass filter following the mixer attenuates unwanted harmonics. In the last step of synthesis, the selected harmonic is upconverted to its original value, mf_r.

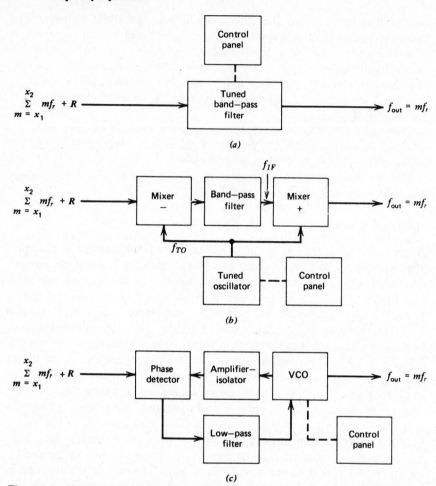

Figure 1-4.   Harmonic synthesis. (a) Passive filter; (b) Active filter, double-mix approach; (c) Active filter, phase-locked loop approach.

The advantages of this approach as compared to passive filtering are numerous. The band-pass filter which attenuates unwanted harmonics operates at a single frequency, and manual or remote tuning is provided by a tuned oscillator. By downconverting input frequencies, spacing is effectively increased manyfold, making the filtering of unwanted harmonics feasible. The tuning of such a filter can be extended to an octave band.

The effects of the tuned oscillator on the stability, accuracy and phase noise of the output signal are canceled by the double-mix techniques; it is assumed that the oscillator is tuned close to the required value of f_{TO} so that f_{IF} falls within the narrow passband of the filter. This can be demonstrated by the following set of equations:

$$f_{IF} = mf_r - f_{TO},$$

$$f_{out} = f_{IF} + f_{TO} = mf_r - f_{TO} + f_{TO} = mf_r.$$

As was expected, f_{TO} is absent from the expression for f_{out}.

Phase-noise degradation of the signal passing through this filter is possible only if the noise level of the input signal is very low or if mixers and associated circuitry are not optimized.

The problem of spurious outputs is as severe here as it is in any other approach utilizing heterodyning and should be carefully investigated.

The phase-locked loop (PLL) approach, depicted in Fig. 1-4c, is most useful when spacing between operating frequencies is so small that neither passive nor double-mix filters can attenuate unwanted harmonics. The approach is inherently free of spurious outputs because it does not require mixers, multipliers, and dividers, but it suffers from leakage problems. Leakage of unwanted harmonics through the phase detector via the amplifier-isolator to the output of the filter should be estimated or measured in order to determine the degree of isolation (or backward gain) expected from the amplifier. Stability analysis should be performed to establish the degree of attenuation that PLL can provide.

Since PLL is a second-order system, the frequency stability and accuracy of the input signal are not modified by the filter as the signal passes through it; however, phase noise is. The subject of phase noise in PLLs is treated in Section 1-3 and in Chapters 4 and 5. Here it will be said that the effect of a PLL filter on the phase-noise spectrum associated with the input signal is to replace this spectrum with the noise spectrum of the voltage-controlled oscillator for all offset-from-signal frequencies above the loop bandwidth.

The techniques are extensively used in synthesizers to generate auxiliary frequencies.

Double-Mix Approach

The principle of double conversion with drift cancellation has been used in radio repeaters for some time (Ref. 8). It was described in the section on harmonic synthesis and will not be elaborated on here. The technique is straightforward and has the advantage of utilizing circuitry that can be implemented at microwave frequencies at relatively low cost.

Triple-Mix Approach

One of the design goals in frequency synthesis has been to develop an approach that utilizes a configuration of standard building blocks repeatedly. Such an approach has many advantages. It reduces the cost and time of design and development as well as the production cost, and it simplifies the operation, alignment, and repair of synthesizers. Triple-mix synthesis was designed to meet this requirement.

Figure 1-5 is a functional block diagram of a typical configuration, a decade, which can be used repeatedly to generate any desired number of frequency increments. It is a modification of the double-mix approach; hence it operates on the same principle of drift cancellation. An additional sum mixer is introduced at the output of the fixed-frequency narrowband filter to inject the next-lower-order frequency increments in the signal path. The selected frequency at the output of the left-hand-side mixer is

$$f_{IF1} = f_{in} + (\Delta f_{0-9})_1 - f_{TO}.$$

The output of the following sum mixer is

$$f_{IF2} = f_{in} + (\Delta f_{0-9})_1 - f_{TO} + f + \frac{(\Delta f_{0-9})_2}{10}.$$

The offset output frequency is

$$f'_{out} = f_{in} + f + (\Delta f_{0-9})_1 + \frac{(\Delta f_{0-9})_2}{10}. \tag{1-2}$$

Notice that f_{TO} disappeared from the expression for f_{out}, as was expected.

The notation $\Sigma f_{in} + (\Delta f_{0-9})_1$ indicates that $f_{in} + (\Delta f_0)_1, f_{in} + (\Delta f_1)_1, \ldots, f_{in} + (\Delta f_9)_1$ are present at the input simultaneously.

In some applications (Refs. 9 to 12) the presence of an offset, f, may not be objectionable, but whenever such decades are used in cascade, the accumulated offset can modify the expression of the output frequency to such an extent that it will be impractical to utilize it in further synthesis. For this reason another difference mixer with a filter (in dotted blocks) is added to the diagram. The function of this mixer is to remove the offset.

The output frequency without the offset is

$$f_{out} = f_{in} + (\Delta f_{0-9})_1 + \frac{(\Delta f_{0-9})_2}{10}. \tag{1-3}$$

Figure 1-6 shows how the basic configuration of the triple-mix approach can be used to generate n frequency increments. Notice that, although a signal with the same frequency spectrum, $\Sigma f_{in} + (\Delta f_{0-9})$, is applied to the inputs of all triple-mix decades simultaneously, frequency-increment generation is an independent process for each decade; that is, the subscripts 1, 2, 3,..., n outside parentheses indicate that the setting of $(\Delta f_0)_1$,

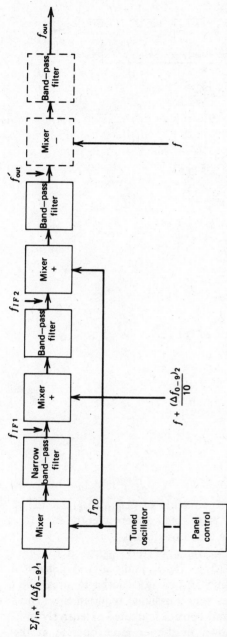

Figure 1-5. Triple-mix synthesis.

13

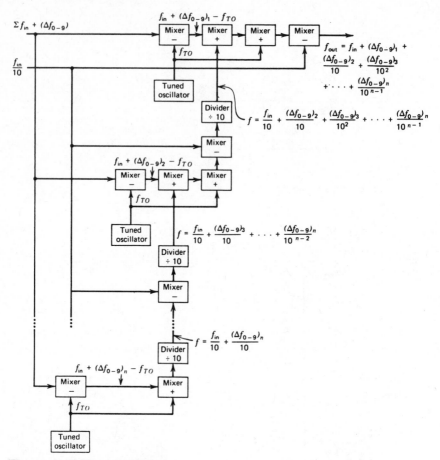

Figure 1-6. Triple-mix synthesizer.

$(\Delta f_1)_1, \ldots,$ $(\Delta f_9)_1$ is selected independently of $(\Delta f_0)_2$, $(\Delta f_1)_2, \ldots, (\Delta f_9)_2$. Similarly, the increment $(\Delta f_{0-9})_2$ is selected independently of $(\Delta f_{0-9})_3$ by appropriately tuning the decade tuned oscillators.

Figure 1-7 is a numerical example of this approach for $n=3$. The frequency-increment setting is $(\Delta f_4)_1$, $(\Delta f_3)_2$, and $(\Delta f_5)_3$. A numerical value for the tuned oscillator frequency, f_{TO}, is not chosen to stress the dependence of f_{TO} on the synthesizer spurious-signal requirement. (The subject of spurious-signal generation and control is treated extensively in Chapter 2.) All triple-mix decades shown in the example operate at the same frequency. This feature allows repeated use of one circuit configuration throughout the system.

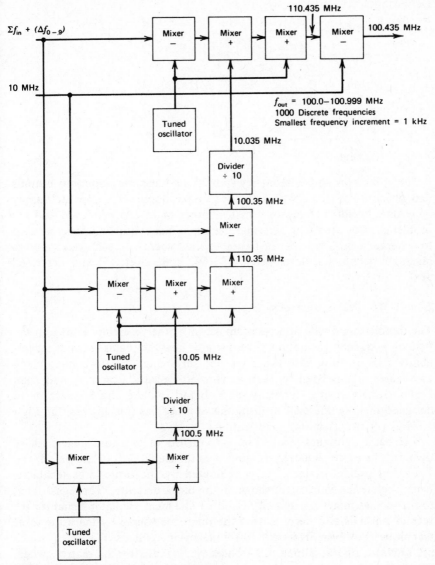

Figure 1-7. Triple-mix synthesis: an example.

Definitions:

1. Δf_{0-9}

$$\left.\begin{array}{l} \Delta f_0 = 0.0 \\ \Delta f_1 = 0.1 \\ \Delta f_2 = 0.2 \\ \quad\vdots \\ \Delta f_9 = 0.9 \end{array}\right] \text{MHz}$$

2. $f_{in} = 100$ MHz.

The comments about frequency stability and accuracy, spurious outputs and phase noise made with reference to the double-mix approach apply here also. Because of the very high number of mixers employed and the problems associated with filtering, the triple-mix synthesis is seldom used to generate a large number of frequency increments. In such cases a more ingenious technique, developed by V. W. Bolie (Ref. 13) and described next, is preferred.

Double-Mix-Divide Approach

The double-mix-divide approach, unlike any other presently known in the field of frequency generation, demonstrates how the requirement of repeatability can be fully satisfied. It is not surprising, therefore, that many companies in this country, such as Hewlett-Packard Company and John Fluke Manufacturing Company, and abroad (Rohde and Schwarz) have designed their synthesizers utilizing this approach as it is described in Figs. 1-8 and 1-9 or with minor modifications.

The basic configuration in Fig. 1-8 consists of two sum mixers and a divider. The input signal, f_{in}, is applied to the signal port of the first mixer. The local oscillator signal, f_1, is generated by the brute-force technique from a reference source (not shown in the block diagram). The signal input to the second mixer is a sum of f_{in} and f_1. The local oscillator signal for the second mixer is also derived from the reference source by the brute-force technique; however, it carries 10 frequency increments, $f_2 + \Delta f_{0-9}$, which are selected, one at a time, for frequency synthesis by the matrix switch. These three frequencies, f_{in}, f_1, and f_2, are chosen so that the spurious-output requirement is satisfied and

$$f_{in} = \frac{f_{in} + f_1 + f_2}{10} = \frac{f_1 + f_2}{9}. \tag{1-4}$$

The output of the second mixer is divided by 10.

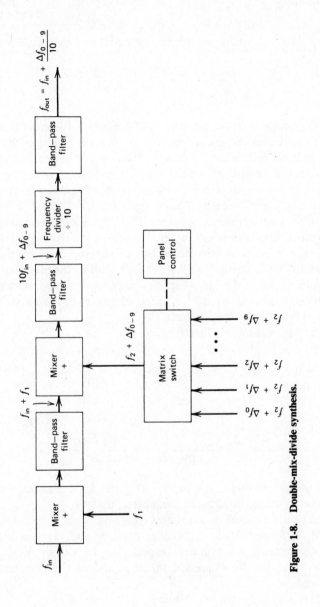

Figure 1-8. Double-mix-divide synthesis.

17

Under the condition expressed in Eq. 1-4, the input frequency appears at the output of the decade unmodified except by addition of the injected increment divided by 10:

$$f_{out} = f_{in} + \frac{\Delta f_{0-9}}{10}.$$ (1-5)

Figure 1-9 demonstrates how the generation of frequency increments is accomplished when n decades are connected in cascade. Notice that frequency increments are selected by rf switching and not by the tuning of n oscillators, as in the triple-mix approach. This is a significant advantage whenever high switching speeds are desired. Another advantage is that theoretically infinitely small frequency increments can be generated by cascading decades, and increments as small as 0.01 Hz have been available in commercial synthesizers. Microwave mixers have been in existence for a long time. Frequency division at microwave can be accomplished by the phase-locked loop divider described in Chapter 6. It is up to the manufacturers of rf switches to develop a large-capacity microwave matrix switch with interchannel isolation of 80 dB or more to make this approach attractive for use at microwave frequencies.

Figure 1-9 shows two available outputs:

$$(f_{out})_1 = 10 f_{in} + (\Delta f_{0-9})_1 + \frac{(\Delta f_{0-9})_2}{10} + \cdots + \frac{(\Delta f_{0-9})_n}{10^{n-1}}$$ (1-6)

and

$$(f_{out})_2 = f_{in} + \frac{(\Delta f_{0-9})_1}{10} + \frac{(\Delta f_{0-9})_2}{10^2} + \cdots + \frac{(\Delta f_{0-9})_n}{10^n}.$$ (1-7)

Equation 1-6 is used more often than Eq. 1-7 because $(f_{out})_1$ is generated over a wider frequency band, in terms of megahertz, than $(f_{out})_2$.

A general expression for the output frequency for any division ratio has been derived by R. R. Stone and H. F. Hastings (Refs. 14 and 15). It is not given in this book because all practical aspects of the design of the double-mix-divide approach are expressed in the equations defining one decade, Eqs. 1-4 and 1-5.

There are various modifications of this approach. Instead of two sum mixers, a sum and a difference mixer may be used. One mixer may be removed from the decade. The division ratio can be any suitable number (Ref. 16).

Relatively few mixers are used in the double-mix-divide synthesis. Frequency dividers such as the locked oscillator described in Chapter 6

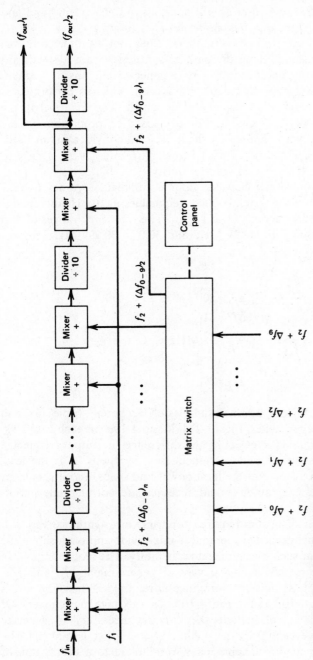

Figure 1-9. Double-mix-divide synthesizer.

provide 10 to 15 dB attenuation even if spurious outputs are in band and close to the signal. This simplifies the problem of filtering.

The phase-noise spectrum of the signal does not degrade with the propagation of the signal along the path of synthesis, again a significant advantage when many decades are used to generate millions of frequencies. This effect is achieved by frequency division, which is performed at the output of each decade. When one deals with a low noise signal, the phase-noise power of the signal is not decreased in proportion to the division ratio but rather is limited by the noise of the divider circuit itself. Nevertheless it is significantly reduced. Signals at f_1 and $f_2 + \Delta f_{0-9}$ have to be synthesized with the required spectral purity.

Figure 1-10a demonstrates how the double-mix-divide synthesis works in practice. Three decades generate 1000 frequencies in 10 kHz increments at $(f_{\text{out}})_1$ and in 1 kHz increments at $(f_{\text{out}})_2$. The frequency ranges of two output signals are 300.0 to 309.99 MHz and 30.0 to 30.999 MHz, respectively.

Δf_{0-9} defined as:

$$\left.\begin{array}{l} \Delta f_0 = 20 \\ \Delta f_1 = 21 \\ \Delta f_2 = 22 \\ \\ \vdots \\ \\ \Delta f_9 = 29 \end{array}\right\} \text{MHz}$$

It is enlightening to compare the examples of incoherent, triple-mix, and double-mix-divide approaches, Figs. 1-2, 1-7, and 1-10a. In each case 1000 frequencies are generated. Neglecting the differences in output frequency, one easily observes the simplicity of incoherent synthesis. It is the least expensive approach, it requires the least power and space, and it should be used whenever frequency stability and accuracy are not of prime importance.

The large number of mixers required for triple-mix synthesis, Fig. 1-7, may lead to the conclusion that the double-mix-divide approach should be used in all cases, but such a generalization is faulty. The block diagram in Fig. 1-10b demonstrates a practical way of generating f_{in}, f_1, and $f_2 + (\Delta f_{0-9})$ when low phase-noise performance is required. When Fig. 1-7 is compared with Fig. 1-10a and b, Fig. 1-10 is seen to have more mixers. Of course, Fig. 1-7 does not include the circuitry necessary to generate $\Sigma f_{\text{in}} + (\Delta f_{0-9})$. However, this is the point that the writer is making. The discussion of coherent direct approaches was limited to a description of frequency synthesis techniques, leaving out most of the auxiliary circuitry,

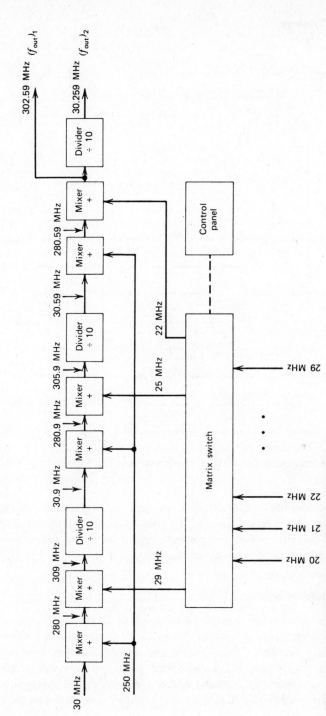

Figure 1-10a. Double-mix-divide synthesis: an example.

21

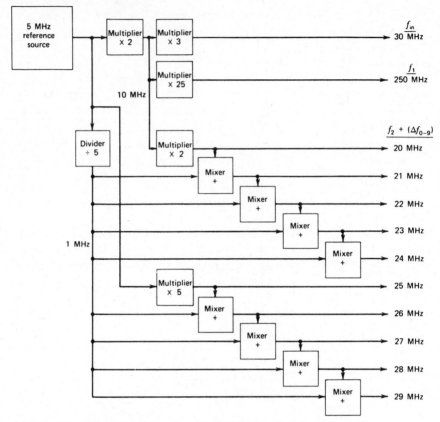

Figure 1-10b.   Fixed-frequency section for the synthesizer shown in Fig. 1-10a.

which should be included in evaluation. In general, preference for one approach over another is determined by requirements, state-of-the-art technology, and current cost of components, and is based on comparison of complete frequency generation systems.

1-3   Coherent Indirect Synthesis

Indirect synthesis utilizes the principle of feedback in generating frequency increments. The technique, known as phase locking, differs from direct synthesis in many respects. The system analysis of indirect synthesis centers on an investigation of phase-locked loop (PLL) stability and acquisition, not on spurious outputs. Mixers, multipliers, dividers, and filters are used in synthesis, but so also are voltage-controlled oscillators

(VCOs), programmable dividers, phase detectors, and frequency discriminators. Phase noise, switching speed, frequency increments, and environmental performance differ from those displayed in direct synthesis. The problems associated with the indirect synthesis techniques are of a dynamic nature (loop stability and acquisition). Providing small-size, lightweight equipment that requires low dc power, these techniques exhibit many advantages not offered by direct synthesis. They will be discussed in the following sections of the text.

Analog Phase-Locked Loop

Figure 1-11a is a block diagram of an analog PLL decade. The frequency of the VCO, f'_{out}, is downconverted and compared to the reference frequency, $f_{in}+(\Delta f_{0-9})_1$. When the difference between these two frequencies is small, the phase detector generates a slowly varying ac voltage, which is passed by the low-pass filter, and pulls the VCO in lock. Under locked condition the output of the phase detector is a dc voltage whose amplitude and polarity are determined by the amount and direction of phase displacement between the reference and downconverted VCO signals. The low-pass filter (often a combination of a low-pass filter and lag-lead network) changes the amplitude and phase of individual signals passing through it as functions of the working frequency to achieve stable loop performance. When the difference between compared frequencies is large, the phase-detector output is a fast-varying ac signal, which is

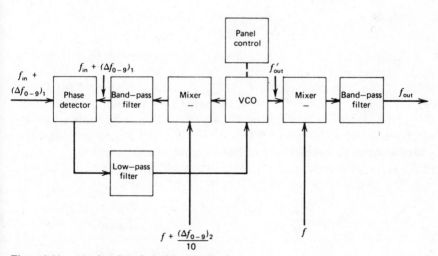

Figure 1-11a.　Analog phase-locked loop synthesis.

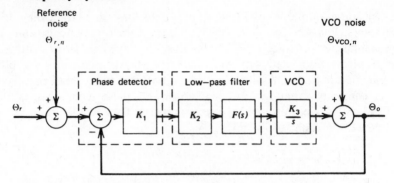

Figure 1-11b. Equivalent block diagram of an analog phase-locked loop.

attenuated by the low-pass filter, preventing locking. When this is expected, other means of locking, discussed in Chapter 4, are employed.

The difference mixer and band-pass filter following the VCO are used to remove the offset, f.

The decade is tuned by selecting appropriate $(\Delta f_{0-9})_1$ and $(\Delta f_{0-9})_2$ and by setting the VCO frequency close to the required frequency to make locking possible.

It can be shown that, for locking to take place,

$$f'_{out} = f_{in} + f + (\Delta f_{0-9})_1 + \frac{(\Delta f_{0-9})_2}{10} \qquad (1-8)$$

and

$$f_{out} = f_{in} + (\Delta f_{0-9})_1 + \frac{(\Delta f_{0-9})_2}{10}. \qquad (1-9)$$

A phase-locked loop is commonly used as a phase-noise filter. It will be demonstrated that a PLL is effectively a low-pass filter with respect to noise of the reference signal, $f_{in} + (\Delta f_{0-9})_1$, and a high-pass filter with respect to noise of the VCO, with the same time constant in both cases. Figure 1-11b is an equivalent block diagram of a PLL. The gain constants of phase detector, filter, and VCO are K_1, K_2, and K_3, respectively. The phase noise contributed by the reference source is $\Theta_{r,n}$, and $\Theta_{VCO,n}$ is the phase noise generated by the VCO. For simplicity assume that $N(s) = K_2 = 1$. The transfer function of the loop with respect to $\Theta_{r,n}$ is

$$\frac{\Theta_0}{\Theta_{r,n}} = \frac{K_1 K_3/s}{1 + K_1 K_3/s} = \frac{K_1 K_3/s}{1 + K_1 K_3/s} \cdot \frac{s/K_1 K_3}{s/K_1 K_3},$$

$$\frac{\Theta_0}{\Theta_{r,n}} = \frac{1}{1 + (1/K_1 K_3)s}. \qquad (1-10)$$

Equation 1-10 is the transfer function of a low-pass filter whose time constant is $1/K_1K_3$.

The transfer function of the loop with respect to $\Theta_{VCO,n}$ is

$$\frac{\Theta_o}{\Theta_{VCO,n}} = \frac{1}{1 + K_1K_3/s},$$

$$\frac{\Theta_o}{\Theta_{VCO,n}} = \frac{1}{1 + 1/(1/K_1K_3)s} \qquad (1\text{-}11)$$

Equation 1-11 is the transfer function of a high-pass filter whose time constant is $1/K_1K_3$.

This characteristic of PLLs plays an important role in vhf and uhf synthesis, which utilizes high multiplication ratios. It indicates that a PLL replaces the noise spectrum of a reference signal with the VCO noise spectrum for all offset-from-signal frequencies above the PLL bandwidth. This results in a significant improvement in phase noise. Chapters 4 and 5 discuss in great detail the subject of phase noise in PLLs.

There are two ways in which the basic PLL configuration can be utilized to generate frequency increments. They are known as parallel and series injection approaches. In the case of parallel injection, all frequency increments are injected into one PLL. Figure 1-12 demonstrates how this is done. Instead of one mixer, numerous mixers are downconverting the VCO frequency. This is disadvantageous from the loop stability point of view when millions of frequencies have to be generated, and further modifications of the approach are necessary to minimize the number of mixers in the loop. On the other hand, the stabilizing, mixing, and dividing circuits are related only indirectly to the final output; hence a high degree of isolation between the circuits generating spurious frequencies and the output is provided, significantly reducing the requirement for filtering.

Assuming that

$$f_{out} > f_{in},$$

$$IF_1 > f_2 + \frac{(\Delta f_{0-9})_2}{10},$$

$$IF_2 > f_3 + \frac{(\Delta f_{0-9})_3}{10^2},$$

$$\vdots$$

$$IF_{n-1} > f_n + \frac{(\Delta f_{0-9})_n}{10^{n-1}},$$

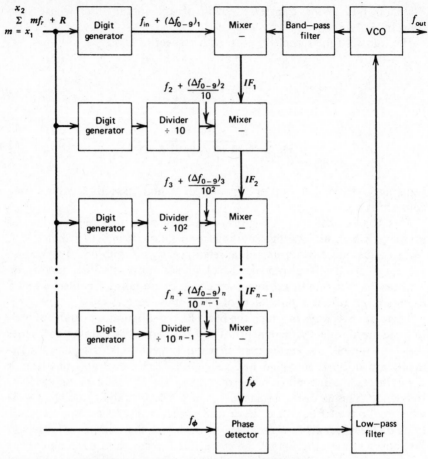

Figure 1-12.   Analog phase-locked loop synthesizer, parallel injection.

and summing up the frequencies injected in the loop, we obtain

$$f_{out} - \left[f_{in} + (\Delta f_{0-9})_1 \right] = IF_1,$$

$$f_{out} = f_{in} + (\Delta f_{0-9})_1 + IF_1,$$

$$IF_1 - \left[f_2 + \frac{(\Delta f_{0-9})_2}{10} \right] = IF_2,$$

$$IF_1 = \left[f_2 + \frac{(\Delta f_{0-9})_2}{10} \right] + IF_2,$$

$$f_{out} = f_{in} + (\Delta f_{0-9})_1 + f_2 + \frac{(\Delta f_{0-9})_2}{10} + IF_2,$$

$$IF_2 - \left[f_3 + \frac{(\Delta f_{0-9})_3}{10^2} \right] = IF_3,$$

$$IF_2 = f_3 + \frac{(\Delta f_{0-9})_3}{10^2} + IF_3,$$

$$f_{out} = f_{in} + f_2 + f_3 + (\Delta f_{0-9})_1 + \frac{(\Delta f_{0-9})_2}{10} + \frac{(\Delta f_{0-9})_3}{10^2} + IF_3;$$

similarly

$$IF_{n-1} - \left[f_n + \frac{(\Delta f_{0-9})_n}{10^{n-1}} \right] = f_\phi,$$

$$IF_{n-1} = f_n + \frac{(\Delta f_{0-9})_n}{10^{n-1}} + f_\phi,$$

and finally

$$f_{out} = (f_{in} + f_2 + f_3 + \cdots + f_n + f_\phi)$$

$$+ \left[(\Delta f_{0-9})_1 + \frac{(\Delta f_{0-9})_2}{10} + \frac{(\Delta f_{0-9})_3}{10^2} + \cdots + \frac{(\Delta f_{0-9})_n}{10^{n-1}} \right]. \quad (1\text{-}12)$$

The frequencies injected into the loop, $f_{in}, f_2, f_3, \ldots, f_n$, and f_ϕ, are selected so that a required output frequency range is obtained, and spurious outputs resulting from each mixing operation are minimized. The digit generators can be of any form described in Fig. 1-4.

There are various modifications of this approach. Instead of following the digit generators, frequency dividers can precede them (Refs. 7, 19, and 20). Some of the mixing can be done outside the loop to eliminate loop stability problems (Ref. 21). Output frequency can be divided by a variable-ratio divider to obtain a larger frequency range and smaller increments (Ref. 21).

The important difference between the approaches previously discussed and PLL is that in the former only a part of a synthesizer system was demonstrated (generation of frequency increments), and more circuitry had to be added to upconvert or downconvert or to multiply or expand generated frequencies until final results were obtained, whereas the system shown in Fig. 1-12 could be designed so that the output frequency, f_{out}, was the final frequency, providing a full frequency range and the increments required for an application. Depending on the application, this characteristic of the PLL approach may (but not always does) save hardware.

An examination of Eq. 1-12 indicates that the output frequency, f_{out}, is in all practical cases much larger than the frequency at which phase comparison is accomplished, f_ϕ. This is undesirable because a change in VCO frequency that is relatively small percentagewise becomes very large when it reaches the phase-detector input and falls outside the phase-detector range. Means for expanding the range have to be provided, in order to ensure locking, by employing a frequency discriminator in parallel with the phase detector, by utilizing sweep circuits, or by both methods.

Figure 1-13 is an example of parallel injection synthesis. Assume that the required output frequency range is 300.0 to 309.99 MHz in 10 kHz increments. To generate 10^3 frequencies, $n=3$. To generate 10 kHz increments,

$$\frac{(\Delta f_1)_n}{10^{n-1}} = \frac{(\Delta f_1)_3}{10^2} = 10^4$$

or

$$\Delta f_1 = 1.0 \text{MHz} \quad \text{and} \quad f_r = \Delta f_1 = 1.0 \text{MHz}.$$

Hence

$$\left.\begin{array}{l} \Delta f_0 = 0.0 \\ \Delta f_1 = 1.0 \\ \Delta f_2 = 2.0 \\ \vdots \\ \Delta f_9 = 9.0 \end{array}\right\} \text{MHz}.$$

Assume that there are no requirements for spurious outputs. Let

$$f_{in} = 277 \text{ MHz}.$$

Then

$$x_1 = 277, \quad x_2 = x_1 + 9 = 286$$

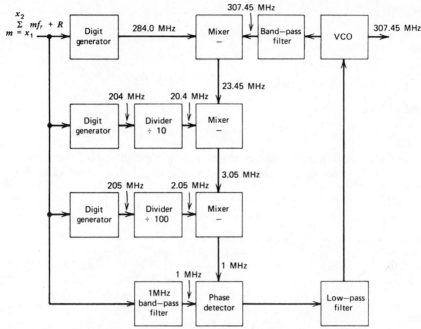

Figure 1-13.   Analog phase-locked loop synthesizer, parallel injection: an example.

and R = all harmonics of f_r except the 277th through 286th. Let

$$f_2 = 20 \text{ MHz},$$

$$f_3 = 2 \text{ MHz},$$

$$f_\phi = 1 \text{ MHz}.$$

Setting (Δf_{0-9})'s to $(\Delta f_0)_1 = (\Delta f_0)_2 = (\Delta f_0)_3 = 0$ MHz and using Eq. 1-12, one computes the minimum output frequency:

$$(f_{\text{out}})_{\text{min}} = 277 + 20 + 2 + 1 = 300 \text{ MHz}.$$

Similarly, the maximum output frequency is obtained by setting (Δf_{0-9})'s to $(\Delta f_9)_1 = (\Delta f_9)_2 = (\Delta f_9)_3 = 9$ MHz:

$$(f_{\text{out}})_{\text{max}} = 286 + 20.9 + 2.09 + 1 = 309.99 \text{ MHz}.$$

If, for example, 307.45 MHz was required, one would set (Δf_{0-9})'s as follows:

$$(\Delta f_{0-9})_1 \quad \text{to} \quad (\Delta f_7)_1 = 7 \text{ MHz},$$

$$(\Delta f_{0-9})_2 \quad \text{to} \quad (\Delta f_4)_2 = 4 \text{ MHz},$$

$$(\Delta f_{0-9})_3 \quad \text{to} \quad (\Delta f_5)_3 = 5 \text{ MHz},$$

and obtain

$$f_{\text{out}} = 284 + 20.4 + 2.05 + 1 = 307.45 \text{ MHz}.$$

The series injection approach of generating n increments is described in Fig. 1-14. It consists of n basic decades (see Fig. 1-11a). Instead of digit

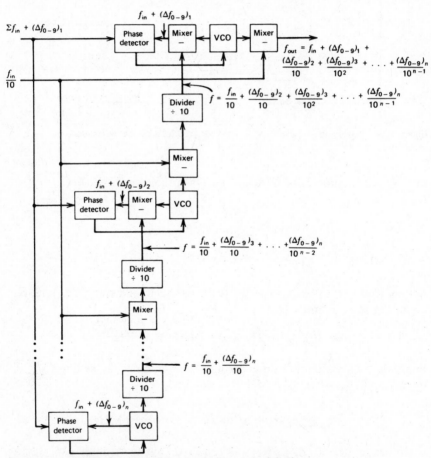

Figure 1-14.   Analog phase-locked loop synthesizer, series injection.

generators producing $f_n + (\Delta f_{0-9})_n$, as in Fig. 1-12, it is assumed that 10 frequencies, $\Sigma f_{\text{in}} + (\Delta f_{0-9})_1$, are available for use in synthesis simultaneously.

Series injection is more adaptable than parallel injection to the generation of large numbers of increments, but to the best of the writer's knowledge neither has been used successfully to generate millions of frequencies as, for example, is possible with the double-mix-divide approach.

The work on analog PLL synthesis dates back to the early days of synthesizers. The emergence of digital synthesis, unjustly, though perhaps only temporarily, puts analog PLL and all other approaches out of favor.

Digital Phase-Locked Loop

No other approach has attracted so much attention, aroused so many unrealized expectations, and been so extensively discussed in print as digital phase-locked loop synthesis. The numerous references given at the end of this chapter (Refs. 22 to 37) are only a few of the publications that the interested student will find on this subject.

The basic form of a digital PLL is shown in Fig. 1-15. The loop consists of a VCO, variable-ratio frequency divider, phase comparator, and low-pass filter. The VCO output is divided and compared with a stable reference. Error voltages derived from the phase comparator maintain the

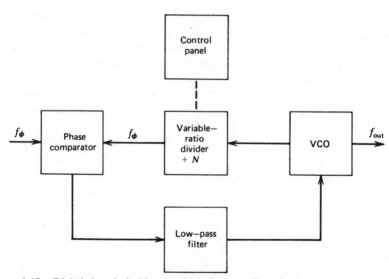

Figure 1-15.   Digital phase-locked loop synthesis: basic configuration.

VCO on frequency. Frequency selection is accomplished by a channel selector (the control panel), which varies the division ratio of the frequency divider.

For locking to occur,

$$f_{out} = Nf_\phi. \tag{1-13}$$

Equation 1-13 indicates that the smallest frequency increment generated by the loop is equal to the phase-comparator frequency, f_ϕ. This is a very convenient feature, particularly when the loop is followed by a frequency multiplier whose multiplication ratio is not to the base 10. Other important characteristics offered by a digital PLL are small size and low dc power consumption. Digital PLLs require few filters, so that integrated circuits and MSI circuits can be used extensively, resulting in miniature equipment (Refs. 22 to 27). At low frequencies, when slow-speed integrated circuits are used, dc power drain is very small, making digital synthesizers suitable for battery operation.

As in the case of analog PLLs, the close-in noise of a digital PLL output signal is determined by the noise of the reference signal at f_ϕ and internal-to-the-loop circuitry, such as the variable-ratio divider. Far-out noise is effectively the VCO noise.

Digital synthesizers seldom are that simple, though. The upper frequency limit of variable-ratio dividers is lagging behind the requirement for VCO frequencies. Present requirements for output frequencies exceed 10 GHz, whereas the variable-ratio dividers available on the market do not operate above 50 MHz and those designed from the high-speed logic operate below 1 GHz. This presents a problem. Small frequency increments lead to a low phase-comparator frequency. This frequency determines the rate at which frequency corrections are made to the VCO. A slow-rate PLL, for example, will not remove vibrational effects, which show up in the VCO as incidental FM noise, at frequencies above this rate (Ref. 28). Conversion of the phase error to dc control voltage is one of the most important processes in digital PLL. This voltage controls the frequency of the VCO. Any variation in control voltage will lead to a VCO phase modulation. It is important, therefore, to design for f_ϕ high enough so that it can be filtered out before it reaches the control input of the VCO (Ref. 22). Acquisition time is a function of loop bandwidth. Since the bandwidth is always smaller than f_ϕ, faster acquisition times require higher f_ϕ (Ref. 22).

It is in the designer's interest, therefore, to raise the phase-comparator frequency to the maximum possible value in order to improve short-term stability, spurious outputs, and acquisition time of a digital synthesizer with a wide band loop. How can these requirements be satisfied simultaneously?

Figure 1-16 shows how the upper frequency limit of the variable-ratio divider can be raised to allow the VCO to operate at microwave frequencies. If the operating frequency range is narrow, the difference mixer alone may downconvert the VCO frequency to within the range of the variable-ratio divider, but 8 to 20 % frequency bands require the addition of a N_2 divider.

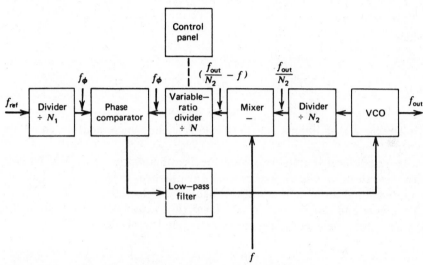

Figure 1-16.　Digital phase-locked loop synthesis, main phase-locked loop.

Usually both the difference mixer and N_2 divider are used. The N_1 divider is shown to indicate that for stability reasons the frequency of the reference signal, f_{ref}, is higher than the phase-comparator frequency (see the discussion of reference sources, Chapter 8).

For locking to occur,

$$f_{out} = (Nf_\phi + f)N_2,\qquad(1\text{-}14)$$

where

$$f_\phi = \frac{f_{ref}}{N_1}.$$

Notice that the smallest frequency increment is no longer f_ϕ but $N_2 f_\phi$. This means that high N_2 ratios are not desirable.

To solve the problem of low sampling rates, a two-loop approach is often used. Figure 1-17 shows one two-loop system, and Refs. 30 and 31 describe two more designs based on the same principle. In Fig. 1-17 frequency synthesis is divided into a main and an auxiliary PLL. The sampling frequency in the main loop, $f_{\phi 1}$, is kept high. The auxiliary loop operates at significantly lower frequency ($f_{out} > f$). For locking to occur,

$$f_{out} = (N f_{\phi 1} + N_3 f_{\phi 2}) N_2, \tag{1-15}$$

where

$$f_{\phi 1} = \frac{f_{ref}}{N_1} \quad \text{and} \quad f_{\phi 2} = \frac{f_{ref}}{N_4}.$$

Under these conditions, the smallest frequency increment is equal to $N_2 f_{\phi 2}$.

An example of double-loop frequency synthesis is given in Fig. 1-18 to demonstrate how this approach works in practice. It is assumed that the required final output frequency range is 7.0 to 7.9 GHz in 1 MHz increments and that a times-4 multiplier (not shown) is used to multiply the output frequency of the synthesizer, f_{out}, to shf. Hence

$$f_{out} = 1.750 \quad \text{to} \quad 1.975 \text{ GHz}$$

in 250 kHz increments. A divide-by-16 circuit (which can be a four-stage parametric divider, described in Chapter 6, with each stage dividing by 2) reduces f_{out} to the frequency range over which the variable-ratio divider operates. The two-loop approach relaxes the dynamic requirements of the main loop, which operates over a 13 % frequency band. The auxiliary loop operates at a significantly lower frequency and over only a 3 % band. This advantage permits the designer to control leakages of the sampling frequency, $f_{\phi 2}$, to the input of the vhf VCO without any difficulty. The synthesis of the system is conducted in such a manner as to achieve the variation of the final output frequency in 100 MHz increments by varying the ratio of the divider in the main loop from 40 to 49. Increments of 1 and 10 MHz are produced by changing the ratio of the divider in the auxiliary loop from 3000 to 3099.

A more detailed discussion of digital phase-locked loops is given in Chapter 5.

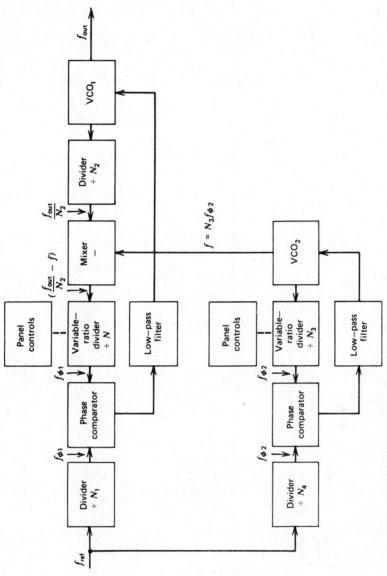

Figure 1-17. Digital phase-locked loop synthesis, two-loop approach.

35

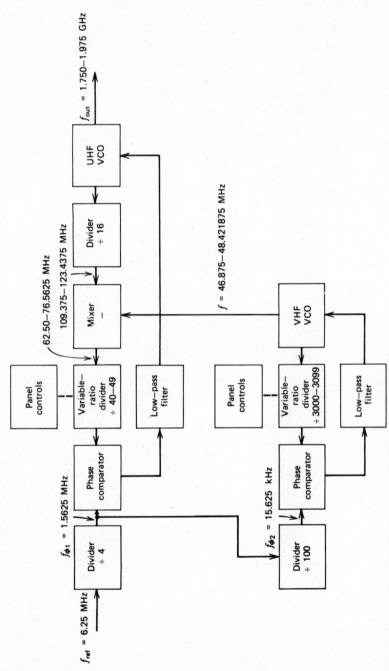

Figure 1-18. Double digital phase-locked loop synthesis: an example.

1-4 Coherent Direct Digital Synthesis

Little practical information concerning direct digital synthesizers has been made available. With few exceptions the use of these systems is confined to the generation of low-frequency signals in very small frequency increments in vhf and uhf synthesizers. This is accomplished at lower cost and less complexity than are possible with other known synthesis techniques. Rapid developments in vhf digital integrated circuits, however, lead one to believe that it may be possible to build direct digital synthesizers generating relatively pure signals in the hf range in the not-too-far-distant future.

As presently defined, direct digital synthesis is a technique by which a signal is generated in the form of a series of digital numbers and converted into analog form by a digital-to-analog converter. Figure 1-19 is a functional block diagram of a direct digital synthesizer showing the five basic building blocks utilized in implementing the approach.

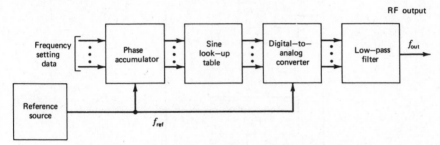

Figure 1-19. Direct digital synthesizer; simplified block diagram.

The reference source is a stable crystal-controlled oscillator used to synchronize the constituent parts of the synthesizer. The accumulator is a device that converts frequency setting data into phase samples which determine the magnitude of the synthesizer output waveform at a given sample time. When addressed with such data, the sine look-up table (a read-only memory) transforms each phase sample into a digital amplitude sample of a sine wave that is converted into the desired analog signal by the digital-to-analog converter. The low-pass filter attenuates the unwanted sampling components, as well as other out-of-band spurious signals.

Figure 1-20 is a block diagram demonstrating in some detail the basic building blocks of a direct digital synthesis approach. The frequency-setting data which represent the desired frequency of the synthesized output signal, in this discussion called "the synthesized signal," are provided by either the front panel switches (manually controlled) or a control unit such as a computer (remotely controlled) directly or by way of a BCD-to-binary converter, not shown in Fig. 1-20. The binary register

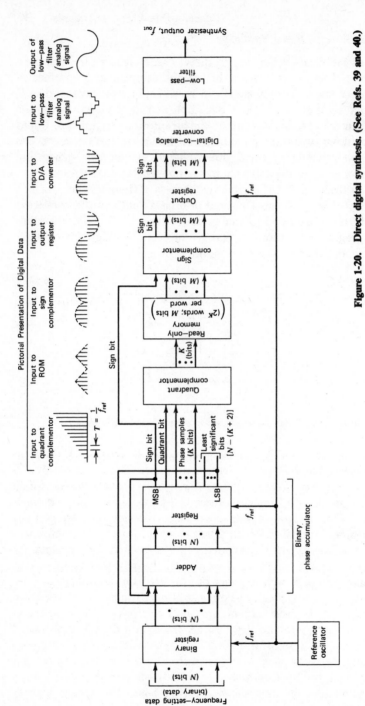

Figure 1-20. Direct digital synthesis. (See Refs. 39 and 40.)

38

accepts these digital data and presents them to the adder of the phase accumulator until a new frequency instruction is received.

The adder adds the digital signals from the binary register to the value of the accumulator register and updates the accumulator register with the most recent sum. The function of the accumulator register is to transfer the updated digital data from the output of the adder to its input at every clock pulse so as to make the modulo 2^N accumulator overflow periodically with a period determined by the frequency setting—the higher the synthesized frequency, the shorter the accumulator period. This cycle of 2^N represents one cycle of the synthesized signal. The accumulator register also presents the updated digital data to the quadrant complementor for further processing. Because of the limited capacity of the read-only memory (ROM) only $K+2$ phase samples are used in further frequency synthesis.

As a consequence, the output of the phase accumulator is the linearly increasing phase value of the synthesized signal, generated in the form of a $(K+2)$-bit digital number, with the magnitude of the phase increment determined by the frequency of the synthesized signal as

$$\text{phase increment} = 2\pi\left(\frac{f_{\text{out}}}{f_{\text{ref}}}\right)\text{rad}, \tag{1-16}$$

where f_{out} = synthesized frequency,
      f_{ref} = clock frequency.

The phase increment, therefore, is small for low synthesized frequencies and large for high frequencies.

When the synthesizer is set to a new frequency, the binary register provides the adder with the data for a new phase increment at the next pulse of the clock, at which time the period of the accumulator changes accordingly.

Under these conditions the most significant bit of the phase accumulator output is a square wave with the repetition rate equal to the repetition rate of the synthesized signal. It is used as a control signal to complement the ROM address, that is, to generate the negative half-cycle of the synthesized signal waveform, by way of the sign complementor, and for that reason it is called the "sign bit." The next-to-the-most significant bit of the accumulator, termed the "quadrant bit," is a square wave with a rate equal to half of the synthesized signal rate. The quadrant bit is used as the control signal in selecting the quadrants of the synthesized waveform by way of the quadrant complementor. The quadrant complementor provides a K-bit address to the ROM.

The ROM is a sine function table of a 2^K word, M bit per word capacity that converts the phase information provided by the phase accumulator into digital amplitude samples of the synthesized waveform. At each consecutive phase value generated by the phase accumulator the ROM provides the sign complementor with the corresponding digital number representing the magnitude of the synthesized waveform at that sample time. The sign complementor operates on these digital data so as to produce a symmetrical sine wave.

All of the bits for each phase sample should appear at the input of the digital-to-analog (D/A) converter at exactly the same time. Otherwise, the different operation times of the D/A converter bit switches will produce a disorganized transition from one sample to the next, affecting the spectral purity of the synthesized signal. The synchronization of the bits with respect to each other is done in the output register, which stores the data supplied by the sign complementor and transfers all $M + 1$ bits to the D/A converter simultaneously at the next clock pulse. In this manner the D/A converter is updated at a rate equal to $1/f_{ref}$, with the result that the envelope of the D/A converter output is a $(\sin X)/X$ function with $X = \pi(f_{out}/f_{ref})$.

As mentioned before, the low-pass filter (LPF) attenuates all out-of-band spurious signals generated in the process of frequency synthesis by the required amount.

Figure 1-20 gives a pictorial presentation of the digital data and analog waveforms at various steps of synthesis.

The smallest frequency increment synthesized by using such a technique is determined by the number of phase-accumulator bits, N, and is equal to

$$\Delta f_{min} = \frac{f_{ref}}{2^N}, \tag{1-17}$$

which is also the lowest synthesized frequency. The highest synthesized frequency is determined by the spurious outputs requirement. This topic is considered next in the order following the direction of the path of frequency synthesis.

The number of ROM words is usually less than 2^N because of the relatively high cost of the ROM. This means that the $N - (K + 2)$ least significant bits of the phase-accumulator output are not used in frequency synthesis, giving rise to a phase quantization error. As a consequence of the quantization error a pair of PM sidebands symmetrically located about the synthesized signal exists at the synthesized frequencies that are not direct submultiples of the reference frequency. The level of these sidebands depends on the ratio of f_{out} to f_{ref} and the ROM capacity. The spurious sidebands cannot be filtered out because of their close proximity to the synthesized signal at some synthesized frequencies and, therefore, set an upper limit to f_{out}.

The digital data from the sign complementor are clocked into the output register so as to hold each value constant between the pulses of the reference clock before it is applied to the D/A converter input. Hence the D/A converter output waveform is a staircase approximation of a sine wave with a spectrum containing the synthesized signal at f_{out} and pairs of spurious sidebands, known as aliased components, symmetrically located about the fundamental component of the sampling frequency, f_{ref}, and its harmonics, as shown in Fig. 1-21. The presence of a high-level spurious signal at f_{spur}, which is the left-hand member of the first pair of the aliased components, effectively limits the upper end of the synthesized frequency range. The spurious sideband-to-signal power ratio can easily be derived as

$$\frac{\text{power of spurious signal at } f_{spur}}{\text{power of signal at } f_{out}} = \frac{\left[E_p \sin\left(\frac{\pi f_{spur}}{f_{ref}}\right) \middle/ \frac{\pi f_{spur}}{f_{ref}} \right]^2 \middle/ R_L}{\left[E_p \sin\left(\frac{\pi f_{out}}{f_{ref}}\right) \middle/ \frac{\pi f_{out}}{f_{ref}} \right]^2 \middle/ R_L}$$

$$= \left[\frac{\sin\left(\frac{\pi f_{spur}}{f_{ref}}\right)}{\sin\left(\frac{\pi f_{out}}{f_{ref}}\right)} \left(\frac{f_{out}}{f_{spur}}\right) \right]^2$$

or (in dB)

$$\frac{\text{power of spurious signal at } f_{spur}}{\text{power of signal at } f_{out}} = 20\log_{10}\left[\frac{\sin\left(\frac{\pi f_{spur}}{f_{ref}}\right)}{\sin\left(\frac{\pi f_{out}}{f_{ref}}\right)} \left(\frac{f_{out}}{f_{spur}}\right) \right],$$

$$(1\text{-}18)$$

where E_p = a constant for a given D/A converter,
$\quad f_{spur} = f_{ref} - f_{out}$,
$\quad R_L$ = a given load.

As f_{out} is increased, the spurious signal at f_{spur} eventually moves so close to the synthesized signal as to make filtering impracticable.

Equation 1-18 is useful in defining the attenuation characteristics of the LPF, particularly in cases where the highest synthesized frequency is

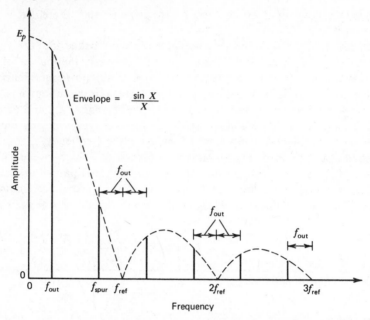

Figure 1-21. Frequency spectrum of a D/A converter output signal. (See Ref. 39.)

significantly smaller than the reference frequency, that is, where the level of the synthesized signal is much higher than the level of the aliased component at f_{spur}, and one can save hardware by knowing the level of the spurious signal. As f_{out} approaches $f_{ref}/2$, the difference between the levels of the two signals diminishes to 1 or 2 dB; this does not affect the filter design significantly, because in such a case the attenuation that the filter has to provide is equal to the requirement for the level of spurious outputs.

The highest synthesized frequency is also limited by the D/A converter switching characteristics. At high frequencies there are few samples per cycle of the synthesized signal with large amplitude differences between adjacent samples. To prevent the D/A converter from introducing a high-level harmonic distortion and generating spurious outputs, the settling time associated with switching from one sample to another should be very short compared to the sampling interval, $1/f_{ref}$, and the amplitude over-shoots and undershoots should be small compared to the sampled value of the synthesized sinusoid.

It is evident that direct digital synthesis exhibits several important design limitations. Among the limitations mentioned here, poor D/A converter switching characteristics predominate at present. Equally important is the requirement for a reference source that is operating at a higher frequency than the synthesized frequency because of (*a*) a relatively low frequency of operation of the state-of-the-art digital circuits utilized in the synthesis, and (*b*) the complexity of the converter required to lock such a synthesizer to an external 1, 5, or 10 MHz reference source, which increases as the synthesized frequency is increased. As the switching characteristics of D/A converters improve, another factor, presently easily controllable because of the relatively high f_{out}-to-f_{ref} ratio, will limit the upper end of the synthesized frequency range: the aliased components. Such are the major disadvantages associated with direct digital synthesis, which limit the use of this approach to synthesizers operating below 1 MHz for levels of spurious outputs lower than -70 dB referred to the level of the synthesized signal.

The advantages of this approach, on the other hand, are numerous and important enough to justify further efforts directed toward improving the technique and simplifying its implementation. First of all, the tuning time of a direct digital synthesizer is an order of magnitude shorter than that of any other synthesizer presently in use (less than 2 μsec). Second, any desired frequency resolution can be obtained at low cost, small volume, low dc power, and light weight by simply increasing the number of accumulator bits. Third, the phase of the synthesized signal can be made continuous as the synthesized frequency is changed or can be remotely controlled—an essential consideration in some applications. A fourth advantage is relaxed shielding requirements; little shielding is required, as is demonstrated in Section 7-4, where a direct digital synthesizer operating between dc and 2 MHz is described. Finally, the circuitry of direct digital synthesizers requires no alignment, except for the D/A converter and low-pass filter. This simplifies the manufacture of the equipment significantly and also reduces cost.

In the synthesis approach described above, a modulo 2^N phase accumulator is used. However, a modulo 10^N accumulator can be employed in frequency synthesis with equal success. A direct digital synthesizer utilizing a modulo 10^N accumulator is described in Chapter 7.

1-5 Fractional-N Phase-Locked Loop

The fractional-N phase-locked loop technique utilizes the digital PLL design and direct digital synthesis to arrive at a phase-locked loop with a programmable divider whose division ratio is the sum of an integer and a fraction of an integer.

The constituent parts of such a PLL are a VCO, a cycle swallower, a variable-ratio divider whose division ratio, N_1, is an integer, a phase comparator, and an accumulator; see Fig. 1-22. With the cycle swallower

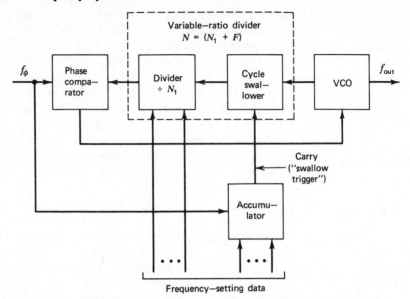

Figure 1-22.   Basic building blocks of a fractional-N phase-locked loop. (See Ref. 42.)

and accumulator removed, the PLL takes the familiar form of a digital PLL, described on pp. 31 and 32, generating a signal at a frequency that is an integer multiple of the reference frequency, f_ϕ. To lock the VCO to a fractional multiple of the reference frequency requires a circuit that divides by a number which has a fractional component, F. This is achieved by designing the divider so that it divides by N_1 for a number of cycles of the VCO signal and then, upon an external command, momentarily divides by $N_1 - 1$ (or $N_1 + 1$). The fractional N is the average of N_1 and $N_1 - 1$. The division by $N_1 - 1$ is required at the moment the phase of the VCO signal differs by exactly -360 degrees from the phase of the signal under locked conditions. Setting the divider to divide by $N_1 - 1$ is equivalent to advancing the output of the variable-ratio divider by one VCO cycle.

For example, assume that a VCO tuned to 990 kHz is to be locked to a 100 kHz reference oscillator, that is, it is desired that $N = 9.9$. Assume also that $N_1 = 10$. Under these conditions, when the reference signal goes through 1 cycle, the VCO signal goes through 9.9 cycles, so that the VCO is 0.1 cycle from being locked. Similarly, when the reference signal goes through 2 cycles, the VCO signal goes through 19.8 cycles; hence the VCO is 0.2 cycle from being locked. The numbers 0.1, 0.2, etc., represent the fractional portion of the VCO signal. Finally, after 10 reference cycles the VCO has gone through 99 cycles, and the phase difference is 1 cycle, or the VCO is -360 degrees from being locked. At this instant the variable-ratio divider is programmed to divide by $N_1 - 1$, which effectively advances the phase of the VCO signal by 1 cycle, and the process of the phase increment growth is repeated again.

The circuit that generates the signal instructing the variable-ratio divider to divide by $N_1 - 1$ is an accumulator similar in design to the phase accumulator described in Section 1-4. When properly programmed, it generates a carry signal at exactly the time when the division ratio has to change from N_1 to $N_1 - 1$. One way of implementing this change is to utilize a cycle swallower, such as the circuit shown in Fig. 1-23a, which removes one VCO pulse, as demonstrated in Fig. 1-23b, each time the division by $N_1 - 1$ is required.

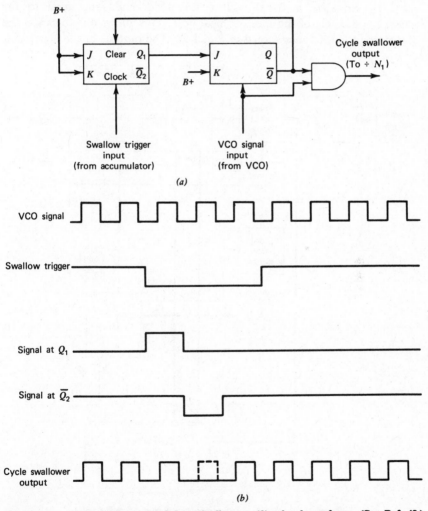

Figure 1-23. Cycle swallower. (a) Schematic diagram; (b) related waveforms. (See Ref. 42.)

Notice that this method of frequency synthesis sets the PPL condition whereby the effective VCO phase not only varies with the VCO mistuning and temperature drift, as is the case in a standard PLL, but also changes periodically, with the period determined by the magnitude of F, the fractional component of N. This results in a spurious beat note generated by the phase comparator whose amplitude is proportional to the comparator gain and whose rate is equal to the rate of $N_1 - 1$ division, that is, the rate with which the accumulator is set to overflow and generate a carry signal.

Fortunately, the accumulator is in possession of the information that is used to synthesize a signal with a waveform and rate equal to the waveform and rate of the spurious beat note. This allows one to cancel the beat note in the manner shown in Fig. 1-24. The outputs of the accumula-

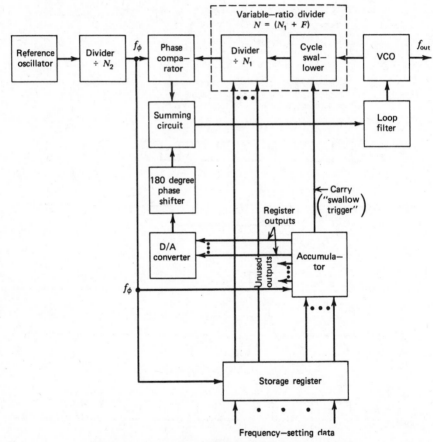

Figure 1-24.   Fractional-N phase-locked loop. (See Ref. 41.)

tor register are used to drive a D/A converter which converts these digital data into an analog signal similar in shape and identical in frequency to the spurious beat note and in phase with it. This signal is shifted by 180 degrees before it is added to the beat note in a sum circuit. The degree of cancellation of the beat note depends on the magnitude of the error made in generating the cancellation signal, that is, on the number of less significant bits of the accumulator discarded because of a limited D/A converter capacity, the resolution and stability of the phase shifter, and, to a lesser extent, the accuracy and stability of the D/A converter output signal.

The major advantage of this technique is obvious: it permits synthesis of small frequency increments at a reference frequency that is orders of

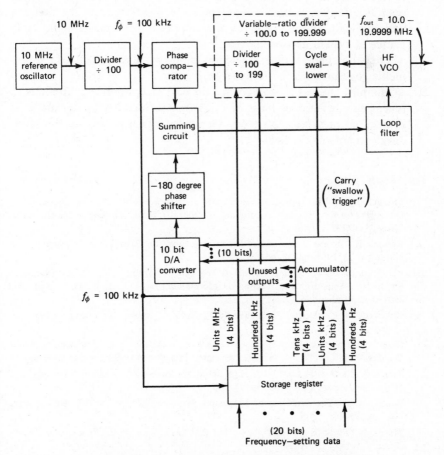

Figure 1-25. An example of a fractional-*N* phase-locked loop.

magnitude higher than the reference frequency required for a standard digital PLL, thus facilitating faster tuning time. The disadvantage is the relatively high FM spurious signals associated with the fractional-N division as it is presently implemented.

Figure 1-25 is a numerical example of a hf fractional-N PLL generating 100 Hz increments at 10 MHz and locked to a 100 kHz reference. (For a practical example of fractional-N PLLs see Section 7-5.)

1-6 Conclusions

Three basic points should be apparent from the survey of methods and basic problems associated with frequency synthesis:

1. A tightening of synthesizer requirements, which has followed shifts of operating ranges to higher frequencies, makes the job of the designer very difficult and emphasizes the importance of making the right choice of system approach.
2. There is no single synthesis technique that satisfies all sets of requirements governing synthesizer design.
3. To simplify the design and reduce the cost of synthesizers, attempts to invent new techniques must be (and are continuously being) made. It is essential in synthesizer work that the designer follow up the latest developments in the field.

References

1. Lindholm, C. and S. Johnson. *A New Method of Frequency Synthesis* (Washington, D. C.: Clearinghouse for Federal Scientific and Technical Information, Department of Commerce, AD 413376, July 1963).
2. Ross, G. F. "Binary Generation of Frequencies Saves on Hardware," *Electronic Design*, November 23, 1964, pp. 38–46.
3. Baltas, M. *Survey of Frequency Synthesis Techniques* (Washington, D. C.: Clearinghouse for Federal Scientific and Technical Information, Department of Commerce, AD 298130, September 1962).
4. Hekimian, N. C. "Digital Frequency Synthesizers," *Frequency*, July/August 1967, pp. 30–34.
5. Chomet, M. and R. Watterson. *An Impedance Probe for Radio Astronomy Explorer Satellite* (Plainview, N. Y.: Sanders Associates, Inc., Geospace Electronics Division).
6. Rasch, P. J. and J. F. Duval. "A High Speed Microwave Frequency Synthesizer," *The Microwave Journal*, June 1966, pp. 97–100.
7. Finder, H. J. "The Problems of Frequency Synthesis," *Journal British IRE*, January 1961, pp. 95–103.
8. Trevor, B. "Radio Repeaters," Patent No. 2,369,268 (Washington, D. C.: U. S. Patent Office, February 13, 1945).
9. Flicker, H. "Triple-Mix Frequency Synthesis," *Frequency*, January/February 1964, pp. 22–27.

10. Wicker, R. G. "Frequency Synthesizers," *G. E. C. Journal*, Vol. 32, No. 2 (1965), pp. 73–78.

11. Barlow, C. E. "Understanding Microwave Frequency Synthesizers," *The Electronic Engineer*, November 1967, pp. 36–38.

12. Saunders, J. T. "A Comparison of Frequency Synthesis Techniques," *Telecommunications*, July 1970, pp. 17–19.

13. Bolie, V. W. "Digital Frequency Synthesizer System," Patent No. 2,829,255 (Washington, D. C.: U. S. Patent Office, April 1, 1958).

14. Stone, R. R., Jr., and H. F. Hastings. "Frequency Synthesizing Techniques Permitting Direct Control and Rapid Switching," *Proceedings of 17th Annual Symposium on Frequency Control*, May 1963, pp. 587–601.

15. Stone, R. R., Jr., and H. F. Hastings, "A Novel Approach to Frequency Synthesis," *Frequency*, September/October 1963, pp. 24–27.

16. Oropeza, F. and J. P. Schoenberg. "Binary Frequency Synthesis: Spectral Purity with Economic Simplicity," *Frequency*, September/October 1966, pp. 14–17.

17. Thompson, G. D., Jr., and R. L. Sydnor. "Programmed Oscillator for Doppler Radar Systems," *Frequency*, July/August 1966, pp. 22–29.

18. McAleer, H. T. "A New Look at the Phase-Locked Oscillator," *Proceedings of the IRE*, June 1959, pp. 1137–1143.

19. Hargreaves, T. F., J. H. Gifford, and G. E. Smythe. "An Airborne Frequency Generating Unit for the HF Communication Band," *Journal British IRE*, February 1961, pp. 129–136.

20. Muller, J. J. and J. Lisimaque. "Portable Single-Sideband High-Frequency Transceiver with Military Applications," *Electrical Communications*, Vol. 43, No. 4 (1968), pp. 360–368.

21. Colodner, A. "Frequency Synthesis Adds Versatility to Stability," *Electronic Design*, September 27, 1963, pp. 124–129.

22. Ulicki, E. M. "Cubic Inch Frequency Synthesizers," *IEEE Mohawk Valley Communications Symposium, Proceedings* (NATCOM), October 1965.

23. Renschler, E. and B. Welling. "An Integrated Circuit Phase-Locked Loop Digital Frequency Synthesizer," *Application Note* AN-463, Motorola Semiconductor Products, Inc., March 1969.

24. Hartley, R. L. "Chopping Costs of Frequency Synthesizers with IC's," *Electronic Products*, July 1968, pp. 46–61.

25. Robin, N. A. "Phase-Locked Frequency Multiplier Cuts Cost," *Electronic Design News*, November 15, 1969, pp. 73–75.

26. Nichols, J. L. "MSI Moves into Transceiver Frequency Selection," *Electronic Design News*, March 1, 1969, pp. 63–65.

27. Gill, W. L. "Use IC's in Your Phase-Locked Loop," *Electronic Design*, April 8, 1968, pp. 76–80.

28. Westwood, D. H. "Study of Trade-offs for Synthesizers," *Electronic Design*, March 1, 1967, pp. 88–90.

29. Blachowicz, L. F. "Dial Any Channel to 500 MHz," *Electronics*, May 2, 1966, pp. 60–69.

30. Thomas, T. C. *Research and Development Investigation of a Precision Digital Frequency Synthesizer for SSB* (Washington, D. C.: Clearinghouse for Federal Scientific and Technical Information, Department of Commerce):
    (a) Quarterly Report No. 1, AD 457162, February 1965;
    (b) Quarterly Report No. 2, AD 462497, March 1965;

(c) Quarterly Report No. 3, AD 621033, August 1965;

(d) Peterson, M. E. Final Engineering Report, AD 483563, January 1966.

31. Breiding, R. J. and C. Vammem. "RADA Frequency Synthesizer," *Frequency*, September/October 1967, pp. 25–32.

32. Evers, A. F. "A Versatile Digital Frequency Synthesizer for Use in Mobile Radio Communication Sets," *Electronic Engineer*, May 1966, pp. 296–303.

33. Ulicki, E. M. "A Microminiature VHF Transceiver," *SSD/CDE*, September 1965, pp. 15–20.

34. Hughes, R. J. and R. J. Sacha. "The Lohap Frequency Synthesizer," *Frequency*, August 1968, pp. 12–21.

35. Editorial. "Synthesizer Reflects Impact of Microcircuits on SSB Mobile Equipment Design," *Communications Designer's Digest*, January 1968, pp. 38–39.

36. Sepe, R. B. "A Frequency Modulation System Utilizing a Digital Control Loop," *Computer Design*, May 1968, pp. 54–63.

37. Gillette, G. C. "The Digiphase Synthesizer," *Frequency Technology*, August 1969, pp. 25–29.

38. Tierney, J., C. M. Rader, and B. Gold. "A Digital Frequency Synthesizer," *IEEE Transactions on Audio and Electroacoustics*, Vol. AU-19, No. 1, March 1971, pp. 48–57.

39. Hosking, R. H. "Direct Digital Frequency Synthesis," *1973 IEEE Intercon Technical Papers*, Paper 34/1.

40. Hosking, R. H. "A Unique Synthesizer and Its Advantages," *ELECTRO 76 Professional Program*, Boston, May 11–14, 1976, Paper 19/3.

41. Correspondence with Brian D. Unter, Engineering Section Manager, Hewlett-Packard Company.

42. Kingsford-Smith, C. A. Patent No. 3,928,813 (Washington, D. C.: U. S. Patent Office, December 23, 1975).

43. Jackson, Leland B. Patent No. 3,735,269 (Washington, D. C.: U. S. Patent Office, May 22, 1973).

2 System Analysis

Six requirements have a great bearing on the choice of a synthesis approach and determine the cost, size, weight, and power consumption of a synthesizer: (a) output frequency range, (b) frequency increments, (c) frequency stability, (d) spurious outputs, (e) phase noise, and (f) switching time. An example of such dependence is illustrated in Figs. 2-1, 2-2, and 2-3, where it is shown that, as new requirements are added or existing requirements are made more stringent, the complexity of the system grows at a rapid rate. It is of great importance, therefore, to be able to select the least expensive approach that is expected to satisfy the requirements and to produce a system design verifying such expectations. This chapter and the following ones describe performance and design techniques for circuits used in frequency synthesis and consider various phases of synthesizer design.

Requirements:

1. Frequency range = 220.0 to 299.999 MHz
2. Frequency increments = 1 kHz
3. Frequency stability = 1 part in 10^8/day
4. Spurious outputs = -60 dB

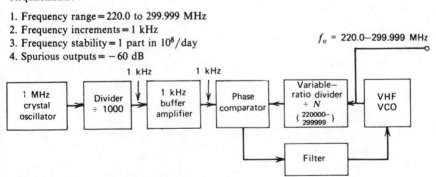

Figure 2-1.   Single-loop digital synthesizer.

2-1   Spurious Outputs

The term spurious output (or spurious signal) was extensively used in comparing the advantages and limitations of the frequency synthesis approaches described in Chapter 1. The term denotes any undesirable

Requirements:

1 to 4. Same as Fig. 2-1
5. Switching time = 5 msec

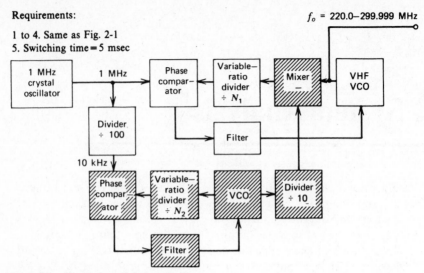

Figure 2-2. Double-loop digital synthesizer.

Requirements:

1. Same as Fig. 2-1
2. Frequency increments = 100 Hz
3. Frequency stability = 1 part in 10^{12}/month
4. Spurious outputs = -90 dB
5. Same as Fig. 2-2

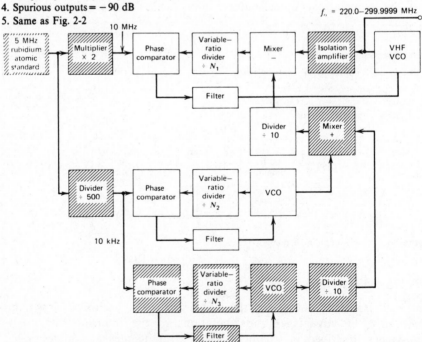

Figure 2-3. Multiloop digital synthesizer.

52

signal present at the output of a frequency generating system or any portion of it. This signal can be present at a single frequency somewhere in the operating band of the equipment, or it may occur in the form of amplitude or frequency modulation. It can be (*a*) radiated externally to and picked up by the system, with possible frequency conversion into the operating band, (*b*) caused by radiation internal to the system (due to a presence of signals at various frequencies used in synthesis), or (*c*) generated in the process of synthesis.

Chapter 3 considers spurious outputs that result from electromagnetic radiation. This section of Chapter 2 describes the mechanism of spurious-output generation in frequency synthesis.

Amplitude Modulation

Amplitude modulation (AM) is a process of producing a wave whose amplitude varies as a function of the instantaneous value of another wave. An amplitude-modulated signal can be described as

$$e(t) = \left[E_c + k_a e_m(t) \right] \cos \omega_c t, \tag{2-1}$$

where

$$e_c(t) = E_c \cos \omega_c t \tag{2-2}$$

is an unmodulated carrier, and k_a is a proportionality factor that determines the maximum variation in carrier amplitude for a given amplitude of modulating signal. For a sine-wave modulation, the modulating signal is

$$e_m(t) = E_m \cos \omega_m t; \tag{2-3}$$

hence

$$e(t) = (E_c + k_a E_m \cos \omega_m t) \cos \omega_c t \tag{2-4}$$

or

$$e(t) = E_c \left(1 + \frac{k_a E_m}{E_c} \cos \omega_m t \right) \cos \omega_c t, \tag{2-5}$$

where

$$m_a = \frac{k_a E_m}{E_c} \tag{2-6}$$

is the modulation index such that

$$e(t) = E_c (1 + m_a \cos \omega_m t) \cos \omega_c t. \tag{2-7}$$

Figure 2-4a through d describes the process of amplitude modulation in graphical form: a sinusoidal modulating signal, the modulated carrier, a vector representation, and the frequency spectrum of an AM wave, respectively. Of a particular interest to the designer of synthesizers is the frequency spectrum of an AM wave. In most cases of spurious AM, the modulation index is much less than unity. To evaluate a very-low-level AM, the frequency spectrum of the desired signal is examined and levels of AM sidebands are measured utilizing sensitive frequency-selective test equipment such as spectrum analyzers.

In general, the modulating signal is a complex wave. This wave can be represented by a Fourier series if it is periodic or by a Fourier integral if it is nonperiodic. Under these conditions each frequency in the modulating signal produces a pair of side frequencies in the frequency spectrum of the modulated signal, so that a modulating signal with frequencies in the band $g(\omega)$ will result in a frequency spectrum with two sidebands:

$$\frac{m_a}{2} g(\omega_c + \omega) \qquad \text{and} \qquad \frac{m_a}{2} g(\omega_c - \omega),$$

as shown in Fig. 2-4e.

It is the unwanted (spurious) amplitude modulation, either in a single-tone form, such as a 60 Hz hum, or in a complex form, such as noise, that plays an important role in synthesizer design. The presence of unwanted AM is undesirable not only because it results in AM spurious outputs, but also because AM is converted to PM in nonlinear devices, used in frequency synthesis. The subject of AM-to-PM conversion is treated later in this chapter.

Single-Tone Frequency and Phase Modulation

Frequency modulation (FM) is a process of producing a wave whose instantaneous frequency varies as a function of the instantaneous amplitude of a modulating wave at a rate given by the frequency of the modulating source. A frequency-modulated signal can be described as follows:

$$e(t) = E_c \cos\left(\omega_c t + k_f \frac{E_m}{\omega_m} \sin \omega_m t\right), \tag{2-8}$$

where the modulating wave is

$$e_m(t) = E_m \cos \omega_m t,$$

and where k_f is a proportionality factor that determines the maximum

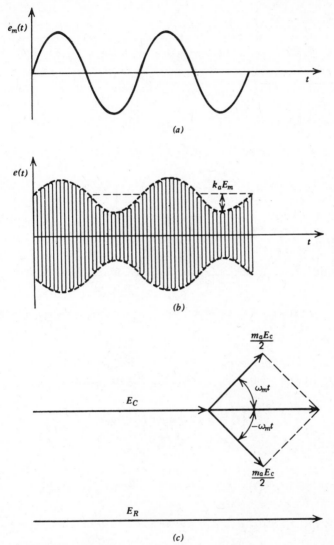

Figure 2-4. Amplitude modulation of a carrier. (*a*) Modulating signal; (*b*) modulated carrier; (*c*) vector representation of AM wave.

variation in frequency for a given modulating signal amplitude, E_m. The modulation index (deviation ratio) is

$$\beta = k_f \frac{E_m}{\omega_m} = \frac{\Delta f_{\text{peak}}}{f_m} \text{ rad,} \qquad (2\text{-}9)$$

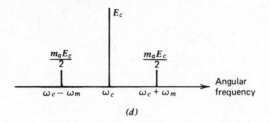

(d)

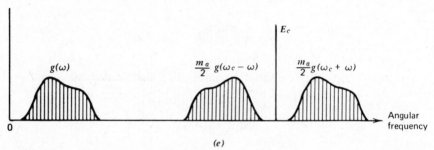

(e)

Figure 2-4. (d) **frequency spectrum of AM wave;** (e) **frequency spectrum of a complex AM wave.**

where Δf_{peak} is the frequency deviation; the maximum swing of the carrier frequency from its mean value, $f_c = \omega_c/2\pi$. Expressed in these terms, a FM wave is

$$e(t) = E_c \cos(\omega_c t + \beta \sin \omega_m t) \qquad (2\text{-}10)$$

or

$$e(t) = E_c \cos\left(\omega_c t + \frac{\Delta f_{\text{peak}}}{f_m} \sin \omega_m t\right). \qquad (2\text{-}11)$$

Expanding Eq. 2-10 for $E_c = 1$ gives

$$e(t) = \cos \omega_c t \cos(\beta \sin \omega_m t)$$
$$- \sin \omega_c t \sin(\beta \sin \omega_m t).$$

For a small modulation index (narrowband FM), or for $\beta \ll \pi/2$,

$$\cos(\beta \sin \omega_m t) \cong 1,$$
$$\sin(\beta \sin \omega_m t) \cong \beta \sin \omega_m t;$$

hence

$$e(t) \simeq \cos \omega_c t + \beta \sin \omega_m t \sin \omega_c t,$$

$$e(t) \simeq \cos \omega_c t - \frac{\beta}{2} \cos(\omega_c - \omega_m)t + \frac{\beta}{2} \cos(\omega_c + \omega_m)t. \qquad (2\text{-}12)$$

Equation 2-12 indicates that the spectrum of a narrowband FM wave consists of the carrier plus two components, one on each side of the carrier. In this sense, narrowband FM is equivalent to AM and cannot be distinguished from it by test equipment that is not capable of identifying the phase of a signal, such as spectrum analyzers. However, it is important in system evaluation and in measurement of spurious outputs and noise to keep in mind the distinction between AM and FM waves: in the case of AM, the carrier envelope varies with the modulating signal, whereas the frequency of the AM wave remains unchanged; in FM, the carrier amplitude is constant, and the carrier instantaneous frequency varies with the modulating signal.

Figure 2-5a to d describes the process of frequency modulation in graphical form: a sinusoidal modulating signal, the modulated carrier, a vector representation, and the frequency spectrum of a FM wave, respectively. Figure 2-5e is the frequency spectrum of a complex FM wave. As in the case of AM, each frequency in the modulating signal produces a pair of side frequencies in the spectrum of the modulated signal, so that a modulating signal with frequencies in the band $h(\omega)$ results in a frequency spectrum with two sidebands:

$$-\frac{\beta}{2} h(\omega_c - \omega) \qquad \text{and} \qquad \frac{\beta}{2} h(\omega_c + \omega).$$

Phase modulation (PM) of a signal is achieved by varying the instantaneous phase of the signal by an amount that is proportional to the amplitude of the modulating signal at a rate equal to the modulating frequency. If

$$e_m(t) = E_m \sin \omega_m t$$

is a modulating signal, the phase-modulated wave is

$$e(t) = E_c \sin(\omega_c t + k_p E_m \sin \omega_m t). \qquad (2\text{-}13)$$

The maximum phase deviation is

$$\theta_d = k_p E_m \text{ rad}. \qquad (2\text{-}14)$$

Hence the equation becomes

$$e(t) = E_c \sin(\omega_c t + \theta_d \sin \omega_m t). \qquad (2\text{-}15)$$

Comparison of Eqs. 2-10 and 2-15 indicates that the frequency spectrum of a PM wave having a maximum phase deviation of x rad is identical in form to the spectrum of a FM wave having a deviation ratio of x rad. With

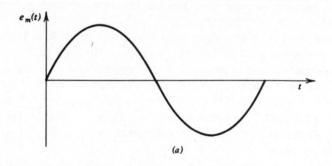

(a)

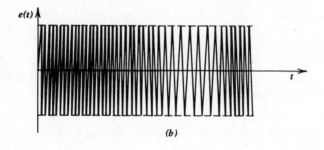

(b)

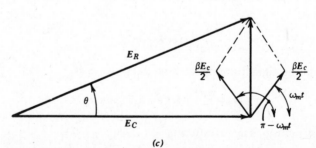

(c)

Figure 2-5. Frequency modulation of a carrier. (a) Modulating signal; (b) modulated carrier; (c) vector representation of FM wave.

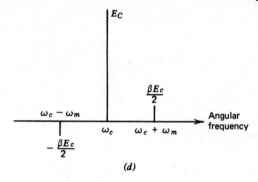

(d)

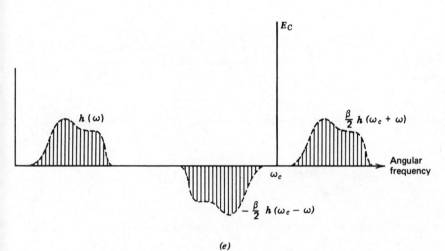

(e)

Figure 2-5. (d) frequency spectrum of FM wave; (e) frequency spectrum of a complex FM wave.

a sinusoidal modulating signal it makes no difference whether one speaks of phase or frequency deviation because the two are related by the rate of modulation as follows:

$$\theta_d = \frac{\Delta f_{\text{peak}}}{f_m} . \qquad (2\text{-}16)$$

A useful way of rating the FM (or PM) of a signal in synthesizer systems is to give the number of decibels by which the level of the one-sided

FM/PM component is below the carrier, or

$$\text{single sideband-to-carrier ratio (dB)} = 10\log_{10}\left(\frac{\text{single sideband power}}{\text{carrier power}}\right).$$

$$(2\text{-}17)$$

A frequency- (or phase-) modulated signal can be expressed by a Bessel function series with argument β, a modulation index:

$$\begin{aligned}
e(t) = E_c \{ & J_0(\beta)\sin\omega_c t + J_1(\beta)\left[\sin(\omega_c + \omega_m)t - \sin(\omega_c - \omega_m)t\right] \\
& + J_2(\beta)\left[\sin(\omega_c + 2\omega_m)t + \sin(\omega_c - 2\omega_m)t\right] \\
& + J_3(\beta)\left[\sin(\omega_c + 3\omega_m)t - \sin(\omega_c - 3\omega_m)t\right] + \cdots \}.
\end{aligned}$$

For a small modulation index, $\beta \ll 1$, $J_0(\beta) \cong 1$, $J_1(\beta) \cong \beta/2$, and $J_2(\beta)$, $J_3(\beta), \ldots, J_n(\beta)$ are approximately zero. Hence the power ratio is as follows:

$$\text{single sideband-to-carrier ratio (dB)} = 10\log_{10}\left[\frac{J_1(\beta)}{J_0(\beta)}\right]^2 = 10\log_{10}\left(\frac{\beta}{2}\right)^2,$$

$$\text{single sideband-to-carrier ratio (dB)} = 10\log\left(\frac{\Delta f_{\text{peak}}}{2f_m}\right)^2,$$

$$\text{single sideband-to-carrier ratio (dB)} = 10\log_{10}\left(\frac{\Delta f_{\text{rms}}}{\sqrt{2}\, f_m}\right)^2. \qquad (2\text{-}18)$$

For convenience, this expression is usually written as

$$\text{single sideband-to-carrier ratio (dB)} = 20\log_{10}\left(\frac{\Delta f_{\text{rms}}}{\sqrt{2}\, f_m}\right). \qquad (2\text{-}19)$$

In terms of the peak phase deviation,

$$\text{single sideband-to-carrier ratio (dB)} = 20\log_{10}\left(\frac{\theta_d}{2}\right). \qquad (2\text{-}20)$$

To illustrate the principles involved, consider the case of a single sideband-to-carrier ratio of -60 dB measured at 100 Hz offset-from-carrier frequency. Using Eq. 2-19, we have

$$-60 = 20 \log_{10} \left(\frac{\Delta f_{rms}}{\sqrt{2} \, f_m} \right)$$

or

$$10^{-3} = \frac{\Delta f_{rms}}{\sqrt{2} \, 100}$$

or

$$\Delta f_{rms} = 0.1414 \text{ Hz.}$$

In terms of peak phase deviation,

$$-60 = 20 \log_{10} \left(\frac{\theta_d}{2} \right),$$

$$10^{-3} = \frac{\theta_d}{2},$$

$$\theta_d = 2 \times 10^{-3} \text{ rad.}$$

The concepts associated with sinusoidal FM and PM waves are utilized in dealing with noise in synthesizer systems. This subject is treated in Section 2-2.

Intermodulation Products in Mixers

Frequency addition or subtraction of two signals is extensively used in frequency generation (see Figs. 1-1 through 1-14 and 1-16 through 1-18). This operation is particularly useful in shifting and expanding the narrow range of a frequency-increment-generation section, such as the one shown in Fig. 1-9, to a required output band of synthesizer frequencies. It is, therefore, of prime importance to the designer of synthesizers to know the frequencies and levels of spurious outputs (intermodulation products) generated in the process of mixing.

The methods of computing spurious outputs generated by a mixer are well known from previous work done in the field of superheterodyne receivers (Refs. 1 to 8). They are briefly described below.

The input-output relationship of a nonlinear device such as a mixer may be expressed by a power series:

$$e_{out} = k_1 e_{in} + k_2 e_{in}^2 + k_3 e_{in}^3 + \cdots + k_n e_{in}^n + \cdots, \qquad (2\text{-}21)$$

where k terms are a function of the nonlinear characteristic of the mixer, e_{in} is the sum of two signals and a dc term:

$$e_{in} = E_0 + A \sin \omega_1 t + B \sin \omega_2 t, \qquad (2\text{-}22)$$

and ω_1, ω_2 are the angular frequencies of the two input signals, respectively. Substituting Eq. 2-22 into Eq. 2-21 and expanding Eq. 2-21, one obtains the frequency components of the resultant output spectrum, Eq. 2-23. For simplicity of presentation, the constants associated with and establishing the level of each term in Eq. 2-23 are omitted. The interested student will find the expressions for these constants in Ref. 2.

$$e_{out} = \text{dc term} + \sin \omega_1 t - \cos 2\omega_1 t + \sin 3\omega_1 t + \left(\begin{array}{c} \text{other harmonic} \\ \text{terms of } \omega_1 \end{array} \right)$$

$$\pm \sin \omega_2 t - \cos 2\omega_2 t \pm \sin 3\omega_2 t + \left(\begin{array}{c} \text{other harmonic} \\ \text{terms of } \omega_2 \end{array} \right)$$

$$\pm \left[\cos(\omega_2 - \omega_1)t - \cos(\omega_2 + \omega_1)t \right]$$

$$\pm \left[\sin(\omega_2 - 2\omega_1)t + \sin(\omega_2 + 2\omega_1)t \right]$$

$$+ \left[\sin(2\omega_2 - \omega_1)t - \sin(2\omega_2 + \omega_1)t \right]$$

$$+ \left[\cos(2\omega_2 - 2\omega_1)t + \cos(2\omega_2 + 2\omega_1)t \right]$$

$$\pm \left[\cos(3\omega_2 - \omega_1)t - \cos(3\omega_2 + \omega_1)t \right]$$

$$\pm \left[\sin(3\omega_2 - 2\omega_1)t + \sin(3\omega_2 + 2\omega_1)t \right] \ldots . \qquad (2\text{-}23)$$

In general, the frequencies of intermodulation products must satisfy the following expression:

$$\omega_{m,n} = m\omega_1 \pm n\omega_2, \qquad (2\text{-}24)$$

where m and n are positive integers, and $\omega_{m,n}$ is the angular frequency of the $m \times n$ intermodulation product. The desired response occurs when $m = n = 1$ and is in the form of $\cos(\omega_2 + \omega_1)$ or $\cos(\omega_2 - \omega_1)$. All other responses are undesirable and must be reduced by filtering to the levels established by system requirements. In most practical cases, however, there are in-band undesirable intermodulation products that cannot be filtered out, and it is the task of the system designer to select mixer input frequencies such that the levels of these products are well within the requirements.

A simple aid often used in system analysis is an intermodulation products chart. There are various forms of this chart (Refs. 5 to 8) serving one main purpose—to guide the designer in the selection of mixer frequencies and in the design of filters following the mixer. Figure 2-6a is an intermodulation products chart that the reader will find very useful in synthesizer work. It is a plot of the mixer output frequency spectrum. Two heavy lines represent sum mixing, $F_2 + F_1$, and difference mixing, $F_2 - F_1$. Figure 2-6a is completely general and gives all intermodulation products up to at least the tenth order ($m + n = 10$) for any two input signals at

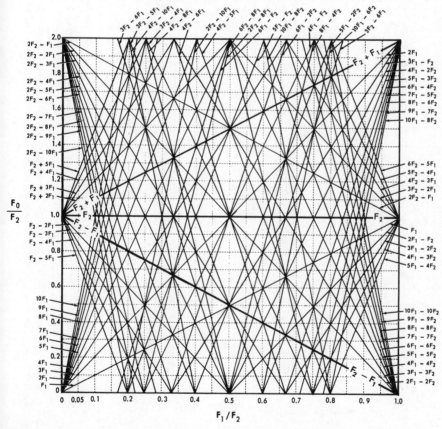

Figure 2-6. (a) Intermodulation products chart. Reprinted by permission, *EDN Magazine*, August 1967.

Figure 2-6 (b). Table of intermodulation product levels for Watkins-Johnson double-balanced mixers. Each cell lists the typical values for models **M1 | M1D | M1E**; the upper line is for the signal at 0 dBm and the lower line is for the signal at −10 dBm.

Harmonics of local oscillator (columns 0–8) × Harmonics of signal (rows 0–7)

Each cell: (top) Signal at 0 dBm — M1, M1D, M1E; (bottom) Signal at −10 dBm — M1, M1D, M1E.

Harmonic of signal	LO 0	LO 1	LO 2	LO 3	LO 4	LO 5	LO 6	LO 7	LO 8
7	79 >99 >99 / >90 >90 >90	69 >99 >99 / >90 >90 >90	80 >99 >99 / >90 >90 >90	74 78 >99 / >90 >90 >90	83 >99 >99 / >90 >90 >90	63 78 >99 / 87 >90 >90	78 >99 >99 / >90 >90 >90	60 81 >99 / >90 >90 >90	71 99 >99 / >90 >90 >90
6	90 >99 >99 / >90 >90 >90	86 >99 >99 / >90 >90 >90	91 >99 >99 / >90 >90 >90	91 >99 97 / >90 >90 >90	90 >99 >99 / >90 >90 >90	84 >99 >90 / >90 >90 >90	93 >99 >99 / >90 >90 >90	84 >99 >99 / >90 >90 >90	88 >99 98 / >90 >90 >90
5	72 93 >99 / >90 >90 >90	70 73 96 / >90 >90 >90	71 87 >99 / >90 >90 >90	52 72 95 / >90 >90 >90	77 88 >99 / >90 >90 >90	46 66 >99 / 68 >90 >90	75 85 >99 / >90 >90 >90	45 64 90 / >90 >90 >90	73 82 >99 / 88 >90 >90
4	80 96 88 / 86 >90 >90	79 80 91 / >90 >90 >90	82 96 >99 / >90 >90 >90	77 80 92 / >90 >90 >90	82 95 90 / >90 >90 >90	76 82 95 / 85 >90 >90	77 98 87 / 86 >90 >90	72 78 94 / 85 >90 >90	77 90 87 / >90 >90 >90
3	51 63 81 / 67 87 >90	49 58 73 / 64 77 >90	53 65 85 / 69 87 >90	51 60 69 / 78 >90 >90	55 65 85 / 77 >90 >90	48 55 68 / 75 >90 >90	54 64 85 / 74 85 >90	53 54 64 / 44 77 89	58 74 87 / 75 88 >90
2	69 68 64 / 73 86 73	72 67 71 / 73 75 83	79 76 62 / 74 84 75	67 67 70 / 75 79 75	75 80 63 / 71 86 80	66 66 70 / 64 74 80	77 82 61 / 69 87 77	68 66 62 / 64 74 82	75 69 64 / 69 84 79
1	25 25 24 / 24 23 24	0 0 0 / 0 0 0	39 39 35 / 35 39 34	13 11 11 / 13 11 11	45 50 42 / 40 46 42	22 16 19 / 24 14 18	54 59 50 / 45 62 49	37 19 39 / 28 19 37	59 59 49 / 49 53 49
0	(shaded)	36 39 29 / 26 27 18	45 42 20 / 35 31 10	52 46 32 / 39 36 23	63 58 24 / 50 47 14	45 37 29 / 41 36 19	60 65 27 / 53 51 17	71 49 30 / 49 37 21	64 75 29 / 51 63 19

Legend (example cell format):

36	39	29
26	27	18

Signal at 0 dBm; local oscillator at +7, +17, and +27 dBm for models WJ—M1, WJ—M1D, and WJ—M1E respectively.

Signal at −10 dBm; local oscillator at +7, +17, and +27 dBm for models WJ—M1, WJ—M1D, and WJ—M1E respectively.

Figure 2-6. (b) Levels of intermodulation products, Watkins-Johnson double-balanced mixers, models **WJ-M1, WJ-M1D, and WJ-M11** Watkins-Johnson Company; reprinted from *Microwaves*, November 1973 : "Selecting Mixers for Best Intermod Performance," p. 49.

frequencies F_1 and F_2 applied to a nonlinear device. The in-band spurious outputs are represented by the products that cross or are tangental to the locus defined by the vertical and horizontal lines corresponding to the normalized input and output frequencies, F_1/F_2 and F_o/F_2, respectively. In cases where F_1 and F_2 are not defined by the system requirement, the designer should choose these frequencies so as to exclude all low-order intermodulation products from the locus of F_1/F_2 and F_o/F_2.

The levels of intermodulation products for a particular mixer may be obtained from the manufacturer of the device. Figure 2-6b provides such information for Watkins-Johnson double-balanced mixers, models WJ-M1, WJ-M1D, and WJ-M1E.

Two examples illustrate the use of the intermodulation products chart.

Example 1. Consider a case of two fixed frequencies, $F_1 = 70\,\mathrm{MHz}$ and $F_2 = 90$ MHz, which are added in a nonlinear device, Fig. 2-7a. The two normalized frequencies are

$$\frac{F_1}{F_2} = \frac{70}{90} \cong 0.778,$$

$$\frac{F_o}{F_2} = \frac{160}{90} \cong 1.778.$$

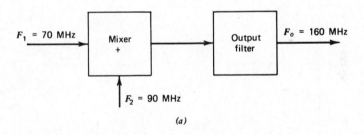

(a)

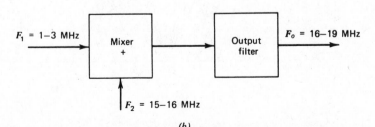

(b)

Figure 2-7. Frequency mixing. (a) Two single frequencies; (b) two variable frequencies.

When vertical and horizontal lines, corresponding to F_1/F_2 and F_o/F_2, respectively, are drawn on the intermodulation products chart, the locus formed is a point on the desired response line, $F_1 + F_2$, Fig. 2-8a. Therefore the in-band intermodulation products are those that cross the locus point. Figure 2-8b, a blowup of the area around the locus point, indicates that there are no such products of low order.

Also of great interest to the designer are out-of-band spurious outputs, which have to be suppressed by the output filter to the level specified in the system requirements. To achieve this, intermodulation products adjacent to the locus point are analyzed. These are tabulated in Table 2-1 for the Watkins-Johnson mixer, model WJ-M1D, operating at a signal level of

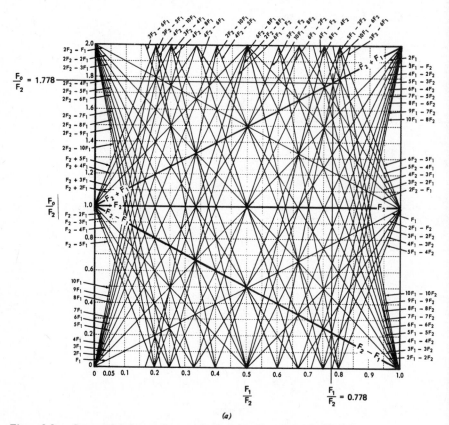

(a)

Figure 2-8a.   Intermodulation products analysis of the mixer shown in Fig. 2-7a.

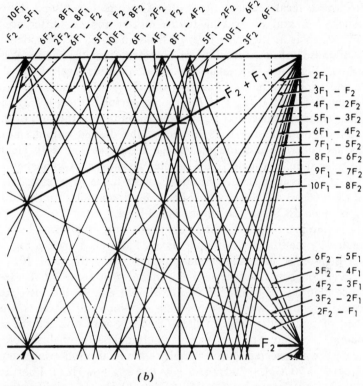

(b)

Figure 2-8b. Blowup of the area around the locus point shown in Fig. 2-8a.

Table 2-1. Levels of In- and Out-of-Band Spurious Outputs for Mixer Input Frequencies of 70 and 90 MHz (Watkins-Johnson Mixer, Model WJ-M1D)

Intermodulation product	Frequency (MHz)	Level below the output (dB)	Attenuation required to satisfy 90 dB specification (dB)
$F_2 - F_1$	20	0	90
F_1	70	-25	65
F_2	90	-39	51
$3F_2 - 6F_1$	150	< -99	0
$5F_2 - 4F_1$	170	-82	8
$5F_1 - 2F_2$	170	-87	3
$8F_1 - 8F_2$	160	< -90	0
$10F_1 - 6F_2$	160	< -90	0
$9F_1 - 9F_2$	180	< -99	0

0 dBm and a local oscillator level of $+17$ dBm. In addition, two input frequencies, F_1 and F_2, leaking to the output and the difference term, $F_2 - F_1$, which is at the same level as the sum term, are considered. The levels of these spurious outputs are obtained from the model WJ-M1D data sheet, Fig. 2-6b. To obtain the lowest possible level of the close-in intermodulation products, $5F_1 - 2F_2$ and $5F_2 - 4F_1$, which coincide in frequency at 170 MHz, F_1 and F_2 are chosen for the low-level and local oscillator signals, respectively. Under these conditions, the sum of the two intermodulation products is 78 dB below the desired output since both signals are coherent and add on voltage basis (see Fig. 2-9). Hence the output filter should provide 12 dB of attenuation at that frequency. In the event that the assignments of F_1 and F_2 were made in reverse, $5F_1 - 2F_2$ would be -66 dB, $5F_2 - 4F_1$ would be -88 dB, and their sum at 170 MHz would be -66 dB below the output signal level.

If a requirement of -90 dB is assumed, only five terms have to be attenuated by the filter following the mixer. These are shown in Fig. 2-10.

Two high-order intermodulation products, $8F_1 - 8F_2$ and $10F_1 - 6F_2$, occurring at the sum frequency are also included in Table 2-1. These terms are more than 90 dB below the desired output and do not interfere with the mixer operation, whether the system of which this mixer is a part is coherent or not. If the order of these products were low and the system incoherent, a severe problem would result unless another set of mixer frequencies was selected. For example, assume the following frequency accuracy: $F_1 = 70$ MHz $- 70$ Hz and $F_2 = 90$ MHz $+ 90$ Hz (± 1 part in 10^6). Then the output frequency $F_o = 160$ MHz $+ 20$ Hz, and the frequency of the first product $8F_2 - 8F_1 = 160$ MHz $+ 1.28$ kHz. The intermodulation product is no longer coincident with the desired output but is 1.26 kHz above it. To attenuate this spurious signal a narrowband phase-locked loop would have to be used. A much less expensive way of eliminating this type of spurious output is to select the mixer frequencies so that all low-order intermodulation products are far from the desired output and can be filtered out by ordinary means.

Example 2.   Consider a case of two signals at variable frequencies, $F_1 = 1$ to 3 MHz and $F_2 = 15$ to 16 MHz, which are added in a nonlinear device, Fig. 2-7b. The limits of the two normalized frequencies are

$$\left(\frac{F_1}{F_2}\right)' = 0.0666 \quad \text{and} \quad \left(\frac{F_1}{F_2}\right)'' = 0.1875,$$

$$\left(\frac{F_o}{F_2}\right)' = 1.068 \quad \text{and} \quad \left(\frac{F_o}{F_2}\right)'' = 1.189.$$

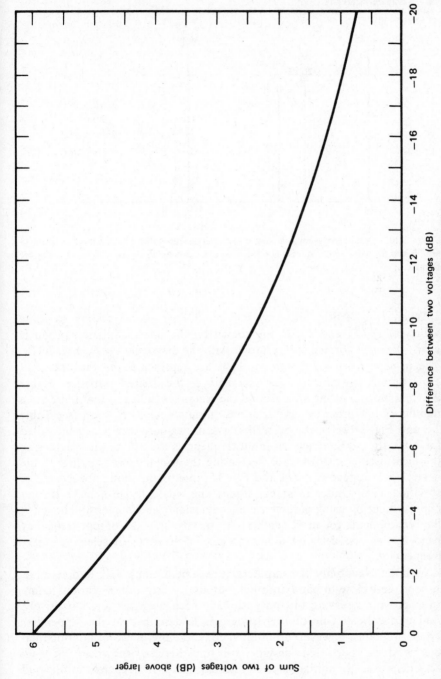

Figure 2-9. Sum of two voltages expressed in decibels. Courtesy of ITT Defense Communication Division.

69

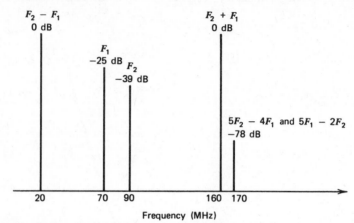

Figure 2-10. Output frequency spectrum of a double-balanced diode mixer; input frequencies are 70 and 90 MHz (only intermodulation products determining the design of the filter are shown).

When two vertical and two horizontal lines corresponding to these values of F_1/F_2 and F_o/F_2 are drawn on the intermodulation products chart, the locus formed is a rectangle with the diagonal $F_1 + F_2$, Fig. 2-11a. The in-band intermodulation products are represented by the lines that cross or are tangential to the rectangle of the desired response. Figure 2-11b, a blowup of the area around the rectangle, indicates that there are a number of low-order in-band spurious outputs, such as $F_2 + 2F_1$ (see Table 2-2 and Fig. 2-12). In the event that a spurious-output requirement called for -90 dB and no choice of input frequencies existed, neither decreasing the local oscillator power nor decreasing the signal power applied to the mixer would reduce $F_2 + 2F_1$ and $F_2 + 3F_1$ intermodulation products to the specified level. Under these conditions one would try to achieve further improvement by using parametric upconversion (see Chapter 6). However, the requirements for most synthesizers specify only the output frequency band, giving the designer a high degree of freedom in selecting mixer frequencies.

With the availability of computers, it became a simple task to determine in-band and close-to-band spurious outputs of any order. Various computer programs serving this purpose have been prepared and extensively used in the field. One of such programs is described in the Appendix. These programs, however, do not invalidate the usefulness of intermodulation products charts. The designer will find the chart shown in Fig. 2-6a very helpful in the initial stage of selecting mixer frequencies, and he will find the computer program very helpful in the verification of final results.

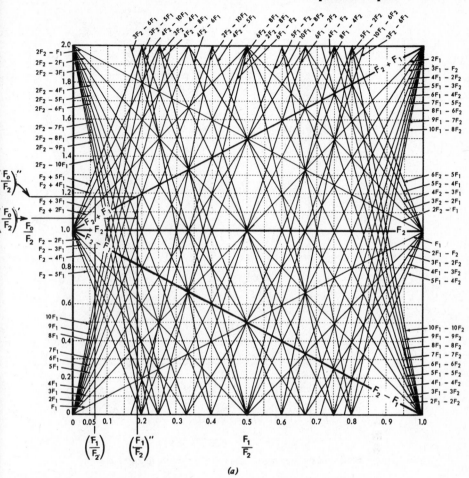

Figure 2-11a. Intermodulation products analysis of the mixer shown in Fig. 2-7b.

Spurious Signals in Frequency Multipliers

A functional block diagram of a frequency multiplier is shown in Fig. 2-13. It consists of a multiplier and a filter that selects the desired output signal. Often a limiter precedes the multiplier. The general equation of the multiplier output is an infinite series (Ref. 9)·

$$e_o'(t) = k_1 e_{in}(t) + k_2 e_{in}^2(t) + \cdots + k_n e_{in}^n(t) + \cdots, \qquad (2\text{-}25)$$

where the nth harmonic term of the series is

$$e_{o,n}'(t) = k_n e_{in}^n(t). \qquad (2\text{-}26)$$

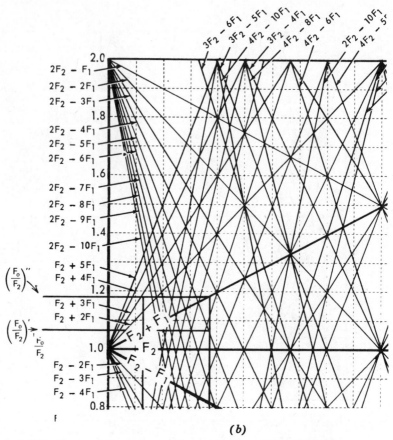

Figure 2-11b.   **Blowup of the area around the rectangle shown in Fig. 2-11a.**

Consider the case of a desired and a spurious signal, $E_1 \cos \omega_1 t$ and $E_2 \cos \omega_2 t$, respectively, applied to the input of such a device, Fig. 2-14a. The signal-to-interference ratio at the input of the multiplier is $20 \log_{10}(E_1/E_2)$. The output of the multiplier is

$$e'_o(t) = k_1(E_1 \sin \omega_1 t + E_2 \sin \omega_2 t)$$

$$+ k_2 \left[\frac{E_1^2 + E_2^2}{2} - \frac{E_1^2}{2} \cos 2\omega_1 t - \frac{E_2^2}{2} \cos 2\omega_2 t \right.$$

$$\left. + E_1 E_2 \cos(\omega_1 + \omega_2)t + E_1 E_2 \cos(\omega_1 - \omega_2)t \right]$$

$+$ third- and higher-order terms.

Table 2-2. Levels of In-band Spurious Outputs for Mixer Input Frequencies of 1 to 3 and 15 to 16 MHz (Watkins-Johnson Mixer, Model WJ-M1D)

Inband intermodulation product	Level below the output (dB)
$F_2 + 2F_1$	-67
$F_2 + 3F_1$	-58
$2F_2 - 5F_1$	-87
$2F_2 - 6F_1$	< -99
$2F_2 - 7F_1$	< -99
$6F_1$	< -99
$7F_1$	< -99

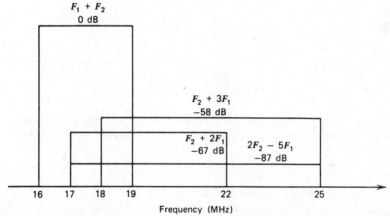

Figure 2-12. Output frequency spectrum of a double-balanced diode mixer; input frequencies are 1 to 3 and 15 to 16 MHz (only low-order intermodulation products are shown).

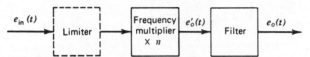

Figure 2-13. Frequency multiplier: functional block diagram.

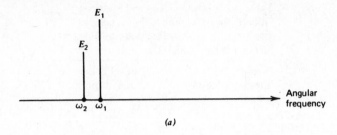

(a)

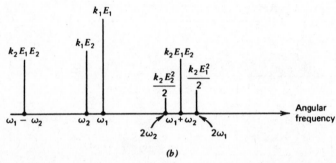

(b)

Figure 2-14. Input and output frequency spectra of a doubler. (a) Input frequency spectrum; (b) output frequency spectrum, only $k_1e_{in}(t)$ and $k_2e_{in}^2(t)$ components of the output signal considered.

For a doubler ($n=2$) the desired output term is $(k_2E_1^2/2)\cos 2\omega_1 t$. If the output term is compared to the sum or the difference term, $E_1E_2\cos(\omega_1 + \omega_2)t$ or $E_1E_2\cos(\omega_1 - \omega_2)t$, respectively (Fig. 2-14b), the signal-to-interference ratio at the output of the multiplier is

$$20\log_{10}\left(\frac{k_2E_1^2/2}{k_2E_1E_2}\right) = 20\log_{10}\left(\frac{E_1}{2E_2}\right).$$

This is a degradation in the signal-to-interference ratio of 6 dB.

For a tripler ($n=3$) the desired output term is $(k_3E_1^3/4)\sin 3\omega_1 t$. When the desired signal is compared to the term

$$\frac{3k_2E_1^2E_2}{4}\sin(2\omega_1 + \omega_2)t,$$

which is a part of the $k_3 e_{in}^3(t)$ term of the series, the signal-to-interference ratio is

$$20\log_{10}\left(\frac{E_1^3/4}{3E_1^2E_2/4}\right)=20\log_{10}\left(\frac{E_1}{3E_2}\right).$$

In general (Ref. 9), as the desired and spurious signals pass a multiplier, the degradation of signal-to-interference ratio is equal to the multiplication factor n or $20\log_{10}n$ dB and is a result of the multiplier being more efficient as a mixer than as a frequency multiplier.

Of special interest to the designer of synthesizers are spurious outputs that fall in or close to the band of the output filter and cannot be attenuated by the filter to the degree specified in the system requirements. For a doubler, if $|\omega_1-\omega_2|\leqslant\frac{1}{2}$ (3dB bandwidth), the $(k_2E_2^2/2)\cos 2\omega_2t$ and $k_2E_1E_2\cos(\omega_1+\omega_2)t$ terms will be in band and of relatively high level. For example, assume that the desired signal at $f_1=5.000,000$ MHz and a spurious signal at $f_2=4.999,999$ MHz are applied to the input of a doubler and that the level of the spurious signal is 80 dB below the desired signal. This situation occurs in synthesizers that operate on an external reference source with an internal reference source still energized but disconnected from the system by a rf switch. Then the spectrum of the doubler input and output, after the unwanted spurious outputs are filtered out, is as shown in Fig. 2-15. At the input of the doubler the signal-to-interference ratio is 80 dB, because of 80 dB isolation provided by the rf switch. The signal-to-interference ratio at the output of the doubler is degraded by 6 dB ($n=2$) and is 74 dB. If further frequency multiplication of the signal is expected, greater degradation of the signal-to-interference ratio will result.

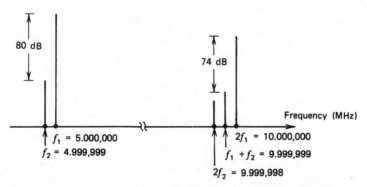

Figure 2-15.   Filtered-out frequency spectra of a doubler before and after multiplication.

When a FM wave

$$e_{in}(t) = E_c \cos(\omega_c t + \beta \sin \omega_m t)$$

is applied to a frequency multiplier, the levels of FM sidebands are enhanced by the multiplication factor so that the nth harmonic term, $e_{o,n}(t)$, is

$$e_{o,n}(t) = k_n E_c^n \cos(n\omega_c t + n\beta \sin \omega_m t).$$

The frequency of the modulating signal, ω_m, is not altered in the multiplication process, so that the frequency spacing between the carrier and FM sidebands is preserved. The level of the sidebands relative to the desired output, however, is enhanced by the multiplication factor, n, or by $20\log_{10} n$ dB.

When an AM wave

$$e_{in}(t) = E_c(1 + m_a \cos \omega_m t)\cos \omega_c t$$

is applied to a frequency multiplier input, AM-to-PM conversion may occur with AM removed by the limiter if one is utilized. The subject of AM-to-PM conversion is discussed on p. 83 of this chapter.

Spurious Signals in Frequency Dividers

The subject of spurious signals generated in the process of frequency division is treated in conjunction with a description of various divider circuits in Chapter 6 because of the dependence of this type of spurious outputs on a particular circuit configuration.

This section of the text deals with leakage of spurious signals, present at the input, to the output of a divider. What happens under these conditions is best illustrated by a practical example.

Consider a section of a frequency synthesizer, shown in Fig. 2-16. It consists of a vhf digital phase-locked loop (PLL) operating at 93 to 112 MHz. The output signal at f_o is down converted to 75 to 93 MHz by an injection signal at f_i, which is varied over the 18 to 19 MHz frequency range in 10 Hz increments, and divided down to 500 kHz. Phase comparison is accomplished at 500 kHz in a sampling-type phase comparator.

The principle of operation of the digital PLL, described in Chapter 5, is not considered here because it is irrelevant to the discussion. It suffices to say that, when the loop is locked, the frequency of both signals at the input of the phase comparator is 500 kHz and that, as the injection frequency, f_i, is varied from 18 to 19 MHz, the loop follows this secondary reference without breaking lock.

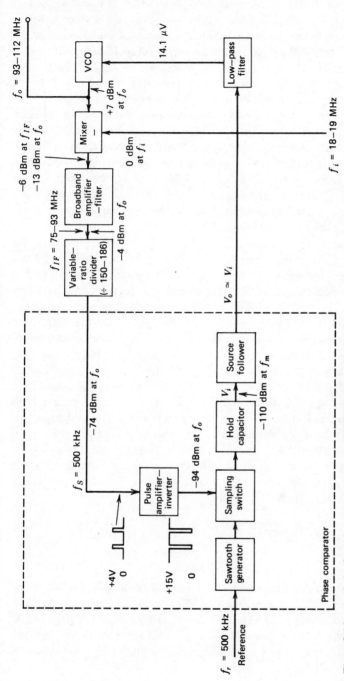

Figure 2-16. Analysis of spurious outputs associated with frequency dividers.

77

Assume that the voltage-controlled oscillator (VCO) is tuned to lock at 93 MHz and that the division ratio of the variable-ratio divider is set to 150. The frequency of the desired signal at the output of the mixer after the loop locks is 75 MHz if the injection frequency is set to 18 MHz. The input levels of the signals at f_o and f_i into the mixer are +7 dBm and 0 dBm respectively, and a 20 dB balancing of f_o is provided by the mixer; hence the level of the signal at f_o appearing at the output of the mixer is −13 dBm. Since this signal is an in-band spurious signal, it is not attenuated by the filter following the mixer. Instead, it is amplified to −4 dBm by the broadband amplifier. An isolation of 70 dB and an attenuation of 20 dB, provided by the variable-ratio divider and the pulse amplifier-inverter, respectively, reduce the signal at f_o to −94 dBm at the input to the sampling switch. This spurious signal is present simultaneously with a 500 kHz pulse, which is used to sample a sawtooth reference. At these exact frequencies no problem is discovered.

Assume now that f_i is changed to 18.01 MHz. The frequency of the output signal at f_o, after this change without breaking lock, becomes 93.01 MHz. Hence the two signals at the input of the sampling switch are 500 kHz and its harmonics and 93.01 MHz. A semiconductor switch is effectively a mixer. The low-level signal at 93.01 MHz mixes with the 186th harmonic of 500 kHz, generating a 10 kHz beat note, the difference product, at the output of the switch. The sum and higher-order intermodulation products are generated also, but they are attenuated by the low-pass filter and appear at the control input to the VCO at a significantly lower level than the beat note at the difference frequency. A conversion loss of 16 dB is assumed for the switch. It is relatively high because both mixing signals are at low power levels. The reactance of the hold capacitor is high at 10 kHz, so that a signal at 10 kHz, whose average power is −110 dBm, appears unattenuated across the input impedance of the source follower. This impedance is assumed to be 20 kΩ to permit one to compute the voltage into (and also out of) the source follower, V_i. The input power is −110 dBm or 10^{-11} mW. Therefore

$$V_i = \sqrt{PR} = \sqrt{10^{-14} \times 2 \times 10^4} \simeq 14.1 \ \mu V \ \text{rms},$$
$$V_o \simeq V_i \simeq 14.1 \ \mu V \ \text{rms}.$$

The loop reacts to this beat note as if it entered the loop at the reference input. Hence the PLL acts as a low-pass filter with respect to it. Assume that the loop bandwidth is 10 kHz and the VCO gain constant $K_{VCO} = 500$ kHz/V. Then the same voltage of 14.1 μV rms appears at the control input of the VCO, unaltered by the loop, and results in Δf_{rms} deviation of

VCO average frequency, where in this particular case

$$\Delta f_{\text{rms}} \cong V_o K_{\text{VCO}} \cong 14.1 \times 10^{-6} \times 500 \times 10^3 \cong 7.1 \text{ Hz rms.}$$

Equation 2-19 is

$$\text{single sideband-to-carrier ratio (dB)} = 20 \log_{10}\left(\frac{\Delta f_{\text{rms}}}{\sqrt{2} \, f_m}\right).$$

Substituting 7.1 Hz for Δf_{rms} and 10 kHz for f_m, one obtains

$$\text{single sideband-to-carrier-ratio} = 20 \log\left(\frac{7.1}{1.414 \times 10^4}\right) \cong -66 \text{ dB.}$$

In view of the fact that the system requirement for spurious outputs is likely to be -80 dB (or more stringent), this is, indeed, a very high level of FM sidebands.

The problem described above can be solved by inserting a rf low-pass filter between the pulse amplifier-inverter and the variable-ratio divider. However, if the problem also resulted from inadequate shielding (see Chapter 3 for rf shielding practices), it would be difficult to identify and eliminate it, and probably a mechanical redesign of the system would be required. This example illustrates a major group of problems associated with frequency dividers (and other circuits as well) and effectively demonstrates the importance of the phase in system analysis during which levels of spurious signals at the output of each building block in frequency synthesis are estimated, based on the designer's knowledge of circuitry, and the effect of these signals on the overall performance of the synthesizer is established.

When a FM wave is applied to a frequency divider, the level of FM sidebands in relation to the desired signal is reduced by the division ratio N or by $20 \log_{10}(1/N)$ dB.

Spurious Signals in an Ideal Limiter

A limiter is a device that truncates the amplitude of a signal at a given value. In terms of the input-output voltage response, an ideal limiter is approached when parameter ε in Fig. 2-17 approaches zero and the actual response curve approaches its asymptotes.

When a sinusoidal wave

$$e_{\text{in}}(t) = E \sin \omega_1 t$$

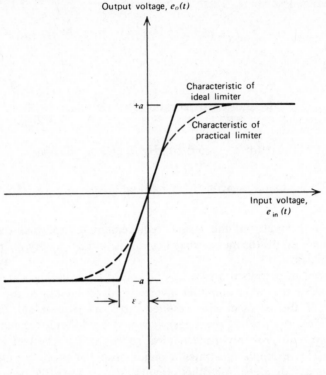

Figure 2-17. Output voltage versus input voltage characteristic of a limiter. Courtesy of the Naval Research Laboratory.

is applied to the input of such a device, the output of the limiter is a series expansion corresponding to the harmonic content of a square wave (Ref. 10), as follows:

$$e_o(t) = \frac{4a}{\pi} \sum_{m=1}^{\infty} \frac{1}{m} \sin\left(\frac{m\pi}{2}\right) \cos(m\omega_1 t). \qquad (2\text{-}27)$$

The relative power content of each term in the series, based on a clipping level of unity, versus frequency is plotted in Fig. 2-18. As was expected, the even-order harmonics are absent from the spectrum of the output wave.

When the input signal of a limiter consists of two sinusoidal waves of equal amplitude and at different frequencies

$$e_{in}(t) = E\left(\cos\omega_1 t + \cos\omega_2 t\right),$$

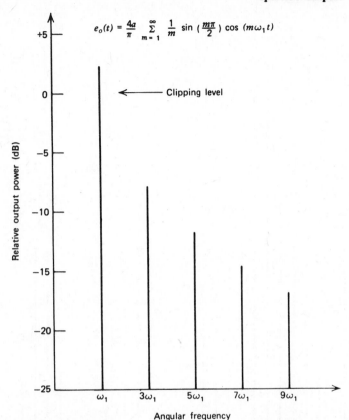

Figure 2-18. Output power spectrum of ideal limiter, single-tone sinusoidal input (for $a=1$). Data from S. F. George and J. W. Wood, *Ideal Limiting*, Part 1 (Washington, D.C.: U.S. Naval Research Laboratory, AD 266069, October 2, 1961).

the output of the limiter is (Ref. 10)

$$e_o(t) = \frac{a}{\pi^2} \sum_{m=0}^{\infty} \sum_{n=0}^{\infty} \varepsilon_m \varepsilon_n \left[1 - (-1)^{m+n} \right] \frac{1}{m^2 - n^2} \sin\left[\frac{(m+n)\pi}{2} \right]$$

$$\times \sin\left[\frac{(m-n)\pi}{2} \right] \left[\cos(m\omega_1 + n\omega_2)t + \cos(m\omega_1 - n\omega_2)t \right], \quad (2\text{-}28)$$

where ε_m = the Neumann factor ($\varepsilon_0 = 1, \varepsilon_m = 2$ for $m > 0$).

Terms of this series exist only when $m + n$ is an odd integer. The power spectrum of $e_o(t)$, Eq. 2-28, for $f_1 = 1$ kHz, $f_2 = 1.0001$ kHz, $a = 1$, and $m, n = 0$ to 5 is shown in Fig. 2-19. Of importance to the designer of

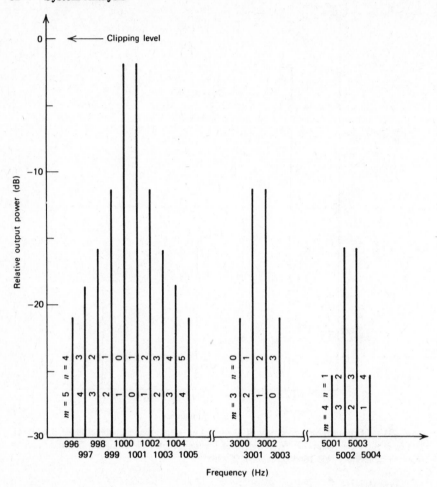

Figure 2-19. Output power spectrum of ideal limiter; input signal is two sinusoids of equal amplitude. Data from S. F. George and J. W. Wood, *Ideal Limiting*, Part 1 (Washington, D.C.: U.S. Naval Research Laboratory, AD 266069, October 2, 1961).

synthesizers is the case in which the frequency spacing between two signals is small, so that some of the spurious signals generated in the limiter with relatively high levels fall in band and cannot be filtered out.

In synthesizer work the presence of two signals of equal amplitude is rare. Usually, the desired signal is higher than all spurious signals. This situation is discussed next.

When the amplitudes of two sinusoidal signals, E_1 and E_2, are not equal to each other, the output of the limiter is (Ref. 10)

$$e_o(t) = \frac{a}{4\pi} \sum_{m=0}^{\infty} \sum_{n=0}^{\infty} \varepsilon_m \varepsilon_n [1 - (-1)^{m+n}] \left\{ \frac{(E_2/E_1)^n \Gamma[(m+n)/2]}{\Gamma(n+1)\Gamma[1+(m-n)/2]} \right\}$$

$$\times \sin\left[\frac{(m+n)\pi}{2}\right]_2F_1\left[\frac{m+n}{2}; \frac{n-m}{2}; n+1; \left(\frac{E_2}{E_1}\right)^2\right]$$

$$\times [\cos(m\omega_1 + n\omega_2)t + \cos(m\omega_1 - n\omega_2)t], \qquad (2\text{-}29)$$

where

$$_2F_1(a; b; c; z) = 1 + \frac{ab}{c}\frac{z}{1!} + \frac{a(a+1)b(b+1)}{c(c+1)}\frac{z^2}{2!} + \cdots$$

= the confluent hypergeometric function

and

$$\Gamma(x) = \int_0^{\infty} e^{-t}t^{x-1}\,dt$$

$$=\ \text{the gamma function for real values}$$
of x greater than zero (Ref. 10a).

Notice that, except for the coefficients of various terms, Eq. 2-29 consists of the same combinations of $\omega_{m,n} = m\omega_1 \pm n\omega_2$ terms as was the case with a frequency mixer, Eq. 2-23. Hence an ideal limiter is effectively a mixer.

The power spectrum of $e_o(t)$, Eq. 2-29, for $f_1 = 1.000$ kHz, $f_2 = 1.001$ kHz, $E_1 = 2E_2$, $a = 1$, and a partial set of values of m and n is shown in Fig. 2-20. In addition to a small number of spurious outputs (as compared to the case when $E_1 = E_2 = E$), a strong attenuation effect of the ideal limiter on the weaker of two signals should be noticed.

Amplitude-Modulation-to-Phase-Modulation Conversion

Conversion from AM to PM occurs in circuits whose phase characteristic (or delay) depends on the instantaneous amplitude of the input signal. Such circuits as limiters, mixers, voltage-tuned filters, and varactor multipliers convert envelope variations at their input to phase variations at their output. If small envelope perturbations are assumed, these circuits can be

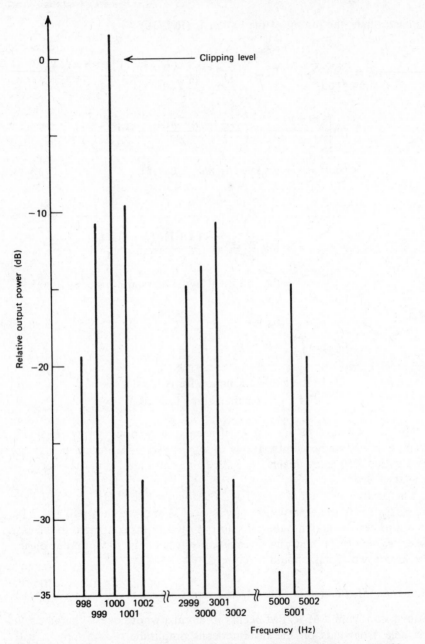

Figure 2-20. Output power spectrum of ideal limiter; input signal is two sinusoids, and amplitude ratio is 2 to 1. Data from S. F. George and J. W. Wood, *Ideal Limiting*, Part 1 (Washington, D.C.: U.S. Naval Research Laboratory, AD 266069, October 2, 1961).

characterized by a device with a conversion constant of K degrees/dB, Fig. 2-21a. As an AM wave passes the converter, K degrees of peak change in the phase of the output signal are produced for every decibel of variation in the envelope of the input signal. This process can be expressed mathematically in the following manner.

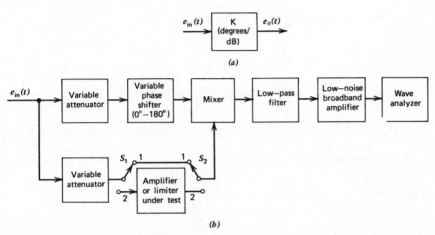

(a)

(b)

Figure 2-21. AM-to-PM conversion. (a) AM-to-PM converter: functional block diagram; (b) test setup for measuring K in amplifiers and limiters.

Let $e_{in}(t)$ be the input voltage to an AM/PM converter, where

$$e_{in}(t) = a(t)E_c \cos \omega_c t$$

and where

$$a(t) = \text{an amplitude modulation function; } a(t) \ll 1.$$

Then the output voltage is

$$e_o(t) = a_1(t)E_c \cos[\omega_c t + ka(t)], \qquad (2\text{-}30)$$

where $a_1(t) = a(t)$, in general;
 $\quad k = $ AM-to-PM conversion coefficient: phase modulation index (rad) divided by amplitude modulation index;
 $ka(t) = $ phase distortion function due to the envelope variations.

One can express k as a function of K by converting radians to degrees and fractional changes to decibels with the following final result (Ref. 11):

$$k = 0.1516K \text{ rad.} \qquad (2\text{-}31)$$

A special case of the $a(t)$ function is a sinusoid. Under these conditions the signal applied to the AM-to-PM converter is (Eq. 2-7)

$$e_{in}(t) = E_c (1 + m_a \cos \omega_m t) \cos \omega_c t,$$

where

$$m_a = \text{modulation index} \left(= \frac{k_a E_m}{E_c} \ll 1 \right).$$

The output of the converter is amplitude and phase modulated (Ref. 12) as

$$e_o(t) = A E_c [1 + m_a (1 - c) \cos \omega_m t] \cos(\omega_c t + k m_a \cos \omega_m t), \qquad (2\text{-}32)$$

where $A = $ gain of AM-to-PM converter,
$c = $ compression factor.

In general, k may be a function of a number of quantities. It may depend on the input signal power, frequency, bias levels, and so forth. However, for small envelope fluctuations in the input signal, k is essentially a constant and may be measured utilizing techniques described at the end of this section.

As an example, consider a diode mixer. The wave applied to the signal port of the mixer is

$$e_s(t) = E_1 \sin \omega_1 t,$$

and the local oscillator wave, amplitude modulated, is

$$e_{LO}(t) = E_2(1 + m_a \sin \omega_m t) \sin \omega_2 t.$$

The sum and the difference terms at the output of the mixer, including the amplitude constants, are

$$e_o(t)_{\pm} = E_3 [1 + m_a (1 - c) \cos \omega_m t] \cos[(\omega_2 - \omega_1) t + k m_a \cos \omega_m t]$$
$$+ E_3 [1 + m_a (1 - c) \cos \omega_m t] \cos[(\omega_2 + \omega_1) t + k m_a \cos \omega_m t].$$

The local oscillator signal switches the mixer diodes on and off. The mixer operates as a limiter with respect to the local oscillator signal and the amplitude modulation is practically removed at the output of the mixer ($c \cong 1$), so that the sum and the difference terms are approximately

$$e_o(t)_{\pm} \cong E_3 \cos\left[(\omega_2 - \omega_1)t + km_a \cos\omega_m t\right]$$
$$+ E_3 \cos\left[(\omega_2 + \omega_1)t + km_a \cos\omega_m t\right].$$

Conversion from AM to PM occurs also in circuits utilizing voltage-variable reactance such as a varicap (in automatically tuned filters) or varactor (in multipliers). A noise voltage modulates the variable reactance of the tuned circuit, varying the resonant frequency as shown in Fig. 2-22. This causes phase modulation in accordance with the variations in noise voltage.

As a numerical example, consider a single-pole, parallel-tuned filter, Fig. 2-23. In this figure

$C =$ voltage variable capacitance (varicap),

$L =$ inductance of the tuned circuit,

$f_0 =$ resonant frequency of the tuned circuit:

$$f_0 = \frac{1}{2\pi\sqrt{LC}}, \tag{2-33}$$

$V =$ dc bias of the varicap,

$C' =$ rf bypass capacitance,

$e(t) =$ rf voltage across the tuned circuit.

Assume that

$$e(t) = E_c(1 + m_a \cos\omega_m t)\cos\omega_c t,$$

where

$$\omega_c = 2\pi f_c, \qquad f_c = f_0.$$

An approximate equation for the varicap capacitance over a small range of bias voltages is

$$C(V) = C_1\sqrt{\frac{V_1}{V}}, \tag{2-34}$$

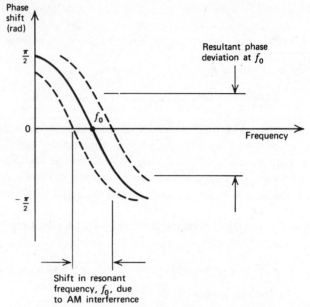

Figure 2-22. AM-to-PM conversion in electronically tuned filters and varactor frequency multipliers.

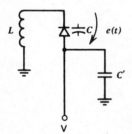

Figure 2-23. Parallel-tuned circuit utilizing voltage-variable capacitance.

where C_1 = capacitance of the varicap, specified at the bias voltage V_1. Differentiating $C(V)$ with respect to V, one obtains

$$\Delta C \cong -\frac{C_1}{2}\sqrt{\frac{V_1}{V^3}} \,\Delta V \qquad \text{for small } \Delta V. \qquad (2\text{-}35)$$

Next consider Eq. 2-33. Expressing C in terms of f_0 and differentiating

$C(f_0)$ with respect to f_0 gives

$$C(f_0) = \frac{1}{(2\pi)^2 L}\left(\frac{1}{f_0^2}\right),$$

$$\Delta C \cong -\frac{5.07 \times 10^{-2}}{Lf_0^3}\Delta f_0 \qquad \text{for small } \Delta f_0. \tag{2-36}$$

Combining Eq. 2-35 with Eq. 2-36 by eliminating ΔC, one arrives at

$$\Delta f_0 \cong 10 L f_0^3 C_1 \sqrt{\frac{V_1}{V^3}}\ \Delta V, \tag{2-37}$$

where $\Delta f_0 =$ the shift in the resonant frequency of the tuned circuit, f_0, due to an amplitude variation of the rf voltage applied to the filter, ΔV.

The phase shift of the signal at frequency f relative to the phase shift at f_0, $\Delta\phi$, in a single-pole resonant circuit is (Ref. 13)

$$\Delta\phi = \tan^{-1}\left(\frac{-2Q\Delta f_0}{f_0}\right),$$

where $\Delta f_0 = f - f_0$,

$Q =$ quality factor of the tuned circuit,

and, for $\Delta\phi \ll 1$,

$$\Delta\phi \cong -\frac{2Q\Delta f_0}{f_0}\ \text{rad},$$

$$\Delta f_0 \cong \frac{-f_0}{2Q}\Delta\phi\ \text{Hz}. \tag{2-38}$$

When Eq. 2-37 is combined with Eq. 2-38 by eliminating Δf_0, a relationship is obtained connecting the phase shift of the output signal of a single-pole resonant circuit utilizing a voltage-variable capacitance to amplitude variations in the input signal:

$$\Delta\phi \cong -20 Q L f_0^2 C_1 \sqrt{\frac{V_1}{V^3}}\ \Delta V\ \text{rad} \qquad \text{for small } \Delta V \tag{2-39}$$

Let

$$E_c = 200 \text{ mV peak,}$$
$$f_c = f_0 = 200 \text{ MHz,}$$

3 dB bandwidth; $BW_{3dB} = 2.5 \text{ MHz}$

$$Q = \frac{f_0}{BW_{3dB}} = 80,$$

varicap: SQ1734, MSI; $C_1 = 22 \text{ pF}$ at $V_1 = -4V$ dc,
varicap operating bias, $V = -5 \text{ V}$ dc.

Then

$$C \cong 22\sqrt{\tfrac{4}{5}} = 19.65 \text{ pF,}$$

$$L = \frac{1}{(2\pi f_0)^2 C} = 0.0323 \ \mu H.$$

Let

$$f_m = \frac{\omega_m}{2\pi} = 60 \text{ Hz,}$$

$$2k_a E_m = E_{c,\max} - E_{c,\min} = E_c - E_{c,\min},$$

$$k_a E_m = \frac{E_c - E_{c,\min}}{2}.$$

For a 1 dB variation in E_c,

$$20\log_{10}\left(\frac{E_c}{E_{c,\min}}\right) = 1 \text{ dB,}$$

$$\frac{E_c}{E_{c,\min}} = 1.122,$$

$$E_{c,\min} = 178.26 \text{ mV peak.}$$

Hence

$$k_a E_m = 10.87 \text{ mV peak.}$$

And in terms of the single sideband-to-carrier ratio:

$$\text{single sideband-to-carrier ratio (dB)} = 20\log_{10}\left(\frac{m_a}{2}\right).$$

But in Eq. 2-6

$$m_a = \frac{k_a E_m}{E_c}.$$

Hence

$$\text{single sideband-to-carrier ratio (dB)} = 20 \log_{10} \left(\frac{k_a E_m}{2 E_c} \right)$$

$$= 20 \log \left(\frac{10.87}{2 \times 200} \right) = -31.3,$$

This is a very high level of 60 Hz sidebands. One would use such levels mainly in measurements of the conversion constant K (degrees/dB). Using a -60 dB single sideband-to-carrier ratio, a requirement for many communications systems, one proceeds as follows:

$$20 \log_{10} \left(\frac{m_a}{2} \right) = -60 \text{ dB},$$

$$k_a E_m = \Delta V = 400 \ \mu V \text{ peak}.$$

This voltage causes the varicap bias to change by the same amount at a 60 Hz rate. Substituting the 400 μV peak for ΔV in Eq. 2-39, one arrives at

$$\Delta \phi \cong -20 \times 80 \times 0.0323 \times 10^{-6} (200 \times 10^6)^2 \times 22 \times 10^{-12} \sqrt{\frac{4}{(5)^3}} \times 400 \times 10^{-6}$$

$$\cong -3.26 \times 10^{-3} \text{ rad}.$$

In this case $\Delta \phi$ is equal to θ_d, the total peak deviation due to -60 dB AM sidebands. The single sideband-phase-interference-to-carrier ratio at 60 Hz offset-from-carrier frequency, Eq. 2-20, is

$$\text{single sideband-phase-interference-to-carrier ratio (dB)} = 20 \log_{10} \left(\frac{-3.26 \times 10^{-3}}{2} \right)$$

$$= -67.8.$$

The equivalent frequency deviation, Eq. 2-16, is

$$\Delta f_{\text{peak}} = 3.26 \times 10^{-3} \times 60 = 0.1895 \text{ Hz}.$$

Figure 2-21b is a block diagram of the test setup used to measure K in degrees per decibel in amplifiers and limiters. A rf signal is applied through a 90 degree phase shifter to the local oscillator port of a mixer. This signal is also applied to the switches, S_1 and S_2, which initially are in position 1. After the rf voltages at both inputs of the mixer are set to the values recommended by the mixer manufacturer, the phase shifter is tuned to give a dc null at the output of the mixer. Under these conditions the two input signals are in phase quadrature with respect to each other, and the mixer operates in a phase-detector mode so that the output dc voltage is proportional to the phase difference between the two mixer input signals. The low-pass filter attenuates the rf signal leaking to the output of the mixer, preventing saturation of the low-noise broadband amplifier.

To measure K, the switches S_1 and S_2 are set in position 2, the rf voltage levels at the mixer are adjusted to the same values as were used during the nulling operation, the rf signal is amplitude modulated by 1 dB at the rate of interest, f_m, and the signal at f_m at the output of the mixer is measured with a wave analyzer (a frequency-selective voltmeter). Utilizing the phase-detector gain constant in volts per degree, obtained during the calibration of the test equipment, one converts volts into equivalent degrees of phase shift produced in the device under test because of a 1 dB amplitude modulation of the rf signal applied to the device.

Calibration of the test setup is achieved after the nulling operation, with the switches S_1 and S_2 still in position 1, by varying the phase shift of the signal applied to the local oscillator port of the mixer in known phase increments and by recording the dc voltage at the output of the mixer. These data permit one to plot a static characteristic of the mixer operating in a phase-detector mode such as the characteristic shown in Fig. 2-24. The

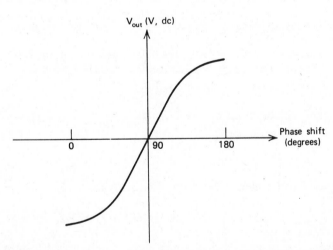

Figure 2-24. Static characteristic of a double-balanced mixer operating in a phase-detector mode.

phase-detector gain constant in volts per degree is obtained from this plot. If the mixer is broadband, the dynamic characteristic will not vary by any significant amount from the static characteristic of the mixer and can be used in dynamic measurements. A way of obtaining the dynamic characteristic of a phase detector is described on p. 132.

Resolution of a Single Spurious Signal into AM and FM Components

A combination of a signal and a single spurious output is commonly encountered in frequency synthesis. This form of interference deserves thorough study not only because it results in spurious outputs at the final synthesizer output but also because it can be transformed into a narrow-band FM somewhere along the path of synthesis (usually, in a mixer). To see this, consider first a single pair of unsymmetrical sidebands,

$$B_1 \cos\left[(\omega_c + \omega_m)t + \theta_1. \right]$$

and

$$B_2 \cos\left[(\omega_c - \omega_m)t - \theta_2 \right],$$

separated from the carrier

$$A \cos \omega_c t$$

by $+\omega_m$ and $-\omega_m$, respectively, Fig. 2-25a. It was shown in Ref. 14, pp. 172–175, that this pair of sidebands can be resolved into a pair of symmetrical and a pair of antisymmetrical sidebands, Fig. 2-25b and c. The amplitude of each symmetrical sideband is

$$A_s = \tfrac{1}{2}\sqrt{B_2^2 + B_1^2 + 2B_1 B_2 \cos(\theta_2 - \theta_1)} \ . \qquad (2\text{-}40)$$

and its phase is

$$\phi_s = \tan^{-1}\left(\frac{B_2 \sin\theta_2 + B_1 \sin\theta_1}{B_2 \cos\theta_2 + B_1 \cos\theta_1} \right). \qquad (2\text{-}41)$$

The amplitude of each antisymmetrical sideband is

$$A_a = \tfrac{1}{2}\sqrt{B_2^2 + B_1^2 - 2B_1 B_2 \cos(\theta_2 - \theta_1)} \ , \qquad (2\text{-}42)$$

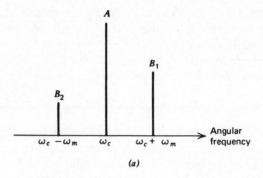

(a)

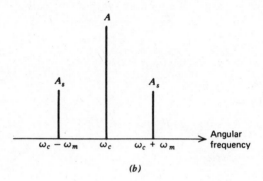

(b)

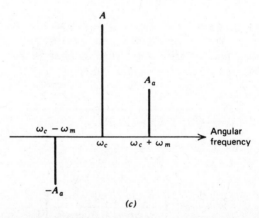

(c)

Figure 2-25. Two unsymmetrical sidebands, resolution into symmetrical and antisymmetrical components. (*a*) Unsymmetrical sidebands and carrier; (*b*) symmetrical sidebands; (*c*) antisymmetrical sidebands.

and its phase is

$$\phi_a = \tan^{-1} \frac{B_1 \sin\theta_1 - B_2 \sin\theta_2}{B_1 \cos\theta_1 - B_2 \cos\theta_2} . \tag{2-43}$$

Consider the case of a desired signal and a single spurious signal, Fig. 2-26a. Let the desired signal be

$$A \cos\omega_c t$$

and the spurious signal be

$$B \cos\left[(\omega_c + \omega_m)t + \psi\right].$$

To find the symmetrical and antisymmetrical sidebands of this configuration, let

$$B_1 = B, \quad \theta_1 = \psi,$$
$$B_2 = 0, \quad \theta_2 = 0.$$

Then from Eqs. 2-40 and 2-41 one has

$$A_s = \frac{B}{2} \quad \text{and} \quad \phi_s = \tan^{-1}\left(\frac{\sin\theta_1}{\cos\theta_1}\right) = \psi,$$

and from Eqs. 2-42 and 2-43

$$A_a = \frac{B}{2} \quad \text{and} \quad \phi_a = \tan^{-1}\left(-\frac{\sin\theta_1}{\cos\theta_1}\right) = \psi.$$

The symmetrical pair of sidebands is equivalent to a component in phase with the carrier, which gives an amplitude-modulated signal with a degree of modulation equal to B/A, Fig. 2-26b. The antisymmetrical pair of sidebands is equivalent to a component in quadrature with the carrier, which for $B \ll A$ gives a frequency-modulated wave with a modulation index equal to B/A and a peak frequency swing of $(B/A)f_m$, where $f_m = \omega_m/2\pi$, Fig. 2-26c. The general expression for the distorted signal is

$$e(t) = A\sqrt{1 + 2(B^2/A^2) - 2(B^2/A^2)\cos\left[2(\omega_m t + \psi - \theta)\right]} \ \cos(\omega_c t + \theta + \varphi),$$

$$\tag{2-44}$$

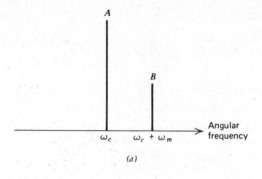

(a)

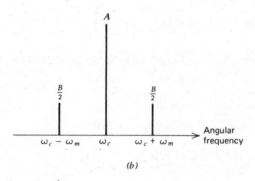

(b)

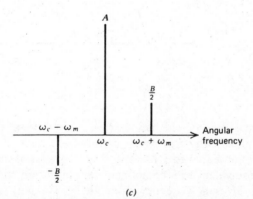

(c)

Figure 2-26. Signal and single-sided spurious output, resolution into AM and FM components. (a) Signal and single-sided spurious outpout; (b) AM components; (c) FM components.

where the modulation angle, φ, is

$$\varphi = \tan^{-1}\left[2\frac{B}{A}\sin(\omega_m t + \psi + \theta)\right].$$ (2-45)

The sidebands at $\omega_c - \omega_m$ in all these cases are a mathematical fiction because there is actually no signal energy at the frequency $\omega_c - \omega_m$; however, conversion from the signal and single-sided spurious output pair to a FM wave takes place in a nonlinear device, for example, when this pair is applied to the local oscillator input of a mixer or to an amplitude limiter (Ref. 15, pp. 470–473). In either case amplitude limiting removes the AM components from the signal, leaving the FM components unaltered. In the presence of stringent FM requirements it is necessary, therefore, to consider such spurious outputs as a possible source of FM in the system analysis of a frequency synthesis approach.

Spurious Outputs at the Power Line Frequency

By "spurious output at the power line frequency" is meant an undesirable amplitude or frequency modulation occurring at a 60 Hz rate, or at a harmonic of 60 Hz, which is caused by an ac ripple at the output of the synthesizer power supply or by radiation from a power line source such as a transformer. The effect of these sources on synthesizer performance is difficult to evaluate theoretically unless a synthesis approach is chosen and the sensitivity of critical circuitry to modulation at the power line frequency is known. By critical circuitry is meant the sections of a synthesizer system that are assigned the most stringent spurious-output requirement. Usually, some testing is performed before system design is completed to estimate the levels of tolerable power supply ripple, and precautions are taken, such as locating the power supply transformer away from critical circuits and enclosing the transformer in a Mu-metal can, to reduce the level of radiation. What is usually done during the system design phase is to assign AM and FM requirements to various sections of a selected synthesis approach and to identify critical circuits. Having this information at his disposal, the designer determines whether or not the specified performance can be realized by this approach in a practical way. An example illustrates how this procedure is followed.

Consider a synthesizer operating over a 220 to 300 MHz band in 1 kHz increments. Assume that the required 60 Hz FM sideband-to-signal ratio is −50 dB. Assume also that the synthesis approach originally chosen is shown in Fig. 2-1; it consists of a 1 MHz reference source, a divide-by-1000 frequency divider, and a digitally controlled phase-locked loop. This PLL is a frequency multiplier with respect to the 1 kHz reference frequency with the highest multiplication ratio of 299,999 or approximately

110 dB. If it is assumed that no attenuation is provided by the PLL at 60 Hz, any FM sidebands at the 60 Hz rate modulating the reference signal are enhanced by 110 dB as the signal is multiplied to the final frequency. The 60 Hz FM requirement imposed on the divide-by-1000 frequency divider is $-50-100=-150$ dB. This is not a practically realizable number. A more practical figure is -110 dB, which means that to meet a -50 dB requirement the multiplication ratio of $110-50=60$ dB (or 1000) is not to be exceeded along any path of synthesis. Figure 2-2 shows an approach satisfying this requirement. Indeed, other requirements such as those involving spurious outputs and phase noise, which for clarity are omitted from this example, should be included in considerations governing the choice of a synthesis approach.

Techniques for Reducing Spurious Outputs

Circuits that are commonly used in frequency synthesis, such as harmonic multipliers and mixers, generate spurious outputs. Techniques for the suppression of these signals utilize passive or active filters. When spurious signals are very close to the band of operation, crystal filters are used if the frequency band is within the bandwidth capabilities of crystal filters. Sometimes spurious outputs fall in band and can be filtered out only by tuned filters. Because of remote-frequency-tuning requirements, these filters have to be electronically tuned. In many cases electronically tuned LC filters are undesirable because of the AM-to-PM conversion process taking place in the filter reactive elements, which are tuned by electronic means. The following approaches either prevent the generation of spurious outputs or eliminate the need for tuned filters.

PHASE-LOCKED LOOP AS A MULTIPLIER. A phase-locked loop (PLL) can be used as a frequency multiplier that generates no spurious outputs at the harmonics of the input frequency, Fig. 2-27. In the event that both the phase comparator and the frequency divider are broadband devices, this PLL replaces a multiplier-tuned filter combination and can be made to operate over an octave band. There are spurious outputs associated with this type of multiplier, such as FM components $\pm f_i$ distance away from the output signal, which result from a leakage of the input signal at f_i through the phase comparator to the VCO control input, but these can easily be reduced to below the required level by proper design procedures. (For PLL design techniques see Chapter 5.)

DIGITAL DIVIDER AS A BROADBAND FILTER. Some digital frequency dividers presently available on the market exhibit a capability of suppressing spurious inputs by not responding to any signal whose power level is below

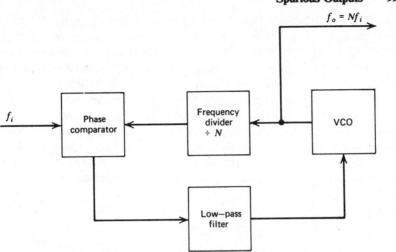

Figure 2-27. Digital phase-locked loop as a frequency multiplier.

the level of the desired input signal by a fixed amount. For example, tests performed by the author on Fairchild variable-ratio dividers designed according to the pulse-swallow technique (see Chapter 6) disclosed that, as long as the difference between the desired and spurious input levels exceeded approximately 200 mV rms into 50 Ω input impedance of the divider, the divider effectively attenuated the spurious signal by more than 70 dB. This attenuation was independent of the division ratio used (which was varied from 150 to 210, an equivalent of 43.5 to 46.5 dB, respectively) but depended on the isolation provided by the circuit layout used in this particular case at 100 MHz. With 600 mV rms input level this would allow spurious signals to be only 10 dB below the desired signal. Investigations performed by others (Ref. 16) indicate that various digital divider circuits possess a capability of suppressing spurious inputs by a factor that is not directly related to the division ratio.

A word of caution is needed about frequency dividers used as filters. In the preceding section of this chapter, which describes spurious signals associated with these circuits, it is pointed out that dividers generate spurious outputs. It is recommended, therefore, that tests be performed on dividers intended for use as filters to permit evaluation of the chosen design and layout before the final choice of a frequency synthesis approach is made.

When the desired input signal of such a device is frequency modulated, the level of FM sidebands is reduced by a factor equal to the division ratio, N, or by $20\log_{10}(1/N)$ dB.

DIGITAL PHASE-LOCKED LOOP AS A MIXER.   Figure 2-28 is a block diagram of a digital phase-locked loop designed to perform up- or down-conversion of the input signal at f_1. At the same time the loop performs the function of a narrowband filter because of the digital divider and low-pass filter present in the loop. In the event that the loop mixer, frequency divider, and phase comparator are broadband devices, the PLL can be electronically tuned over an octave band, preserving the narrowband filter characteristic throughout the frequency band of interest. This approach of mixing two signals is particularly useful and saves hardware in the design of multiloop frequency synthesizers, where a choice of combining outputs from various PLLs either internally or externally to the loops exists.

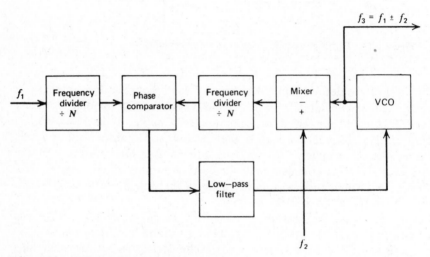

Figure 2-28.   Digital phase-locked loop as a mixer.

Techniques for Measuring Spurious Outputs

Before the days of spectrum analyzers, the techniques for measuring spurious outputs utilized tuned receivers. The frequency band of interest was analyzed by slowly tuning a receiver over it and by investigating every response recorded by the instrument to determine whether the response was part of the frequency spectrum being evaluated or was generated in the receiver itself. This procedure required many engineering hours.

Spectrum analyzers display a wide frequency spectrum of a rf signal on a cathode-ray-tube screen and make the search for spurious signals much easier. Unfortunately for the designer, the requirements in regard to spurious outputs are made more stringent as the synthesizer applications become more complicated and as the synthesizer operating bands shift to higher frequencies. At present, specifications for many vhf synthesizers require that the levels of spurious outputs be -90 to -110 dB below the desired signal, and that the range of measuring spurious outputs be extended to 10 Hz offset-from-signal frequency. State-of-the-art spectrum analyzers do not have a 120 dB dynamic range. Neither can they provide a 10 Hz resolution at vhf and higher frequencies. Because of a lack of equipment capable of performing these measurements, various techniques have been developed to assist the designer of synthesizers in the evaluation of a system.

TUNED FILTER-SPECTRUM ANALYZER APPROACH.   The factor limiting the dynamic range in commercial spectrum analyzers is the noise-to-saturation range of the front-end mixer, which, at present, is less than 90 dB. When saturated by the signal whose spectrum is viewed, the mixer generates high-level intermodulation products (spurious signals internal to the analyzer) and makes identification of synthesizer spurious outputs difficult. To eliminate the problem of saturation, a tunable band-pass filter is used in conjunction with a spectrum analyzer, Fig. 2-29, which attenuates synthesizer fundamental output as the spectrum of this signal is scanned. This approach is particularly attractive at frequencies above 1 GHz because YIG (Yttrium-Iron-Garnet) band-pass filters operate in this range over a number of octaves and can be current-swept at the same rate as the spectrum analyzer sweep rate. Such equipment as the AIL spectrum analyzer, Model 707, provides a sweep voltage output for this purpose. At vhf a number of mechanically tuned filters must be used if the frequency band of interest is wider than an octave.

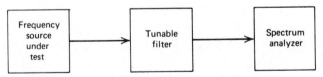

Figure 2-29.   High-dynamic-range test setup for measuring spurious outputs.

AUTOCORRELATION APPROACH.   The autocorrelation method of Tykulsky for measurements of spurious outputs (Ref. 17) provides an effective means of carrier cancellation. Figure 2-30 is a block diagram of the test setup based on this approach. The output of a frequency source to be tested is applied directly to the local oscillator port and, by way of a delay line of d seconds, to the signal port of a double-balanced mixer. The mixer

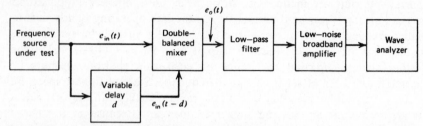

Figure 2-30. Autocorrelation techniques for measuring spurious outputs. Proceedings IEEE, Vol. 54, No. 2 (February 1966). Reprinted by permission.

provides 20 to 40 dB cancellation of the carrier. Further cancellation is obtained from the low-pass filter that follows the mixer. This filter also prevents the carrier from saturating the low-noise broadband amplifier. The operation of this equipment is described in mathematical terms as follows.

The output of the mixer is a product of two input signals:

$$e_o(t) = k e_{in}(t) e_{in}(t-d), \tag{2-46}$$

where $k =$ gain constant of the mixer.

It was demonstrated in this chapter that the combination of a desired signal at ω_c and a single spurious signal at $\omega_c + \omega_m$ has a mathematical equivalent consisting of the sum of two AM and two FM sideband components. If the AM sideband components alone are considered, the input signals to the mixer are

$$e_{in}(t) = E_c \left[\cos \omega_c t + \frac{m}{2} \cos(\omega_c + \omega_m)t + \frac{m}{2} \cos(\omega_c - \omega_m)t \right] \tag{2-47}$$

and

$$e_{in}(t-d) = k_1 E_c \left[\cos \omega_c (t-d) + \frac{m}{2} \cos(\omega_c + \omega_m)(t-d) \right.$$
$$\left. + \frac{m}{2} \cos(\omega_c - \omega_m)(t-d) \right]. \tag{2-48}$$

Substituting Eqs. 2-47 and 2-48 into Eq. 2-46, we have

$$e_o(t) = \frac{k k_1 E_c^2}{2} \left[\cos \omega_c d + 2m \cos(\omega_c d) \cos \left(\frac{\omega_m d}{2} \right) \cos \omega_m \left(t - \frac{d}{2} \right) \right], \tag{2-49}$$

where $m =$ modulation index,
$\quad k_1 =$ loss of the delay line.

Hence the mixer output consists of a dc term and a term at ω_m. The test setup has maximum sensitivity to AM when the dc output is maximum, that is, when

$$\cos \omega_c d = 1$$

or

$$d = \frac{N\pi}{\omega_c}, \qquad (2\text{-}50)$$

where $N =$ integer. This occurs when the delay line is an integral number of even half-wavelengths (including $d = 0$) at ω_c. For d as given in Eq. 2-50, Eq. 2-49 simplifies to

$$e_o(t) = \frac{kk_1 E_c^2}{2} + kk_1 E_c^2 m \cos\left(\frac{\omega_m d}{2}\right) \cos \omega_m \left(t - \frac{d}{2}\right). \qquad (2\text{-}51)$$

The maximum mixer output at ω_m occurs when

$$\cos\left(\frac{\omega_m d}{2}\right) = 1$$

or when

$$d = \frac{M\pi}{\omega_m}, \qquad (2\text{-}52)$$

where $M =$ integer. This occurs when the delay line is an integral number of even half-wavelengths (including $d = 0$) at ω_m.

Tuning for maximum sensitivity to AM is achieved by varying the delay d until a maximum dc level is obtained at the output of the mixer. This condition satisfies Eq. 2-50 and validates Eq. 2-51.

Calibration of the test setup is accomplished either by modulating the frequency source under test at ω_m to produce AM sideband components of easily measurable level (such as -40 db) or by injecting another signal at $\omega_c + \omega_m$ or at $\omega_c - \omega_m$ simultaneously with the signal whose spectrum is viewed, Fig. 2-31, and by measuring this reference level at ω_m with the wave analyzer. The reference level is then linearly extrapolated down to the mixer noise.

Still another carrier suppression technique is described in Ref. 18.

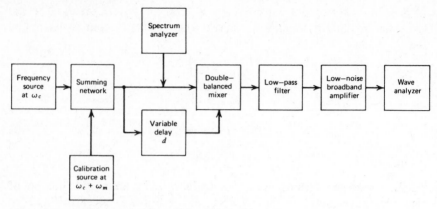

Figure 2-31.   Calibration of the test setup shown in Fig. 2-30.

2-2 Phase Noise

Any electric signal generated by known means is phase (or frequency) modulated in a random fashion. When this signal is passed through electronic circuitry, often the noise spectrum of the signal is modified. This section describes common types of phase noise and estimates the power levels and frequency distributions of phase noise in various circuits used in frequency synthesis.

Stability of Frequency Sources

There are numerous ways of specifying the stability of frequency sources (Refs. 9, 16, 19 to 34). In this text two definitions, one in the frequency domain and another in the time domain, are given and an approximate relationship connecting sinusoidal PM mathematical expressions with measured noise is described.

Consider a frequency source whose instantaneous output signal is

$$e(t) = \left[E_c + A(t) \right] \sin\left[\omega_c t + \theta(t) \right], \tag{2-53}$$

where E_c and ω_c are the nominal amplitude and angular frequency of the output, respectively. Equation 2-53 indicates that there is a concentration of noise power surrounding a rf signal. A typical spectral distribution of such a signal is shown in Fig. 2-32. In general, both AM noise and FM noise, $A(t)$ and $\theta(t)$, respectively, are present, accounting for the lack of symmetry in the envelope of the spectrum. If it is assumed that

$$A(t) \ll 1$$

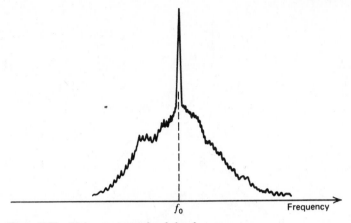

Figure 2-32.   Noise spectrum of a rf signal.

and

$$\frac{d\theta(t)}{dt} = \mathring{\theta}(t) \ll 1$$

for any time t, the fractional instantaneous frequency deviation from nominal frequency may be defined as (Ref. 34)

$$y(t) \equiv \frac{\mathring{\theta}(t)}{\omega_c}. \tag{2-54}$$

A definition for the measure of frequency stability proposed in Ref. 34 is the spectral density, $S_y(f_m)$, of the function $y(t)$, where the spectrum is considered to be one-sided on a per hertz basis. The function $S_y(f_m)$ has the dimensions of $1/\text{Hz}$.

In the following discussion it is assumed that measuring devices are available which permit viewing the AM spectrum separately from the FM spectrum. Such devices are described in the sections of this chapter on measuring techniques. To escape repetition, the discussion here is limited to considerations of FM noise only. However, the method of presentation can be applied to AM noise with equal success.

The FM noise spectrum of a rf signal is shown in Fig. 2-33. Although the noise distribution on each side of the signal is continuous, one can subdivide the spectrum into a large number of strips of width ΔB (the measured passband) located f_m distance away from the signal, and view the energy in ΔB as being caused by a sinusoidal frequency-modulating signal

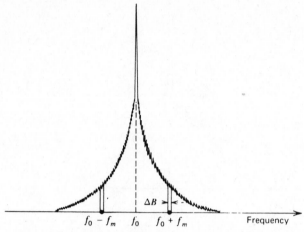

Figure 2-33.   Phase-noise spectrum of a stable-frequency source.

centered in ΔB with a deviation proportional to the amplitude of the spectrum at f_m. It is assumed that ΔB is much smaller than f_m so that the envelope of the noise spectrum is flat within ΔB, though it may have $1/f$ or any other characteristic in the vicinity of f_m. This amounts to treating a continuous noise spectrum as if it consisted of a very large number of sinusoidal FM sideband components symmetrically distributed about the signal so that, when all the equivalent sinusoidal voltages are added on a power or root-sum-square basis, the same total mean power as the actual noise spectrum is produced. Notice that this analogy is based on the assumption that contributions from amplitude modulation to the noise spectrum considered are negligible compared to those from frequency modulation.

It has been customary to characterize the noise performance of signal sources as the ratio of the measured power in one noise sideband component, on a per hertz of bandwidth spectral density basis, to the total signal power (Ref. 32):

$$\alpha(f_m) \equiv \frac{\text{power density (one sideband, phase only)}}{\text{power (total signal)}} \text{ per hertz bandwidth.}$$

$$(2\text{-}55)$$

Equation 2-55, introduced by Dr. Donald Halford of the National Bureau of Standards (Ref. 32), does not have a particular mathematical significance but rather specifies a measure of performance. A comparison of $S_y(f_m)$ with $\alpha(f_m)$ indicates that the two quantities are directly proportional to each other if the same total signal power is assumed in both cases.

Using sinusoidal analogy, Eq. 2-20, $\alpha(f_m)$ in terms of peak phase deviation is

$$\alpha(f_m) \cong \frac{\theta_d}{2} \qquad (2\text{-}56)$$

or, expressed in decibels per hertz bandwidth,

$$\alpha(f_m) \cong 20\log_{10}\left(\frac{\theta_d}{2}\right), \qquad (2\text{-}57)$$

and since, Eq. 2-16,

$$\theta_d = \frac{\Delta f_{\text{peak}}}{f_m},$$

$\alpha(f_m)$ expressed in terms of Δf is

$$\alpha(f_m) \cong 20\log_{10}\left(\frac{\Delta f_{\text{peak}}}{2f_m}\right) \text{ dB/Hz} \qquad (2\text{-}58)$$

or

$$\alpha(f_m) \cong 20\log_{10}\left(\frac{\Delta f_{\text{rms}}}{\sqrt{2}\,f_m}\right) \text{ dB/Hz}. \qquad (2\text{-}59)$$

Having measured $\alpha(f_m)$ for various values of f_m, one can estimate an equivalent peak phase or rms frequency deviation using Eq. 2-57 or Eq. 2-59, respectively. If the equivalent single sideband-to-signal ratio was measured in a bandwidth different from 1 Hz, the conversion from a single sideband-to-signal ratio in x Hz bandwidth, ΔB_x, to $\alpha(f_m)$ is expressed as

$$\alpha(f_m) \cong \left.\frac{\text{single sideband noise power}}{\text{signal power}}\right|_{\Delta B_x} - 10\log_{10}(\Delta B_x) \text{ dB/Hz.} \qquad (2\text{-}60)$$

Throughout this book, whenever the subject of phase noise is introduced, it is assumed, unless stated otherwise, that phase noise has a flat frequency response within the bandwidth of measurement.

In general, the conversion of a single sideband-to-signal ratio (in dB) from a bandwidth ΔB_1 to a bandwidth ΔB_2 is accomplished as

$$\left. \frac{\text{single sideband noise power}}{\text{signal power}} \right|_{\Delta B_2} = \left. \frac{\text{single sideband noise power}}{\text{signal power}} \right|_{\Delta B_1}$$

$$- 10\log_{10}\left(\frac{\Delta B_1}{\Delta B_2} \right). \qquad (2\text{-}61)$$

For the rms frequency deviation (in Hz) this conversion is

$$\Delta f_2 = \Delta f_1 \sqrt{\frac{\Delta B_2}{\Delta B_1}} \qquad (2\text{-}62)$$

Equations 2-60 to 2-62 apply to the noiselike content of the spectrum only. The power of a discrete FM sideband component is independent of the bandwidth of measurement.

To illustrate the technique, consider a single sideband noise-to-signal power ratio of -100 dB measured in a 10 Hz bandwidth at 1 kHz offset-from-signal frequency. Using Eq. 2-60, one has

$$\alpha(10^3) \cong -100 - 10\log_{10}(10) = -110 \text{ dB}.$$

To find an equivalent peak phase deviation, θ_d, in 1 Hz bandwidth, Eq. 2-56 is used:

$$\text{The ratio of } -110 \text{ dB} = 3.16 \times 10^{-6} = \frac{\theta_d}{2},$$

$$\theta_d = 6.32 \times 10^{-6} \text{ rad/Hz}.$$

Similarly, using Eq. 2-59 gives

$$3.16 \times 10^{-6} = \frac{\Delta f_{\text{rms}}}{\sqrt{2}\, f_m},$$

$$\Delta f_{\text{rms}} = 3.16 \times 10^{-6} \times 1.414 \times 10^3 = 4.468 \times 10^{-3} \text{ Hz/Hz bandwidth}.$$

Some applications of frequency sources in communications and radar require phase-noise data to be taken close to the signal. The techniques used in measuring $\alpha(f_m)$ in the frequency domain do not provide accurate measurements of noise at offset-from-signal frequencies below 10 Hz because the narrowest bandwidth of state-of-the-art wave analyzers is 1 Hz. To satisfy these requirements, techniques of measuring phase noise below 10 Hz have been developed. This leads to a definition of frequency stability in the time domain.

Define $\bar{y}_k$, the average fractional frequency offset obtained during the kth measurement interval, as (Ref. 34)

$$\bar{y}_k \equiv \frac{1}{\tau} \int_{t_k}^{t_k+\tau} y(t)\, dt = \frac{\theta(t_k+\tau) - \theta(t_k)}{2\pi f_c \tau}, \qquad (2\text{-}63)$$

where $t_{k+1} = t_k + T$,
$\quad\quad k = 0, 1, 2, \ldots,$
$\quad\quad T$ = repetition interval for measurements of duration τ ($T = \tau$ if there is no dead time between measurements),
$\quad\quad \tau$ = duration of each measurement (averaging time).

Conventional frequency counters measure the number of cycles in a period τ or $f_c \tau (1 + \bar{y}_k)$.

The second definition of frequency stability proposed in Ref. 34 is based on the sample variance of the fractional frequency fluctuations and is expressed in terms of $\bar{y}_k$ as

$$\sigma^2(N, T, \tau) \equiv \left\langle \frac{1}{N-1} \sum_{n=1}^{N} \left(\bar{y}_n - \frac{1}{N} \sum_{k=1}^{N} \bar{y}_k \right)^2 \right\rangle, \qquad (2\text{-}64)$$

where $\langle X \rangle$ = infinite time average of X,
$\quad\quad N$ = positive integer that gives the number of data points used in obtaining a sample variance.

In practice N is a finite number. To improve the compatibility of data, it is important to specify the particular N and T used in determining $\sigma^2(N, T, \tau)$. Reference 34 recommends choosing $N = 2$ and $T = \tau$ (i.e., no dead time between measurements). Let

$$\langle \sigma_y^2 (N = 2, T = \tau, \tau) \rangle = \sigma^2(\tau),$$

where $\sigma^2(\tau)$ is the Allen variance. Then the proposed measure of frequency stability in the time domain is (Ref. 34):

$$\sigma_y^2(\tau) = \left\langle \frac{(\bar{y}_{k+1} - \bar{y}_k)^2}{2} \right\rangle \qquad \text{for } T = \tau. \qquad (2\text{-}65)$$

This definition represents one estimate of the Allen variance of the angular frequency fluctuations for $N = 2$, made from two samples ($\bar{y}_k$ and $\bar{y}_{k+1}$) of the angular frequency averaged over the sample time, τ. The value of the Allen variance depends to some extent on both τ and the bandwidth of the system used in measuring angular frequency fluctuations. These quantities have to be specified, together with N and T, for any given measurement.

An expression that permits computations of the standard deviation of the fractional frequency fluctuations directly from the measured data is given in Ref. 27 and in the section of this chapter on measurement techniques (p. 141).

Types of Noise

Devices such as resistors, capacitors, diodes, and transistors generate internal noise. Noise analysis is relatively easy at frequencies above approximately 5 kHz. In this region shot noise and thermal noise predominate until a decrease in gain of active devices comes into play at high frequencies. At frequencies below 5 kHz the noise observed experimentally exceeds the shot noise and thermal noise. It varies inversely with frequency and is identified as $1/f$ noise.

SHOT NOISE.   In most active devices the current is carried by individual electrons that flow as charged particles. This current fluctuates with time because the number of electrons (and hence the amount of charge) passing a cross section of a semiconductor or leaving the cathode of a vacuum tube varies. Schottky showed in 1918 that the mean square current fluctuation is

$$i_{n,\text{rms}}^2 = 2qI\,\Delta B, \tag{2-66}$$

where q = electron charge (1.6×10^{-19} C),
I = dc current flowing in semiconductor or tube (A),
ΔB = incremental bandwidth (Hz).

The amplitude distribution of the shot noise is gaussian since the noise is made up of a very large number of independent contributors.

THERMAL NOISE.   Because of the random thermal motion of free electrons in a conductor, an ac voltage is developed across the conductor. According to Johnson and Nyquist (Refs. 35 and 36), the mean value of the open-circuit voltage across any conductor is given by

$$e_{n,\text{rms}}^2 = 4kTR\,\Delta B, \tag{2-67}$$

where k = Boltzmann's constant $(1.3805 \times 10^{-23} \text{ J}/°\text{K})$,
  T = absolute temperature of the thermal noise source $(°\text{K})$,
  R = resistance of conductor (Ω),
  ΔB = incremental bandwidth (Hz).

The amplitude distribution of the thermal noise (also called Johnson noise, thermal agitation, and resistance noise) is also gaussian, and the spectrum of the noise is independent of frequency, as was the case with shot noise. It can be shown that a thermal noise source delivers a maximum power of $kT\Delta B$. Expressed in dBm at room temperature, the maximum power is

$$P_{n,\max} = -174 + 10\log_{10} \Delta B. \qquad (2\text{-}68)$$

This expression is very useful in system analysis. For example, if one wishes to determine the lowest power level that gives a desired signal-to-noise ratio at the input of a device, it can be computed as follows:

lowest power level of signal at input of device = $-174 + 10\log_{10}\Delta B$

$$+ NF + \frac{S}{N}(\text{desired}) \text{ dBm},$$

$$(2\text{-}69)$$

where NF = noise figure of the device (amplifier or mixer) (dB),
S/N (desired) = desired signal-to-noise ratio referred to a given bandwidth (dB).
  For an amplifier with a 10 dB noise figure and desired signal-to-noise ratio of 160 dB in 1 Hz bandwidth, one obtains

lowest power level of signal at input of amplifier = $-174 + 10 + 160$

$$= -4 \text{ dBm}.$$

LOW-FREQUENCY $(1/f)$ NOISE. A third commonly encountered type of gaussian distribution noise is $1/f$ noise (also known as contact noise, excess noise, and flicker noise). This noise is associated with contact and surface irregularities in semiconductors and appears to be caused by fluctuations in the conductivity of the medium.
  The expression for the spectral density of $1/f$ noise is given in Ref. 15 as

$$P(f) = \frac{k}{f^{\nu}}\text{W}, \qquad (2\text{-}70)$$

where the exponent ν varies as

$$0.8 \leqslant \nu \leqslant 1.5.$$

The noise was first observed by Schottky in vacuum tubes and named the flicker effect because it seemed to be due to a kind of flickering of the electron emission from the cathode. This noise is also observed in resistors (Ref. 37) and capacitors (Ref. 38). Unlike the shot and thermal effects, the flicker effect is not considered an ultimate source of noise and can be somewhat reduced by proper processing of semiconductor surfaces. A significant reduction of the $1/f$ noise in some circuits can be achieved by using a technique described later in this chapter (pp. 119, 127).

Phase Noise in Oscillators

The phase-noise performance of an oscillator can be described in terms of the simplified model shown in Fig. 2-34. In this model the components of the oscillator feedback loop are a single-tuned resonant circuit and a limiting amplifier. A tuned buffer amplifier follows the oscillator. Phase noise internal to the oscillator is given as a voltage, $V_{n,\text{int}}$, injected into the oscillator feedback loop through a phase modulator which introduces a

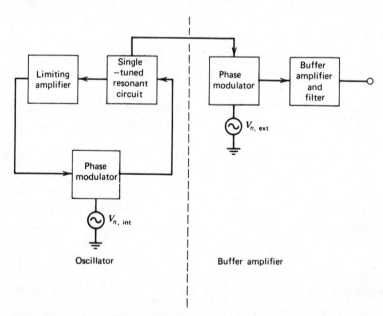

Figure 2-34.   Phase-noise model of an oscillator and a buffer amplifier. Proceedings IEEE, Vol. 53, No. 7 (July 1965). Reprinted by permission.

fluctuating phase shift in the loop. Oscillator frequency follows the phase shift fluctuations, which frequency-modulate the oscillator output frequency because the frequency of oscillations is determined by the phase relationships established in the feedback loop. The phase noise of the buffer amplifier is described as a noise voltage, $V_{n,ext}$, injected at the input of the buffer amplifier into a modulator that phase-modulates the rf signal at that point.

Both noise sources can be subdivided into two predominant components: (*a*) a low-frequency term due to the flicker effect, which is characterized by a narrow frequency spectrum displaying $1/f$ behavior and centered at the frequency of oscillations, and (*b*) an additive term whose phase spectrum is flat, extends widely about the oscillator frequency, and, therefore, is modified by the oscillator and buffer amplifier resonant circuits.

The subject of phase noise in amplifiers is considered in the corresponding section of this chapter (p. 119). Here the oscillator noise is discussed.

It has been demonstrated in Ref. 39 that the two-sided phase-noise spectrum of an oscillator can be described as shown in Fig. 2-35, where $S_\phi(f_m)$ is the two-sided phase spectrum density, expressed in square radians per hertz bandwidth or in decibels relative to 1 rad^2/Hz bandwidth, and f_m is the offset-from-signal frequency in hertz. Both quantities are plotted on

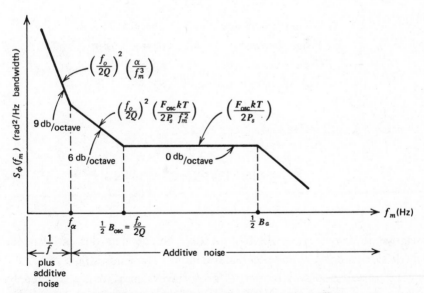

Figure 2-35. An approximation of the phase-noise spectrum of an oscillator.

logarithmic scales. Other parameters used in Fig. 2-35 are defined in the list of symbols on p. 118.

For offset-from-signal frequencies below the value for which flicker noise equals additive noise, that is, below f_α, the flicker noise of the oscillator predominates. The two-sided phase spectral density is related to the two-sided frequency spectral density, $S_{\dot\phi}(f_m)$, as (Ref. 40)

$$S_\phi(f_m) = \frac{1}{f_m^2} S_{\dot\phi}(f_m).$$ (2-71)

Hence this portion of the phase-noise plot displays a $1/f^3$ characteristic. Above f_α, flat additive noise predominates and the slope of the curve changes from 9 to 6 and then to 0 dB/octave accordingly.

It may occur that the oscillator additive noise is attenuated at 6 dB/octave down to the thermal noise level for offset frequencies above half of the oscillator bandwidth, that is, for $f_m > f_0/2Q$. However, subsequent band-limiting filtering and the phase noise of the amplifier circuit following the oscillator always modify $S_\phi(f_m)$ and should be included in phase-noise analysis.

As an example, consider an uhf voltage-controlled oscillator, with the following constants:

Resonant frequency,	$f_0 = 300$ MHz
RF signal power,	$P_S = -3$ dBm
Effective noise figure,	$F_{osc} = 10$ dB
Feedback loop bandwidth,	$B_{osc} = 3$ MHz

Hence $\frac{1}{2} B_{osc} = 1.5$ MHz and

$$\frac{F_{osc} kT}{2P_S} = 10 - 174 - 3 + 3 = -164 \text{ dB}$$

relative to 1 rad^2 in 1 Hz bandwidth. Notice that this is also the single sideband-phase-noise-to-signal ratio in 1 Hz bandwidth.

The curves displaying 6 and 9 dB/octave slopes are constructed by observing in Fig. 2-35 that the 6 dB/octave rise occurs at $f_m = f_\alpha$. The values of f_α for various oscillator types are given in Table 2-3 (Ref. 41). The final plot of the estimated oscillator phase noise is shown in Fig. 2-36.

Table 2-3. Values of f_α for Various Types of Oscillators[a]

Type	f_α(Hz)
5 MHz standard crystal oscillator	10^4
VHF crystal oscillator	10^3
UHF voltage-controlled oscillator	10^5
UHF cavity oscillator	3×10^5
X-Band gunn-effect diode oscillator	3×10^5
Two-cavity klystron	3×10^3
H Maser	10^2

[a]Reprinted by permission of *Microwave Journal*, June 1970.

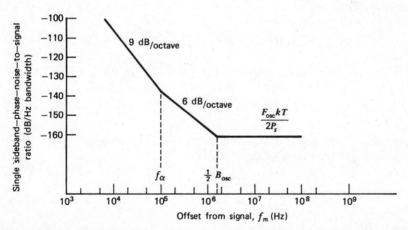

Figure 2-36. Estimated phase-noise spectrum of a uhf voltage-controlled oscillator, noise of reactance control circuit excluded.

When the VCO tuning constant, $K_{T,\text{vco}}$, in hertz per volt, is high (in cases of broadband electronic tuning), another source of noise not considered in the derivation of the curve in Fig. 2-35, that is, FM noise due to the voltage-controlled reactance circuit of the VCO, has to be evaluated. One can assume that this noise occurs in an equivalent resistor, R_{eq}, placed across the control-voltage input to the reactance circuit. The rms frequency deviation due to this noise component in bandwidth, ΔB, is approximated by

$$\Delta f_{\text{rms}} = K_{T,\text{vco}} \left(\sqrt{4kTR_{\text{eq}}\Delta B} + \text{flicker noise component} \right) \text{Hz}. \quad (2\text{-}72)$$

If the $1/f$ component is neglected, the total deviation is determined by the half-bandwidth of the oscillator resonant circuit, $\frac{1}{2}B_{\text{osc}}$, and can be expressed as (Ref. 42)

$$\Delta f_{\text{rms, total}} = K_{T,\text{vco}} \sqrt{2kTR_{\text{eq}}B_{\text{osc}}} \text{ Hz}. \quad (2\text{-}73)$$

In broadly tuned VCOs FM noise due to the reactance-controlled circuit not only predominates but also may be 20 to 40 dB higher than the noise of the oscillator with the reactance-controlled circuit removed. This is demonstrated below.

Assume that $K_{T,\text{vco}}$ is equal to 500 kHz/V and to 5 MHz/V at the upper and lower ends of the band, respectively. The value of R_{eq} measured in the transmission line uhf VCO described in Chapter 6 is 150 kΩ. At room temperature kT is 4×10^{-21} W/Hz. Hence the estimated value of the rms frequency deviation is

$$(\Delta f_{\text{rms}})_{\text{upper}} = 5 \times 10^5 \sqrt{4 \times 4 \times 10^{-21} \times 150 \times 10^3} = 0.0245 \text{ Hz}.$$

in 1 Hz bandwidth at the upper end of the band. At the lower end it is

$$(\Delta f_{\text{rms}})_{\text{lower}} = 5 \times 10^6 \sqrt{4 \times 4 \times 10^{-21} \times 150 \times 10^3} = 0.245 \text{ Hz}.$$

Using Eq. 2-16, one obtains the values of the equivalent peak phase deviations for both ends of the band:

$$(\theta_d)_{\text{upper}} = \frac{0.0346}{f_m} \text{ rad in 1 Hz bandwidth}$$

and

$$(\theta_d)_{\text{lower}} = \frac{0.346}{f_m} \text{ rad in 1 Hz bandwidth}.$$

The plots of the single sideband-phase-noise-to-signal ratio, shown in Fig. 2-37, are constructed using these values of θ_d and Eq. 2-20. Comparison of these plots with the curve in Fig. 2-36 shows a significant degradation of the VCO noise due to the reactance-controlled circuit.

This phenomenon takes place in VCOs utilizing nonlinear tuning elements such as a voltage-variable capacitance. Since YIG tuning displays excellent linearity, YIG-tuned microwave oscillators exhibit an almost

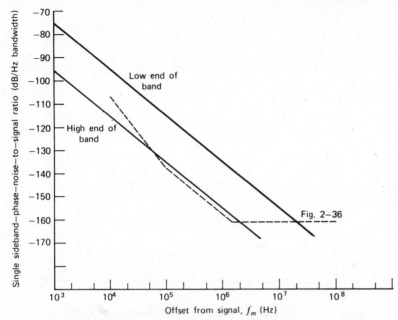

Figure 2-37.   Estimated phase noise due to reactance control circuit.

constant phase-noise spectrum over an octave frequency band. Practical levels of phase noise associated with YIG-tuned oscillators are given in Section 7-3 on microwave synthesizers.

A number of expressions for oscillator noise in the time domain have been derived by various researchers. Again, the predominant components are due to $1/f$ and additive noise. Unfortunately, at present no mathematical expression exists that describes the dependence of flicker noise on oscillator configuration or on parameters of circuit components in the time domain. Hence in this book the flicker noise term is identified by a constant, C. An approximate expression for the standard deviation of rms fractional frequency fluctuations versus averaging time, derived for an oscillator by E. Hafner (Ref. 43), is given as

$$\sigma\left(\frac{\Delta f_{\text{rms}}}{f}\right) = \frac{2\pi}{\tau}\sqrt{\frac{4kTF_{\text{osc}}}{P_{\text{osc}}Qf_0}} + C, \tag{2-74}$$

where τ, F_{osc}, and Q are as defined in the list of symbols on p. 118 and $P_{\text{osc}} =$ total rf power dissipated in the oscillator resonant circuit.

Figure 2-38 graphically describes the dependence of the rms fractional frequency fluctuations in an oscillator on averaging time. In this figure both of these quantities are plotted on the logarithmic scales.

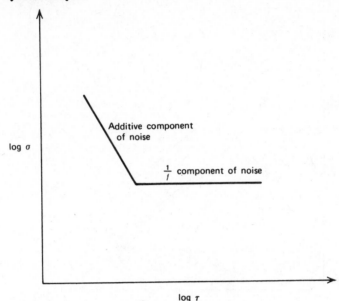

Figure 2-38.   RMS fractional frequency fluctuations in an oscillator.

In cases where a stable crystal oscillator operates at a low rf power, the noise of the buffer amplifier following the oscillator may predominate. The reader is referred to the following section of this chapter on noise in amplifiers for an appropriate expression of the rms fractional frequency fluctuations.

List of Symbols Used in Fig. 2-35

$f_0 =$ average frequency of oscillations

$Q =$ operating, or loaded, quality factor of the oscillator resonant circuit

$\alpha =$ a constant determined by the level of $1/f$ variations

$f_\alpha =$ offset-from-signal frequency for which the flicker noise is equal to the additive noise

$F_{osc} =$ effective noise figure of the limiting amplifier in the oscillator feedback loop

$k =$ Boltzmann constant

$T =$ temperature (°K)

$kT = -174$ dBm/Hz bandwidth at room temperature

$P_S =$ maximum available rf signal power from the oscillator

$B_{osc} =$ operating 3 dB bandwidth of the oscillator

$B_a =$ operating 3 dB bandwidth of the buffer amplifier following the oscillator

Phase Noise in RF Amplifiers and Frequency Multipliers

The first portion of this section is based on findings reported by Dr. D. Halford of the National Bureau of Standards at the 22nd Annual Symposium on Frequency Control, April, 1968 (Ref. 38; also see Refs. 44 and 45).

For convenience a rf amplifier is considered a frequency multiplier with a multiplication factor of unity. Hence the information presented in this section applies to both devices.

RESIDUAL NOISE. It has been demonstrated experimentally that phase noise generated internally in rf amplifiers and frequency multipliers (residual noise) varies as $1/f$ in (variance) spectral density in the 1 Hz to 5 kHz range of offset-from-signal frequency and is due to intrinsic, direct phase modulation of the rf signal by the transistors. Without any rf feedback, a typical value of $\alpha(1)$, that is, $\alpha(f_m)$ at $f_m = 1$ Hz, is -115 dB in 1 Hz bandwidth, referred to the input of the device. The worst and the best cases are -110 and -120 dB, respectively, depending on the transistor used in the circuit. A typical one-sided spectral density of the flicker noise of phase is about $10^{-11.2}/f_m$ rad^2 or $\alpha(f_m) = -(112 + 10\log_{10} f_m)$ dB referred to the input of the device. (See Table 2-4 for the decible-power ratio relationship.)

The maximum spread in $\alpha(1)$ of 10 dB is the result of differences among transistor types and within one type. The following transistor types have been tested: silicon, germanium, high f_T, low f_T, low dc flicker noise, high dc flicker noise, plastic, hermetic can, bipolar, FET, and overlay.

Only local, negative rf feedback (emitter degeneration) helps. When the feedback is utilized, more than 30 dB reduction of the flicker noise of phase in amplifiers and multipliers is achievable, $\alpha(f_m) = -(142 + 10\log_{10} f_m)$ dB. A 40 dB improvement is possible and practical. With a rf negative feedback and silver mica capacitors selected for low noise, the achieved $\alpha(1)$ is better than -155 dB.

Neither signal frequency for frequencies at least between 5 and 100 MHz nor rf drive level for reasonable drive levels has a significant effect on $\alpha(1)$. No dependence of $\alpha(1)$ on either the amplifier class (A, B, or C) or the order of multiplication, including times 1, that is, a buffer amplifier, is observed.

For offset-from-signal frequencies above 5 kHz thermal noise predominates. In this frequency region the single sideband-to-signal ratio is determined by the rf power drive, thermal noise level, and noise figure of the device and is given by

single sideband-to-signal ratio in 1 Hz bandwidth,

$$\text{referred to input of device} = 10\log_{10}\left(\frac{kT}{P_i}\right) + F, \quad (2\text{-}75)$$

where P_i = rf power at the input of the device (W),
$\quad kT = 4 \times 10^{-21}$ (W),
$\quad F$ = noise figure of the device (dB).

For $P_i = 10$ dBm, a noise figure of 15 dB, and $kT = -174$ dBm in 1 Hz bandwidth, the single sideband-noise-to-signal ratio is

$$-10 - 174 + 15 = -169 \text{ dB}.$$

An estimation of the thermal noise floor is important because this noise is often half AM and half PM at the point where it first appears (Ref. 27).

In the time domain the noise consists of two predominant components also: flicker noise and additive noise. An approximate expression for the standard deviation of the rms fractional frequency fluctuations versus averaging time, derived for an amplifier followed by a narrowband single-pole filter by L. S. Cutler and C. L. Searle (Ref. 40), is

$$\sigma\left(\frac{\Delta f_{rms}}{f_0}\right) = \frac{1}{2\pi f_0 \tau}\left[\frac{P_N}{P_i}(1 - e^{-\pi f_1 \tau})\right]^{1/2} + C', \quad (2\text{-}76)$$

where f_0, τ, and P_i are as defined above and

P_N = total noise power at the input of the amplifier:

$$P_N = \frac{\pi}{2}kTf_1 F, \quad (2\text{-}77)$$

C' = flicker noise factor,
f_1 = bandwidth of the filter following the amplifier,
F = noise figure of the amplifier.

EXTERNAL NOISE.   The data cited above were obtained with AM and PM noise of the input signal negligible compared to the noise generated internally to the device. In practice, AM noise of a significant level may be present at the input of the device. If the device is a frequency multiplier

Table 2-4. Decibels Versus Voltage and Power Ratios

A	$20\log_{10}(A)$ dB	A	A	$20\log_{10}(A)$ dB	A	B	$10\log_{10}(B)$ dB	B	B	$10\log_{10}(B)$ dB	B
1.0000	0.00	1.0000	0.5129	5.8	1.950	1.0000	0.00	1.0000	0.2630	5.8	3.802
0.9988	0.01	1.0012	0.5070	5.9	1.972	0.9977	0.01	1.0023	0.2570	5.9	3.890
0.9977	0.02	1.0023	0.5102	6.0	1.995	0.9954	0.02	1.0046	0.2512	6.0	3.931
0.9966	0.03	1.0035	0.4955	6.1	2.018	0.9931	0.03	1.0069	0.2455	6.1	4.074
0.9954	0.04	1.0046	0.4898	6.2	2.042	0.9908	0.04	1.0093	0.2399	6.2	4.169
0.9943	0.05	1.0058	0.4842	6.3	2.065	0.9886	0.05	1.0116	0.2344	6.3	4.266
0.9931	0.06	1.0069	0.4786	6.4	2.089	0.9863	0.06	1.0139	0.2291	6.4	4.365
0.9920	0.07	1.0081	0.4732	6.5	2.113	0.9840	0.07	1.0162	0.2239	6.5	4.467
0.9908	0.08	1.0093	0.4677	6.6	2.138	0.9817	0.08	1.0186	0.2188	6.6	4.571
0.9897	0.09	1.0104	0.4624	6.7	2.163	0.9795	0.09	1.0209	0.2138	6.7	4.677
0.9886	0.1	1.012	0.4571	6.8	2.188	0.9772	0.1	1.023	0.2089	6.8	4.786
0.9772	0.2	1.023	0.4519	6.9	2.213	0.9550	0.2	1.047	0.2042	6.9	4.898
0.9661	0.3	1.035	0.4467	7.0	2.239	0.9333	0.3	1.072	0.1995	7.0	5.012
0.9550	0.4	1.047	0.4416	7.1	2.265	0.9120	0.4	1.096	0.1950	7.1	5.129
0.9441	0.5	1.059	0.4365	7.2	2.291	0.8913	0.5	1.122	0.1905	7.2	5.248
0.9333	0.6	1.072	0.4315	7.3	2.317	0.8710	0.6	1.148	0.1862	7.3	5.370
0.9226	0.7	1.084	0.4266	7.4	2.344	0.8511	0.7	1.175	0.1820	7.4	5.495
0.9120	0.8	1.096	0.4217	7.5	2.371	0.8318	0.8	1.202	0.1778	7.5	5.623
0.9016	0.9	1.109	0.4169	7.6	2.399	0.8128	0.9	1.230	0.1738	7.6	5.754
0.8913	1.0	1.122	0.4121	7.7	2.427	0.7943	1.0	1.259	0.1698	7.7	5.888
0.8810	1.1	1.135	0.4074	7.8	2.455	0.7762	1.1	1.288	0.1660	7.8	6.026
0.8710	1.2	1.148	0.4027	7.9	2.483	0.7586	1.2	1.318	0.1622	7.9	6.166
0.8610	1.3	1.161	0.3981	8.0	2.512	0.7413	1.3	1.349	0.1585	8.0	6.310

Table 2-4 (Continued)

A	$20\log_{10}(A)$ dB	A	A	$20\log_{10}(A)$ dB	A	B	$10\log_{10}(B)$ dB	B	B	$10\log_{10}(B)$ dB	B
0.8511	1.4	1.175	0.3936	8.1	2.541	0.7244	1.4	1.380	0.1549	8.1	6.457
0.8414	1.5	1.189	0.3890	8.2	2.570	0.7079	1.5	1.413	0.1514	8.2	6.607
0.8318	1.6	1.202	0.3846	8.3	2.600	0.6918	1.6	1.445	0.1479	8.3	6.761
0.8222	1.7	1.216	0.3802	8.4	2.630	0.6761	1.7	1.479	0.1445	8.4	6.918
0.8128	1.8	1.230	0.3758	8.5	2.661	0.6607	1.8	1.514	0.1413	8.5	7.079
0.8035	1.9	1.245	0.3715	8.6	2.692	0.6457	1.9	1.549	0.1380	8.6	7.244
0.7943	2.0	1.259	0.3673	8.7	2.723	0.6310	2.0	1.585	0.1349	8.7	7.413
0.7852	2.1	1.274	0.3631	8.8	2.754	0.6166	2.1	1.622	0.1318	8.8	7.586
0.7762	2.2	1.288	0.3589	8.9	2.786	0.6026	2.2	1.660	0.1288	8.9	7.762
0.7674	2.3	1.303	0.3548	9.0	2.818	0.5888	2.3	1.698	0.1259	9.0	7.943
0.7586	2.4	1.318	0.3508	9.1	2.851	0.5754	2.4	1.738	0.1230	9.1	8.128
0.7499	2.5	1.334	0.3467	9.2	2.884	0.5623	2.5	1.778	0.1202	9.2	8.318
0.7413	2.6	1.349	0.3428	9.3	2.917	0.5495	2.6	1.820	0.1175	9.3	8.511
0.7328	2.7	1.365	0.3388	9.4	2.951	0.5370	2.7	1.862	0.1148	9.4	8.710
0.7244	2.8	1.380	0.3350	9.5	2.985	0.5248	2.8	1.905	0.1122	9.5	8.913
0.7161	2.9	1.396	0.3311	9.6	3.020	0.5129	2.9	1.950	0.1096	9.6	9.120
0.7079	3.0	1.413	0.3273	9.7	3.055	0.5012	3.0	1.995	0.1072	9.7	9.333
0.6998	3.1	1.429	0.3236	9.8	3.090	0.4898	3.1	2.042	0.1047	9.8	9.550
0.6918	3.2	1.445	0.3199	9.9	3.126	0.4786	3.2	2.089	0.1023	9.9	9.772
0.6839	3.3	1.462	0.3162	10.0	3.162	0.4677	3.3	2.138	0.1000	10.0	10.000
0.6761	3.4	1.479	0.2985	10.5	3.350	0.4571	3.4	2.188	0.08913	10.5	11.22
0.6683	3.5	1.496	0.2818	11.0	3.548	0.4467	3.5	2.239	0.07943	11.0	12.59

Voltage Ratio

Ratio	dB	Ratio	Ratio	dB	Ratio
0.6607	3.6	1.514	0.2661	11.5	3.758
0.6531	3.7	1.531	0.2512	12.0	3.981
0.6457	3.8	1.549	0.2371	12.5	4.217
0.6383	3.9	1.567	0.2239	13.0	4.467
0.6310	4.0	1.585	0.2113	13.5	4.732
0.6237	4.1	1.603	0.1995	14.0	5.012
0.6166	4.2	1.622	0.1884	14.5	5.309
0.6095	4.3	1.641	0.1778	15.0	5.623
0.6026	4.4	1.660	0.1585	16.0	6.310
0.5957	4.5	1.679	0.1413	17.0	7.079
0.5888	4.6	1.698	0.1259	18.0	7.943
0.5821	4.7	1.718	0.1122	19.0	8.913
0.5754	4.8	1.738	0.1000	20.0	10.000
0.5689	4.9	1.758	0.03162	30.0	31.620
0.5623	5.0	1.778	0.01	40.0	100.00
0.5559	5.1	1.799	0.003162	50.0	316.20
0.5495	5.2	1.820	0.001	60.0	1,000.00
0.5433	5.3	1.841	0.0003162	70.0	3,162.00
0.5370	5.4	1.862	0.0001	80.0	10,000.00
0.5309	5.5	1.884	0.00003162	90.0	31,620.00
0.5243	5.6	1.905	10^{-5}	100.0	10^5
0.5188	5.7	1.928			

Power Ratio

Ratio	dB	Ratio	Ratio	dB	Ratio
0.4365	3.6	2.291	0.07079	11.5	14.13
0.4266	3.7	2.344	0.06310	12.0	15.85
0.4169	3.8	2.399	0.05623	12.5	17.78
0.4074	3.9	2.455	0.05012	13.0	19.95
0.3981	4.0	2.512	0.04467	13.5	22.39
0.3890	4.1	2.570	0.03981	14.0	25.12
0.3802	4.2	2.630	0.03548	14.5	28.18
0.3715	4.3	2.692	0.03162	15.0	31.62
0.3631	4.4	2.754	0.02512	16.0	39.81
0.3548	4.5	2.818	0.01995	17.0	50.12
0.3467	4.6	2.884	0.01585	18.0	63.10
0.3388	4.7	2.951	0.01259	19.0	79.43
0.3311	4.8	3.020	0.01000	20.0	100.00
0.3236	4.9	3.090	0.00100	30.0	1,000.00
0.3162	5.0	3.162	0.0001	40.0	10,000.00
0.3090	5.1	3.236	0.00001	50.0	10^5
0.3020	5.2	3.311	10^{-6}	60.0	10^6
0.2951	5.3	3.388	10^{-7}	70.0	10^7
0.2884	5.4	3.467	10^{-8}	80.0	10^8
0.2818	5.5	3.548	10^{-9}	90.0	10^9
0.2754	5.6	3.631	10^{-10}	100.0	10^{10}
0.2692	5.7	3.715			

driven by a limiter, this noise will be first converted into PM noise and then enhanced by $20\log_{10}n$ dB, where n is the multiplication factor. If the amplifier-driver operates in class A, AM noise rides along the signal by a mixing process and is also enhanced by $20\log_{10} n$, as demonstrated in Ref. 9.

Similarly, PM noise is enhanced by $20\log_{10}n$, as one would expect after reading the discussion on spurious signals in frequency multipliers.

An important point made above is that multipliers, in addition to generating their own internal phase noise, make an important noise contribution by enhancing the noise present at their input.

Noise in Limiters

A limiter was graphically defined in Fig. 2-17 (the section on spurious signals in an ideal limiter, p. 80). For the purpose of this discussion, the performance of a limiter is described as

$$v_{\text{out}}^{+}(t)=\begin{cases} a\big[\,v_{\text{in}}(t)\,\big]^{1/n} & \text{for } v_{\text{in}}(t)\geqslant 0, \\ 0 & \text{for } v_{\text{in}}(t)<0, \end{cases} \qquad (2\text{-}78)$$

where $v_{\text{out}}^{+}(t)=$ positively going limiter output voltage. Hence, for $n=1$ the device is an amplifier, and for $n=\infty$ the device is an ideal limiter.

Let the input signal be a sinusoid plus noise:

$$v_{\text{in}}(t)=E\cos(2\pi f_0 t)+N(t),$$

where $N(t)=$a noise wave gaussian in nature that has a narrowband spectrum centered in the vicinity of the input signal frequency, f_0.

Assume also that the limiter is followed by a band-pass filter, which displays an ideal rectangular passband transfer characteristic centered on f_0, and that the filter is wide enough to pass all of the limiter output spectrum centered about f_0. Then for an ideal symmetrical limiter ($n=\infty$) the following input-output signal-to-noise power ratio relationship has been derived by W. B. Davenport (Ref. 45a):

$$\left(\frac{S}{N}\right)_{\text{out}}\simeq\frac{\pi}{4}\left(\frac{S}{N}\right)_{\text{in}} \qquad \text{for } \left(\frac{S}{N}\right)_{\text{in}}\to 0 \qquad (2\text{-}79)$$

and

$$\left(\frac{S}{N}\right)_{\text{out}}\simeq 2\left(\frac{S}{N}\right)_{\text{in}} \qquad \text{for } \left(\frac{S}{N}\right)_{\text{in}}\to\infty. \qquad (2\text{-}80)$$

Equations 2-79 and 2-80 indicate that the output signal-to-noise ratio, $(S/N)_{out}$, is approximately linearly proportional to the input signal-to-noise ratio, $(S/N)_{in}$, for all values of $(S/N)_{in}$. For the case of the band-pass square-rooter ($n=2$), which is a good approximation of a practically realizable limiter, this relationship is similarly

$$\left(\frac{S}{N}\right)_{out} \simeq 0.96\left(\frac{S}{N}\right)_{in} \quad \text{for } \left(\frac{S}{N}\right)_{in} \rightarrow 0 \tag{2-81}$$

and

$$\left(\frac{S}{N}\right)_{out} \simeq \frac{8}{5}\left(\frac{S}{N}\right)_{in} \quad \text{for } \left(\frac{S}{N}\right)_{in} \rightarrow \infty. \tag{2-82}$$

The case of $(S/N)_{in} \rightarrow 0$ is of no interest to the designer of synthesizers. Throughout a synthesizer system rf voltage levels are selected to give signal-to-noise ratios in excess of 100 dB/Hz bandwidth. The case of $(S/N)_{in} \rightarrow \infty$ is important, and Eqs. 2-81 and 2-82 reveal the fact that the noise spectrum of a rf signal is improved as the signal passes through a limiter if $(S/N)_{in}$ of the device is very high. This statement is not true, however, if the limiter is poorly designed and displays a high noise figure or if the limiter is improperly matched to the adjacent circuitry. Under either or both of these conditions, the signal-to-noise ratio at the input of the device, $(S/N)_{in}$, is low, resulting in a degraded spectrum of the signal passing the limiter as compared to the spectrum of the signal with the limiter not inserted.

Phase Noise in Frequency Dividers

Phase noise generated internally in frequency dividers is considered in Chapter 6 in conjunction with the design of divider circuits, as well as in Appendix B, Fig. B-1, which shows the measured phase noise of some integrated-circuit digital dividers. This section of Chapter 2 describes the changes in the phase (or frequency) spectrum of a rf signal whose frequency is divided by a number N.

Figure 2-39 is a block diagram of a frequency divider followed by an active circuit, which usually is a buffer amplifier but can be a phase comparator, mixer, or frequency multiplier. In general, the noise present at the input of a divider is reduced by the division ratio, N, or by $20\log_{10}(1/N)$ dB, as a rf signal passes the device. When the division ratio is high or when very low phase-noise levels are involved, the $1/f$ noise and the noise figure of the active circuit following the divider determine the lowest limit of noise. For example, consider the frequency synthesis approach shown in Fig. 2-1, part of which, for convenience, is redrawn in Fig. 2-40. The frequency of a 1 MHz reference source consisting of a crystal oscillator, a buffer amplifier, and a narrowband crystal filter, inclosed in a temperature-controlled oven, is divided by 1000. Since $20\log_{10}(10^4) = 80$ dB, one would expect the phase-noise spectrum of the reference source (Fig. 2-40, curve 1) to be reduced by 80 dB, as shown in Fig. 2-40, curve 2. This is not the case even if the internal divider noise is neglected.

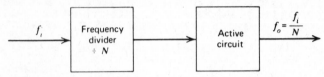

Figure 2-39.   Frequency divider.

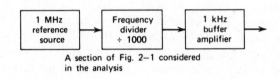

A section of Fig. 2–1 considered
in the analysis

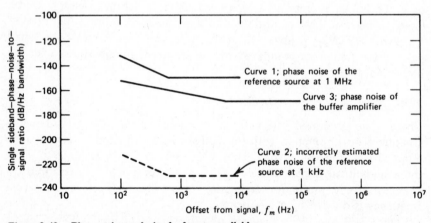

Figure 2-40.   Phase-noise analysis of a frequency divider.

Let us assume that the active circuit following the divider is a buffer amplifier with a 4 dB noise figure designed for low flicker-noise-of-phase operation. Assume also that the power level of the signal applied to the input of the amplifier is 0 dBm. From previous reading the designer knows that, when a sufficient emitter degeneration is used, the flicker noise of phase in amplifiers can be approximated by the expression

$$\text{single sideband-phase-noise-to-signal ratio in 1 Hz bandwidth} = -(132 + 10\log_{10} f_m) \text{ dB}.$$

It is also known that, with a 0 dBm input signal and a 4 dB noise figure, the buffer amplifier thermal noise, referred to the input of the amplifier, is -174 dBm/Hz + 4 dB = -170 dB/Hz. Curve 3, Fig. 2-40, is the noise spectrum of the buffer amplifier referred to the amplifier input. It demonstrates that the amplifier noise is significantly higher than the noise of the reference source divided to 1 kHz and is a limiting factor in the noise analysis.

Phase-Noise Reduction Techniques

Techniques of phase-noise reduction consist mainly of noise optimization of individual circuits and of a system design that results in the least enhancement of noise by frequency synthesis. Circuit optimization is discussed in Chapter 6, and an example of low-phase-noise system design is given at the end of this chapter. When the noise level at the output of a synthesizer exceeds the requirement, often little can be done to reduce it except further circuit optimization or system redesign. However, in some cases circuits such as crystal filters or phase-locked loops can be used to reduce phase noise, either at the output of certain critical sections or at the final output of the synthesizer.

Crystal filters are often used to reduce reference source noise when this noise, multiplied to the final synthesizer frequency, exceeds system requirements and when the residual noise of the synthesizer (the noise of the synthesizer excluding the reference source) is not a limiting factor. If properly matched into the system, a crystal filter will reduce the phase noise of an oscillator outside the filter bandwidth. Close-in noise reduction can be achieved by making the filter bandwidth narrow and by putting the filter in an oven to stabilize bandwidth drifts with temperature. However, bandwidths narrower than 40 Hz at 5 MHz are difficult to realize because of matching problems, so that phase noise at offset-from-signal frequencies below approximately 20 Hz cannot be reduced by any known filtering technique. A typical example of oscillator noise reduction is given in Fig. 2-41. The oscillator frequency is 1 MHz, and the bandwidth of the crystal filter is 40 Hz. In this particular case the crystal filter reduces the phase noise by 12 dB at an offset frequency of 100 Hz. Often crystal filters are used to reduce phase noise in the frequency synthesis portion of a synthesizer. This is not permissible, however, when the application requires high-speed frequency switching or when the reference signal to which a synthesizer is locked is derived from a distant source whose average frequency is varied according to Doppler shifts at constant or variable rates, as in shipboard-to-satellite communications.

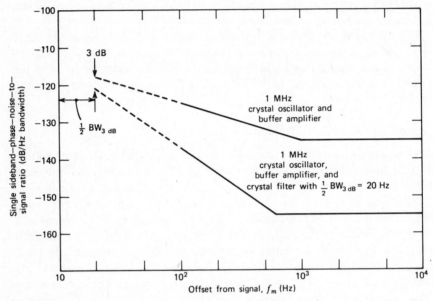

Figure 2-41.   Phase-noise spectrum of a stable crystal oscillator.

Phase-locked loops have been successfully utilized in the reduction of phase noise at offset frequencies above 10 kHz. As is demonstrated in Chapter 1, a PLL acts as a low-pass filter with respect to the noise of the reference source, replacing the reference noise spectrum by the VCO noise spectrum at offset frequencies above the loop bandwidth. Below approximately 10 kHz, VCO noise is higher than the state-of-the-art synthesizer requirements and reduction in loop bandwidth does not result in noise improvement, although such bandwidth reduction may be desirable from the system point of view. An example of the low-pass filter effect of a phase-locked loop is given in the section on system design (p. 151).

Techniques of Measuring Phase Noise

It is not an easy task to measure the phase-noise content of high-stability frequency sources. Although various techniques of phase-noise measurements will be considered, all present interference problems, some of which are briefly discussed after the description of measurement techniques. It is the task of the test engineer, however, to identify sources of interference associated with his test setup and the environment in which he performs the tests and to reduce interference to a tolerable level.

DOUBLE-BALANCED MIXER METHOD. A typical noise spectrum of a frequency source is shown in Fig. 2-32. It consists of both AM and PM noise. In applications where the exact level of phase noise is unimportant as long as it does not exceed a specified value or in cases where AM noise is known to be more than 20 dB below PM noise, the double-balanced mixer test setup, Fig. 2-42, is used. To perform the measurement, two identical frequency sources are utilized. Since there is no phase coherence between the individual noise components of the two sources, their noise adds on power basis. Hence the noise of each source is 3 dB lower than the measured noise. In the event that the two frequency sources do not contribute equally to the measured noise, the noise of the source that is the major contributor is no more than 3 dB greater than the assumed noise.

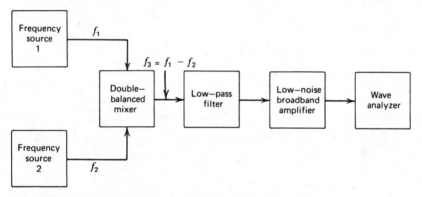

Figure 2-42.   Test setup for measuring total noise of frequency sources.

The frequencies of two sources, f_1 and f_2, are set so that one is higher than the other by a fixed offset, $f_3 = f_1 - f_2$. These two signals are applied to a frequency downconverter (a double-balanced mixer). The purpose of downconversion is to reduce the rf frequencies of both sources to within the range of narrowband wave analyzers. If measurements close to the signal are to be made, where bandwidths of 1 to 3 Hz are required, the two sources can be set to give $f_3 = 1$ kHz. When it is desirable to determine the noise spectrum further out, f_3 can be increased to any desired value.

The output at the difference frequency, f_3, is passed through a low-pass filter that attenuates both signals at f_1 and f_2, leaking to the output of the mixer, the sum at $f_1 + f_2$ and higher-order intermodulation products preventing these signals from saturating the low-noise broadband amplifier that follows the filter. The noise is measured in relation to the level of the difference output at $f_1 - f_2$ with a wave analyzer (a frequency-selective voltmeter) of known bandwidth.

For measurements to be accurate, it is important that the wave analyzer bandwidth be small compared to the offset-from-signal frequency. For example, measurements of noise at 20 Hz offset frequency require a 1 Hz wave analyzer bandwidth.

To increase the dynamic range of the wave analyzer, the frequencies of both sources are multiplied by a factor n, as is shown in Fig. 2-43. In such cases phase noise is enhanced by the same factor or, when expressed in decibels, by $20 \log_{10} n$. Often amplitude limiters are used in conjunction with multipliers. When this is done, AM noise present in the signal is eliminated. Care must be taken in the design of the multipliers not to produce significant AM-to-PM conversion.

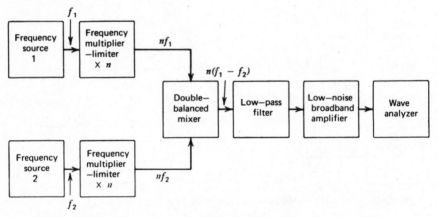

Figure 2-43. **High-sensitivity test setup for measuring phase noise of frequency sources.**

In the event that the noise of the multipliers is of interest, it can be measured utilizing the dual-channel approach shown in Fig. 2-44. In this approach a reference oscillator of known noise spectrum drives both frequency multiplier-limiters. If the mixer provided perfect balancing, the oscillator noise would not appear at the output of the mixer. In practical cases, though, the mixer cancels this noise only partially so that the oscillator noise, enhanced by multiplication, should be at least comparable in magnitude to, and preferably lower than, the expected multiplier noise.

If the input-to-the-mixer voltages are kept constant during the measurements performed, as shown in Figs. 2-43 and 2-44, the measured level of the difference signal at $n(f_1 - f_2)$ can be used as a reference relative to which multiplier-limiter noise is measured. The multiplier-limiter noise spectrum at the output of the mixer in Fig. 2-44 is folded because the frequency of the difference signal is zero. Hence the wave analyzer measures the power of both lower and upper sideband components of phase noise. The corresponding noise components are coherent, so that the associated noise power is proportional to the square of the sideband amplitude. The single-sided noise power is then 6 dB lower than the measured power.

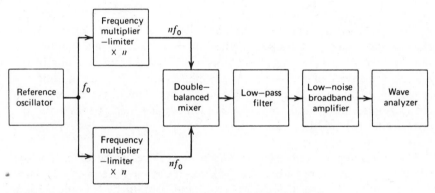

Figure 2-44. Test setup shown in Fig. 2-43, sensitivity measurement.

PHASE-DETECTOR METHOD. This technique of phase-noise measurement utilizes a double-balanced mixer as a phase detector of very high sensitivity (Fig. 2-45). The output of one frequency source under test is applied to the local oscillator port of the mixer. The output of the other source operating at the same frequency is applied to the signal port of the mixer by way of a 180 degree phase shifter. The mixer is a very broad double-balanced type of device utilizing Schottky diodes, which exhibit very low $1/f$ noise. Balancing against both inputs is required so as not to demodulate the amplitude noise of either input signal. The mixer provides a 90 degree phase shift of one input signal in relation to another which is required for the phase-detector mode of operation. The phase shifter is used in calibration and dc nulling of the test setup. The low-pass filter following the mixer prevents all spurious mixer outputs from saturating the low-noise broadband amplifier.

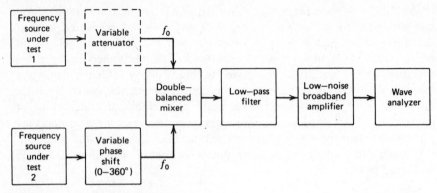

Figure 2-45.   A sensitive phase detector.

The test setup is tuned for the phase-detector mode of operation by shifting the phase of the signal at the signal port of the mixer until a dc null is obtained at the output of the mixer. This condition results in the highest sensitivity of the equipment to phase noise. Factors limiting the sensitivity of this test setup are the noises originating in the mixer and in the low-noise broadband amplifier. To achieve the highest sensitivity, the mixer output should be made as high as possible while the system noise is held to a minimum.

The test setup is calibrated by offsetting the frequency of one of the sources slightly and measuring the peak voltage at the difference frequency at the output of the low-noise broadband amplifier. An appropriate amount of attenuation is provided by the variable attenuator in order not to saturate the amplifier. If it is assumed that the measured peak voltage, appropriately corrected for the setting of the attenuator, is E_m and that a sinusoidal phase detector is employed, the noise voltage measured by the wave analyzer, e_d, is

$$e_d = E_m \sin \theta_n,$$

where θ_n is the phase deviation due to noise.

The following correction factors should be used to obtain $\alpha(f_m)$ from the measured phase-noise data when calibration is performed as described in the preceding paragraph.

1.   The equipment measures the noise of both sources. If equal noise contribution from each frequency source is assumed, the noise of one source is 3 dB lower than the measured noise.
2.   During the measurement the frequency of one source is set to equal the frequency of the other source. Hence a folded (double-sided) noise

spectrum is displayed at the mixer output. The single-sided spectrum is 6 dB lower.

3. Most wave analyzers are average reading devices. They are calibrated to read the rms value of a measured signal for a special case of sinusoidal input. When the input signal is white noise, the wave analyzer meter reads approximately 1 dB lower. To obtain the rms value, the measured noise should be increased by 1 dB.

4. The equivalent noise bandwidth of a wave analyzer is proportional to the 3 dB bandwidth of the analyzer. It can be defined as the band-width of an ideal rectangular band-pass filter having the same max-imum gain and passing the same average power as the actual band-pass filter. The noise bandwidth associated with a particular wave analyzer can be either measured or obtained from the manufacturer of the equipment.

5. It is convenient to express phase noise in 1 Hz bandwidth. If the measurement is made with an analyzer that does not have 1 Hz bandwidth resolution, conversion from x Hz to 1 Hz bandwidth is accomplished with the help of Eq. 2-60. It is assumed that the noise spectrum is flat within the bandwidth of measurement so that the noise power is proportional to the bandwidth of measurement.

The phase-detector method of measuring phase noise is very useful in dual-channel measurements of residual phase noise of phase-locked loops, multipliers, and synthesizers. By "residual phase noise" is meant the phase noise associated with a device, excluding the noise of the reference source driving it. Figure 2-46 shows how dual-channel testing is done. Two

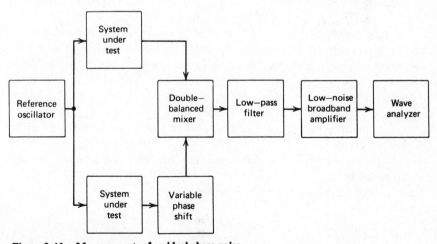

Figure 2-46. **Measurements of residual phase noise.**

systems under test are driven by one reference oscillator. This ensures that the noise of the oscillator does not appear at the output of the mixer. The phase noise measured with the wave analyzer at any offset-from-signal frequency is, therefore, the sum of the individual components of the systems under test.

AUTOCORRELATION METHOD.   The autocorrelation method of Tykulsky (Ref. 17) is a comprehensive measurement technique for analyzing the noise performance of frequency sources of moderate short-term stability. Figure 2-47 is a block diagram of the test setup based on this approach. The output of a frequency source under test is applied directly to the local oscillator port and, by way of a delay of d seconds, to the signal port of a double-balanced mixer.

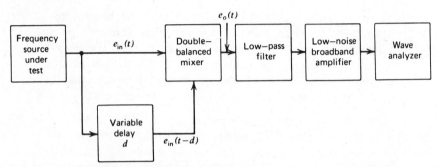

Figure 2-47.   Autocorrelation technique of measuring phase noise. Proceedings IEEE, Vol. 54, No. 2 (February 1966). Reprinted by permission.

The mixer and the delay line constitute a frequency discriminator when the test setup is properly aligned. The low-pass filter, broadband amplifier, and wave analyzer perform the same functions as in the previous cases. The advantage of this method is that only one frequency source to be tested is required for phase-noise measurements and the long-term frequency drift of the source may be any value characteristic of VCOs and other frequency-generating circuits because of a drift cancellation provided by the test setup.

The operation of this test equipment can be described in mathematical terms as follows.

Let $e_{in}(t)$ be a single-tone FM wave:

$$e_{in}(t) = E_c \left[\cos \omega_c t - \frac{\beta}{2} \cos(\omega_c - \omega_m)t + \frac{\beta}{2} \cos(\omega_c + \omega_m)t \right]. \quad (2\text{-}83)$$

Then

$$
e_{in}(t-d) = k_1 E_c \left[\cos \omega_c (t-d) - \frac{\beta}{2} \cos(\omega_c - \omega_m)(t-d) \right.
$$

$$
\left. + \frac{\beta}{2} \cos(\omega_c + \omega_m)(t-d) \right], \qquad (2\text{-}84)
$$

where $k_1 = $ loss in the delay line, a ratio.

The output of the mixer is a product of $e_{in}(t)$ and $e_{in}(t-d)$:

$$
e_o(t) = k_2 e_{in}(t) e_{in}(t-d), \qquad (2\text{-}85)
$$

where $k_2 = $ insertion loss of the mixer, a ratio.

Substituting the expressions for both mixer input signals, Eqs. 2-83 and 2-84, in Eq. 2-85 and eliminating terms centered on $2\omega_c$, which are filtered out by the low-pass filter, yields

$$
e_o(t) = \frac{k_1 k_2 E_c^2}{2} \left\{ \cos \omega_c d \left(1 + \frac{\beta^2}{2} \cos \omega_m d \right) \right.
$$

$$
- 2\beta \sin \omega_c d \sin \frac{\omega_m d}{2} \cos \left[\omega_m \left(t - \frac{d}{2} \right) \right]
$$

$$
\left. - \frac{\beta^2}{2} \cos \left[(\omega_c + \omega_m) d \right] \cos 2\omega_m t \right\}. \qquad (2\text{-}86)
$$

In the case of phase-noise modulation, $\beta \ll 1$. Hence Eq. 2-86 reduces to

$$
e_o(t) = \frac{k_1 k_2 E_c^2}{2} \left\{ \cos \omega_c d - 2\beta \sin \omega_c d \sin \frac{\omega_m d}{2} \cos \left[\omega_m \left(t - \frac{d}{2} \right) \right] \right\}. \qquad (2\text{-}87)
$$

The output consists of a dc term:

$$
E_{o,dc} = \frac{k_1 k_2 E_c^2}{2} \cos \omega_c d
$$

and a term at ω_m that is the desired output:

$$
e_{o,\omega_m}(t) = -k_1 k_2 E_c^2 \beta \sin \omega_c d \sin \frac{\omega_m d}{2} \cos \left[\omega_m \left(t - \frac{d}{2} \right) \right]. \qquad (2\text{-}88)
$$

The system has maximum sensitivity with respect to FM when

$$\sin \omega_c d = \pm 1 \text{ (hence, when } \cos \omega_c d = 0)$$

or when

$$d = \left(\frac{2N+1}{\omega_c}\right)\frac{\pi}{2}, \qquad\qquad (2\text{-}89)$$

where $N =$ an integer. This occurs when the ω_c terms in $e_{in}(t)$ and $e_{in}(t-d)$ are in phase quadrature and the delay line is an integral number of odd quarter-wavelengths at ω_c. For d as given in Eq. 2-89, Eq. 2-88 becomes

$$e_{o,\omega_m}(t) = -k_1 k_2 E_c^2 \beta \sin \frac{\omega_m d}{2} \cos\left[\omega_m\left(t - \frac{d}{2}\right)\right]. \qquad (2\text{-}90)$$

The system has maximum sensitivity to FM when

$$\sin \frac{\omega_m d}{2} = \pm 1$$

or when

$$d = \left(\frac{2M+1}{\omega_m}\right)\pi, \qquad\qquad (2\text{-}91)$$

where $M =$ an integer. This occurs when the delay line is an integral number of odd half-wavelengths at ω_m.

This system is tuned for maximum sensitivity to FM by varying the delay, d, until a dc null is obtained at the output of the mixer.

The equipment can be calibrated by utilizing a test oscillator that is capable of being frequency modulated, such as a voltage-controlled oscillator, operating at ω_c instead of the frequency source under test, Fig. 2-48. Balancing of the system is achieved in this case by varying either ω_c or d to obtain a dc null. The VCO is frequency modulated at ω_m, and the power of the sidebands at $\omega_c \pm \omega_m$ is set to a desired level below the signal at ω_c by observing the combined spectrum on the screen of a spectrum analyzer. Under these conditions the voltage at ω_m read on the wave analyzer following the low-noise amplifier corresponds to a known level of FM sidebands. Since for a low modulation index ($\beta \ll 1$) $e_{o,\omega_m}(t)$ is linearly proportional to β, calibration can be performed with easily observable sideband levels (e.g., 30 to 40 dB below the carrier) and then $e_{o,\omega_m}(t)$ linearly extrapolated down to the mixer noise level.

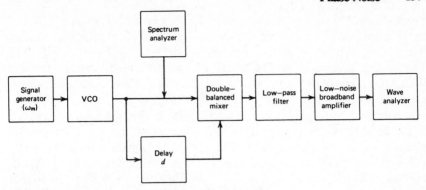

Figure 2-48.   Calibration of the test setup shown in Fig. 2-47.

This test setup is sensitive to changes in the rf power level of the input signal. Calibration should be performed, therefore, at the power level expected from the frequency source under test.

Another way of calibrating the test setup is presented in Fig. 2-49. A signal from a calibration oscillator at $\omega_c + \omega_m$ is added to the output of the frequency source being tested, which operates at ω_c, thus simulating frequency and amplitude modulation. The level of the calibration oscillator is set to a convenient level below the signal at ω_c. Under these conditions the voltage at ω_m read on the wave analyzer corresponds to a level of FM sidebands 6 dB lower than the level of the single-sided spurious signal. (See the section of this chapter on resolution of a signal and a single spurious output into AM and FM components, p. 93.) The short-term stability of the calibration oscillator should be such that the measured tone at ω_m remains in the narrow bandwidth of the wave analyzer during the time of measurement.

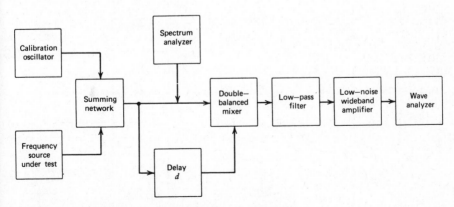

Figure 2-49.   Calibration of the test setup shown in Fig. 2-47.

PHASE-LOCKED LOOP METHOD. Phase-locked loops are successfully utilized in spectrum analysis of low-phase-noise crystal oscillators. A functional block diagram of a PLL phase-noise test setup is shown in Fig. 2-50. The system consists of a low-noise, voltage-controlled crystal oscillator (VCXO), a phase detector (a double-balanced mixer), an operational amplifier with a bandwidth wide enough to pass the highest frequency component of noise, and a loop filter that determines the loop bandwidth. If the leakage of the signal at f_0 to the output of the phase detector is excessive, a low-pass filter is used to provide adequate attenuation at f_0. The system is designed so as to result in very loose phase locking, that is, very small loop bandwidth. A typical VCXO tuning rate is 1 part in 10^9 per volt. Voltage at the output of the phase detector varies as phase in short term but as frequency in long term, so that the VCXO is locked to the crystal oscillator under test. This voltage is proportional to the double-sided phase noises of both oscillators. The double-sided phase noise of each oscillator, therefore, is 3 dB lower than the measured noise.

This equipment is calibrated by replacing the crystal oscillator under test with a VCXO operating at the same frequency, by frequency-modulating the VCXO at the rate equal to f_m until a conveniently measurable sideband level is obtained, and by measuring the voltage at f_m at the output of the low-noise operational amplifier. The test setup can be calibrated also by offsetting the oscillator under test slightly until the phase lock is broken and a beat note at $\Delta f = f_1 - f_2$ appears at the output of the phase detector (see Fig. 2-51). A substitute offset frequency source may be used if the frequency of the oscillator under test cannot be varied by the amount

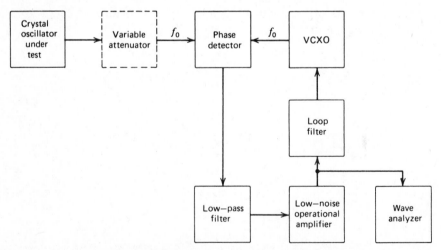

Figure 2-50. **Phase-locked loop technique of measuring phase noise.**

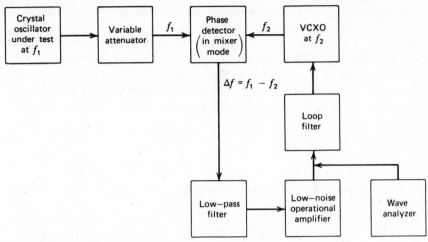

Figure 2-51.   Calibration of the test setup shown in Fig. 2-50.

necessary to break the lock. The variable attenuator is used to reduce the level of the beat note so as to prevent this signal from saturating the low-noise operational amplifier. The level of the beat note, measured with the wave analyzer, and the setting of the attenuator serve as the references relative to which noise is measured.

Frequency multipliers can be used to increase the dynamic range of the spectrum analyzer as shown in Fig. 2-52. This technique is sometimes

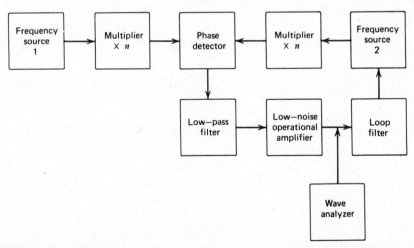

Figure 2-52.   High-dynamic-range test setup for measuring phase noise, phase-locked loop technique.

employed in measuring the phase noise of frequency synthesizers that have provisions for being voltage controlled.

Another way of utilizing phase locking in phase-noise measurements is shown in Fig. 2-53. Two frequency sources are used in conjunction with two multipliers, a mixer, and a phase detector. The frequency of one source is offset so that the difference between the selected nth harmonic of the sources, nf_1 and nf_2, respectively, is equal to the frequency of one of the sources:

$$f_{\text{IF}} = n(f_1 - f_2)$$

or, when the loop is phase locked ($f_{\text{IF}} = f_2$),

$$f_1 - f_2 = \frac{f_2}{n}$$

The output of the phase detector contains the phase noises of both frequency sources enhanced by the multiplication factor, n.

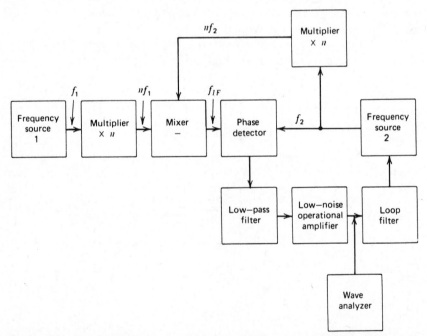

Figure 2-53.   High-dynamic-range test setup for measuring phase noise, phase-locked loop–mixer technique.

FREQUENCY DISCRIMINATOR METHOD. Frequency discriminators find some use in measurements of phase noise at microwave frequencies. In the test setup shown in Fig. 2-54 the signal of a frequency source under test is downconverted to the frequency of an IF discriminator. A band-pass filter and a tuned IF amplifier attenuate all unwanted intermodulation products and amplify the desired mixer output. Phase noise is measured in terms of an equivalent frequency deviation at a given modulation rate, f_m, that is, f_m Hz away from the IF frequency.

This test setup does not possess a large dynamic range capability in the region close to signal, as compared to other methods described above, because of the relatively high level of $1/f$ noise inherent in discriminators.

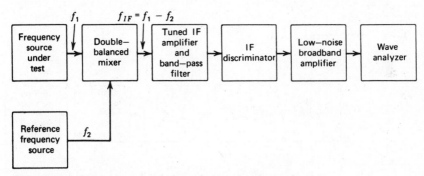

Figure 2-54.   IF discriminator technique of measuring phase noise.

DIGITAL COUNTER METHOD, THE TIME DOMAIN. The digital counter, or the multiple-period measuring test setup used in determining the variance of the fractional frequency fluctuations, is shown in Fig. 2-55. Two signal sources, slightly offset in average frequency, feed a mixer. The difference frequency, f_3, is used to trigger a Schmitt trigger at the zero crossings. The low-pass filter attenuates f_1 and f_2, leaking to the output of the mixer, $f_1 + f_2$, and higher-order intermodulation products and prevents false triggering. The period of the sharp leading edge of the Schmitt trigger is measured by the counter and displayed by the analog recorder, and each measurement is printed out on the digital recorder. The equation used to compute the value of the standard deviation of the fractional frequency fluctuations directly from the measured data for $N = 2$ is given in Ref. 27 as

$$\sigma\left[\frac{\Delta f}{f_1}(2, T, \tau, \Delta B)\right] = \left(\frac{f_3}{f_1}\right)\left(\frac{1}{\tau f_{TB}}\right)\left[\frac{1}{N-1}\sum_{k=1}^{N-1}\frac{(n_{k+1}-n_k)^2}{2}\right]^{1/2}, \quad (2\text{-}92)$$

where $T =$ repetition rate for measurements of duration τ (sec) (often $T = \tau$, or zero dead time between measurements),

$f_1 =$ signal frequency (Hz),

$f_3 =$ frequency offset between two frequency sources (in Hz),

$\tau =$ averaging time of one measurement (sec),

$f_{TB} =$ time base frequency of the counter used for the period measurement (in Hz),

$N =$ positive integer giving the number of data points used in obtaining a sample variance (100 or more),

$n_1, n_2, \ldots, n_N =$ number shown on the output record of the digital printer,

$\Delta B =$ bandwidth of the measuring system (Hz).

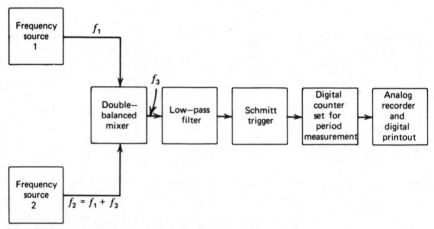

Figure 2-55. Measurement of fractional frequency fluctuations. Proceedings IEEE, Vol. 54, No. 2 (February 1966). Reprinted by permission.

The parameters T, τ, and ΔB must be specified for any given measurement of fractional frequency fluctuations because the value of the Allan variance depends to some extent on these variables.

If it is assumed that the two frequency sources have identical phase-noise contents, the estimated fractional frequency deviation of each source is $1/\sqrt{2}$ (or 3 dB) smaller than the measured deviation.

Long-term frequency drift data should be taken on both sources before making measurements of fractional frequency fluctuations. The long-term frequency drift expected during these measurements should be estimated from the frequency drift data and removed from the estimate of the fractional frequency deviation.

To increase the resolution and sensitivity of the measurement, frequency multipliers can be used in both channels. If carefully designed, the multipliers remove any amplitude changes by limiting, introducing little AM-to-PM conversion.

Measurement of fractional frequency fluctuations in frequency synthesizers is accomplished by utilizing a modified version of this test setup (see Fig. 2-56). The dual-channel approach used in this case (Ref. 27) provides cancellation of the reference oscillator noise and permits the residual phase noises of both synthesizers to be measured. The digital counter is driven by the external reference oscillator, thus eliminating the possibility of phase modulation, which would occur if a small frequency offset between the internal-to-the-counter reference oscillator and the external reference used to drive both synthesizers were present. This modulation could be sufficiently large to invalidate the data.

Basic problems associated with taking measurements in the time domain are similar to the problems arising in the frequency domain. These are discussed next.

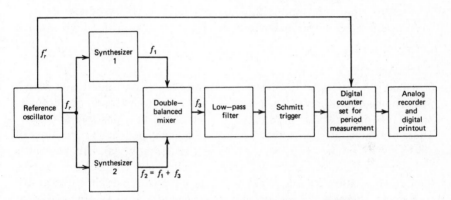

Figure 2-56. Measurement of fractional frequency fluctuations in synthesizers.

INTERFERENCE.   Low-level phase-noise measurements present many problems. The problems that are most often encountered in practice are associated with implementation of the test equipment and with the environment in which tests are performed and must be identified and eliminated by the test engineer himself. Only a few problems tend to present difficulties in practically all cases. These are as follows.

1. Ground loops. Reduction of interference due to instrumentation is one of the most difficult tasks to be performed before accurate phase-noise

measurements can be made. Good isolation transformers and grounding techniques have to be utilized to avoid ground loops due to ac current flowing through the line filter capacitors and primary-to-secondary winding capacitors of the test equipment used in measurements. The interested student will find an extensive discussion of this subject in Ref. 46.

2. Microphonic noise. Rotational devices used in the cooling of test equipment, such as oscilloscopes, frequency counters, and dc power supplies located near the frequency sources under test, result in low-frequency microphonic noise. The level of this noise depends on the sensitivity to mechanical vibration of the circuits undergoing testing and can be 10 to 20 dB higher than that of the noise being measured. Equipment utilizing these devices should be physically isolated from the frequency sources under test.

3. Impedance matching. The data obtained with any one of the phase-noise-measuring test setups described above depend on the rf power level of the signals whose frequency spectra are analyzed. Poor impedance matching anywhere between the frequency sources under test and the test equipment and within the test setup itself introduces calibration errors. Good impedance matching is mandatory for high accuracies.

SUMMARY OF IN-HOUSE EQUIPMENT.   Having been confronted with such a wide choice of phase-noise-measuring equipment in the frequency domain, one may wonder about the failure to specify the best possible approach. There are three basic reasons why such a choice is not made here. First of all, hardware implementation of some test setups at certain frequencies is not as convenient or economical as that of other test setups. For example, the autocorrelation technique cannot be economically implemented at low frequencies because of delay line requirements; it is more usable at frequencies above 100 MHz. At these frequencies it has two advantages, compared to other approaches, that make it very attractive: (a) only one frequency source, that which is tested, is required; and (b) the same test equipment is used over a multioctave frequency band, whereas the wide-dynamic-range mixer-PLL test setup of Fig. 2-53, for example, requires numerous frequency multipliers to cover even an octave band. Second, the nature of the frequency sources to be evaluated often dictates the choice of an approach. Thus VCOs and VCXOs are most easily tested by PLL techniques, Figs. 2-50 to 2-53, whereas synthesizers with no provisions for voltage control cannot be evaluated in this way. Finally, because of the random nature of the quantity measured and the many factors that may (and do) interfere with measurements, it is highly advisable, in order to achieve a high degree of confidence in measured data, to make phase-noise measurements by utilizing two different techniques.

COMMERCIAL EQUIPMENT. Presently available commercial equipment designed to measure phase noise is scarce and is intended for use at microwave frequencies only. The FEL phase jitter analyzer, model 800B, operates between 1 and 12.4 GHz with the sensitivity shown in Fig. 2-57a. When the internal 8 MHz local oscillator is replaced by a 100 MHz external oscillator (see Fig. 2-58), the sensitivity of the phase jitter analyzer is improved as shown in Fig. 2-57b. Spectra Electronics equipment covers 1 to 18 GHz with 10 plug-in units, models SE200 through SE209. The sensitivity of Spectra Electronics equipment at 9.8 GHz is plotted in Fig. 2-59.

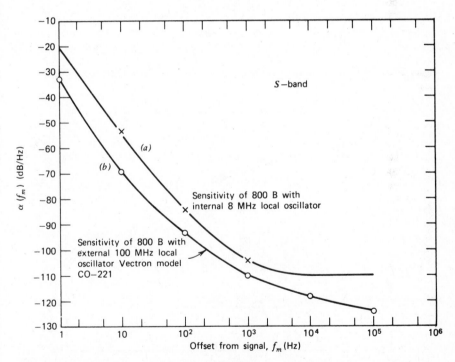

Figure 2-57. Sensitivity of the FEL phase jitter analyzer, model 800B. Courtesy of Frequency Engineering Labs.

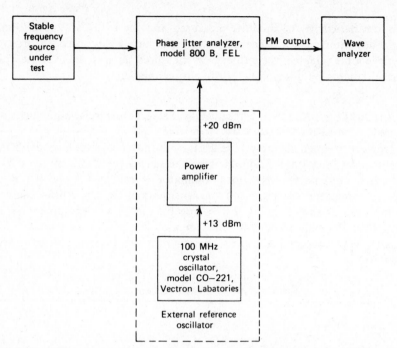

Figure 2-58. **FEL test setup for measuring phase noise. Courtesy of Frequency Engineering Labs.**

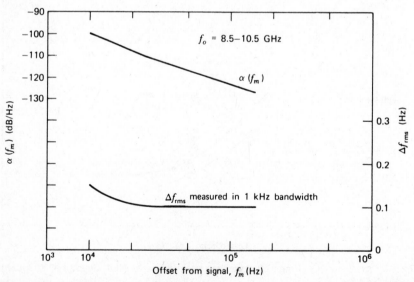

Figure 2-59. **Sensitivity of the Spectra Electronics phase-noise analyzer, model SE-200. Courtesy of Spectra Electronics, Division of Ancom.**

RECOMMENDED TEST EQUIPMENT. The following list of equipment will assist the test engineer in the design of a phase-noise-measuring test setup.

1. Double-balanced mixers
    (a) Hewlett-Packard, model 10514A
         Frequency range: 200 kHz to 500 MHz
         LO power: +7 dBm
    (b) Watkins-Johnson, model WJ-M7A
         Frequency range: 10 to 1200 MHz
         LO power: +7 dBm
    (c) Lorch Electronics Corporation, model FC-235Z
         Frequency range: 2 to 1000 MHz
         LO power: +20 dBm
2. Low-noise wideband amplifier
    Aertech Industries, model V1100
    Voltage gain: 45 dB (typical)
    3 dB bandwidth: 325 Hz to 16 MHz
    Noise: total of 5 μV rms referred to the input in the specified 3 dB bandwidth
3. Delay line
    Fixed delay: 1000 ft of $\frac{1}{2}$ in., 50 Ω foam polyethylene dielectric cable, type FXA 12-50J, manufactured by Cablewave Systems
    Variable delay: to be selected depending on the operating frequency range
4. Wave analyzers (frequency-selective voltmeters)
    (a) Quan-Tech Laboratories, model 304
         Frequency range: 1 Hz to 5 kHz
         Narrowest 3 dB bandwidth: 1 Hz
    (b) Hewlett-Packard, model 3580A signal analyzer
         Frequency range: 5 Hz to 50 KHz
         Narrowest 3 dB bandwidth: 1 Hz
    (c) Philco-Ford, model 128A
         Frequency range: 10 kHz to 15 MHz
         Narrowest 3 dB bandwidth: 250 Hz

2-3 Bandwidth-Switching Speed Considerations

The time needed to switch a synthesizer from one frequency to another is one of the major requirements. All synthesizers use some kind of filtering, active or passive. In many cases the narrowest filter bandwidth affects the switching time of the system. It is important, therefore, to be able to estimate the longest switching time associated with the narrowest bandwidth of the filters chosen either to attenuate certain spurious outputs or to reduce phase noise.

The switching time of analog and digital phase-locked loops is considered in Chapters 4 and 5, respectively. This section of Chapter 2 deals with the switching time of passive filters.

A well-known fundamental relationship between the 3 dB bandwidth of a filter, $\Delta B_{3\,\mathrm{dB}}$, and the rise time, t_r, is

$$ t_r \cong \frac{k}{\Delta B_{3\,\mathrm{dB}}}, \qquad (2\text{-}93) $$

where

$$ 0.3 \leqslant k \leqslant 0.45 $$

and $t_r \overset{\Delta}{=}$ the time required for an impulse to rise from 10% to 90% of its final value.

This simplified expression for the filter switching time does not convey any information with regard to the overshoot and ringing of the filter response, which often are of great importance also. Both the overshoot and the ringing depend on the type of filter, design parameters, and number of poles and are an integral part of filter design, which is a field in its own right and is outside the scope of this book. The reader is referred to the book by A. I. Zverev (Ref. 47, Chapter 7) for an excellent discussion of filter characteristics in the time domain and for the plots of transient response for various types of filters.

2-4 Technique of Measuring Synthesizer Switching Time

Of great importance to the users of remotely controlled synthesizers is the synthesizer switching time. It is defined as the time elapsed between a command to change frequency and the time the synthesizer generates an output signal at the desired frequency with the specified frequency error.

Figure 2-60 is the block diagram of the test setup used to measure the switching time (Ref. 2-49). Two synthesizers driven by the same reference source are set to the frequency f_0 at which the measurement is made. The frequency of synthesizer 1 is kept constant during the measurement. The frequency of synthesizer 2 is varied at a rate that is equal to the repetition rate of the trigger pulse, T. The pulse, provided by the pulse generator, triggers the circuits of the frequency switching network which generate the frequency control inputs to synthesizer 2.

The waveforms associated with the switching time measurement are

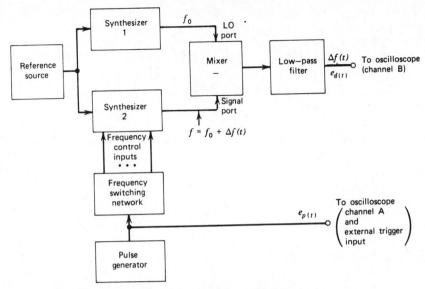

Figure 2-60.   Test setup for measuring synthesizer switching time.

shown in Fig. 2-61. When the frequencies of two synthesizers are equal, the mixer operates in a phase-detector mode and the low-pass filter output (after the switching transient dies out) contains no ac component. Whenever synthesizer 2 is switched to a frequency f different from f_0, a beat note appears at the output of the low-pass filter. The filter attenuates all low-order mixer intermodulation products with the exception of the difference product. The time it takes for the beat note to disappear after the application of a trigger pulse is the synthesizer switching time. The slope of the transient observed during the transition phase is equal to the frequency error at the time of interest.

The test is performed as follows. With the trigger pulse removed from the input of the frequency switching network, both synthesizers are set to f_0. The levels of the signals applied to the mixer inputs are set so that the mixer operates in a linear mode. To calibrate this test setup the frequency of synthesizer 2 is slightly offset to produce a beat note at the low-pass filter output and the peak amplitude of the beat note, E_m, is measured. Synthesizer 2 is set again to f_0 and a trigger pulse is applied to the input of the frequency switching network. The transition phase takes place during the time the mixer operates in the phase-detector mode. Hence the expression of the voltage transient displayed on the oscilloscope in terms of the phase angle between the mixer input signals is

$$e_d(\theta) = E_m \sin\theta \qquad (2\text{-}94)$$

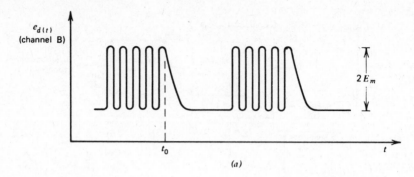

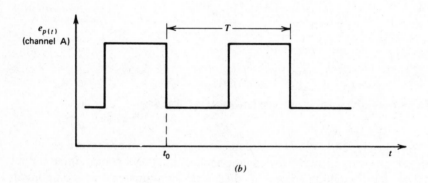

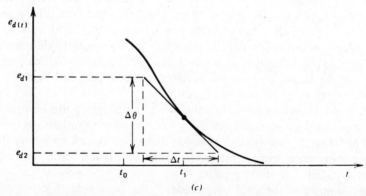

Figure 2-61. Waveforms associated with switching time measurement. (*a*) **Low-pass filter output waveform.** (*b*) **Trigger pulse.** (*c*) **Expanded view of the low-pass filter output waveform during the transition phase.**

and for $\theta \ll 1$

$$e_d(\theta) \cong E_m \theta. \qquad (2\text{-}95)$$

Having measured e_{d1} and e_{d2}, we compute θ_1 and θ_2 using Eq. 2-94 or 2-95. The frequency error at the time t_1 is

$$\Delta f_{\text{error}} = \frac{\Delta\theta}{\Delta t}\bigg|_{t=t_1}, \qquad (2\text{-}96)$$

where $\Delta\theta = \theta_1 - \theta_2$.

Depending on the synthesizer frequency and application, the specified phase and frequency errors may be as small as 0.1 rad and 1 Hz, respectively.

In describing this technique of measuring synthesizer switching time it was assumed that (a) the rise and fall times of the trigger pulse are orders of magnitude shorter than the synthesizer switching time; (b) the time delay introduced by the frequency switching network is negligible compared to the switching time; (c) the repetition rate of the trigger pulse is slow enough to establish the steady-state condition before the transition takes place, and (d) the 3 dB cutoff frequency of the low-pass filter is high enough to pass the beat note unattenuated and to introduce a negligible delay in the transient response of the test setup.

2-5   An Example of System Design

A well-thought-out design procedure is the first major step to the successful completion of a synthesizer project. There are probably as many "best" design procedures as there are experienced synthesizer designers. This section is intended to assist an inexperienced reader in the development of his own design "style."

Given a synthesizer specification, the designer studies the most important requirements, such as frequency range, frequency increments, spurious outputs, phase noise, and switching time, and decides on a synthesis approach that satisfies these requirements and results in the most economical system occupying the smallest possible volume and operating at the lowest dc power. Having done this, the designer may proceed in the following manner.

Step 1.   Frequency plan. On the basis of knowledge associated with such considerations as spurious outputs, phase noise, switching time, and ease of frequency selection, the designer prepares a preliminary frequency plan for the system. For example, if the digital synthesis approach is chosen, the decision is made whether to use a single- or multiloop approach. This plan is modified as the analysis progresses.

Step 2.   Intermodulation products. Intermodulation products generated in mixers used in frequency synthesis are determined from charts or from appropriate computer programs, and the frequency plan is modified so as to shift all high-level spurious outputs out of operating bands.

3. Preliminary filter design. A preliminary filter design is performed to establish the feasibility of reducing out-of-band spurious outputs to the required level by standard filter techniques. Attenuation provided by PLLs and frequency dividers, if such are used, is included in the considerations on which the design is based.

Step 4. Phase noise. A preliminary phase-noise analysis is performed to determine whether or not the phase-noise requirement can be met with the selected approach and commercially available reference sources.

Step 5. Block diagram. A preliminary detailed block diagram is prepared upon successful completion of steps 1 to 4.

Step 6. Radio-frequency power levels. On the basis of his knowledge of circuitry and system requirements, the designer establishes rf power levels at the input and output of each building block on the detailed block diagram.

Step 7. Leakage of rf signals. Expected leakage of rf signals through circuits, such as the local oscillator signal leakage to the output of a mixer, is identified, power levels of leaking signals are estimated, and possible spurious combinations with desired signals are investigated. The block diagram is modified if additional circuits are required to reduce the leakage.

Step 8. Discrete FM sideband components. Based on the synthesizer requirement for discrete FM sideband components, tolerable limits for FM components are established at the output of various sections of the synthesizer and realization of these limits with conventional circuitry is investigated. In cases where there are no readily available answers, tests should be performed to determine the expected levels of FM components before initiation of the hardware design and construction. This investigation concerns the effects of power supply ripple in particular.

Step 9. Schematic diagram. Upon successful completion of steps 1 to 8, a preliminary schematic diagram of the synthesizer circuitry is prepared.

Step 10. Packaging. With a schematic diagram available, the synthesizer circuitry is divided into modules, and the modules are divided in turn into shielded compartments to provide adequate shielding of sensitive-to-pick-up circuits.

Step 11. Schedule. Based on the information provided by preliminary block and schematic diagrams, manpower requirements are defined and a milestone schedule is prepared.

Step 12. Circuit design. Having gone this far, the designer proceeds with a detailed hardware design and construction.

At any time during the development of the equipment further modifications of the preliminary frequency plan may be necessary, in which case some of the steps described above will have to be repeated.

The example given below demonstrates the phase-noise analysis of a digital synthesizer. Only the noise of the voltage-controlled oscillator and reference source is considered. Other sources of noise should be treated in a similar manner.

The specified single sideband-phase-noise-to-signal ratio is -80 dB in 200 Hz bandwidth centered at an offset frequency of 500 Hz. It is assumed that other requirements of the synthesizer specification are satisfied by the choice of the single-loop approach shown in Fig. 2-1.

Having read the section on the oscillator phase noise, one knows that an oscillator displays a $1/f_m^2$ or $1/f_m^3$ performance at offset frequencies close to signal (Fig. 2-35). One is also aware of the fact that under this condition, for phase-noise measurements performed on a synthesizer to be meaningful, the bandwidth of the test equipment should be significantly smaller than the offset frequency at which the noise is measured. Probably at the 500 Hz offset frequency the noise will be measured with a 7 or 10 Hz bandwidth wave analyzer such as Hewlett-Packard model 302A. Hence, to make a comparison between the measured and the specified noise possible, the requirement must be expressed in 1 Hz bandwidth. Equation 2-61 cannot be used in this case because the phase-noise spectrum is not flat in the band of interest.

If the envelope of the predicted noise spectrum in the region of interest is approximated by straight line segments with more than one slope (as in the region of f_m below $\frac{1}{2}B_{osc}$ in Fig. 2-35), the spectrum derivation is achieved by successive approximations. Several curves representing various single sideband-phase-noise-to-signal ratios in 1 Hz bandwidth are plotted, each curve is integrated over the specified frequency band, and the curve that, when integrated, satisfies the total noise requirement is selected as the synthesizer specification. Each line segment is integrated individually, and the sum of noise powers for each curve is computed by using Fig. 2-62.

If the envelope of the predicted spectrum is approximated by one line segment, a straightforward way of arriving at the envelope of the noise spectrum may be employed. One proceeds as follows.

total single sideband-phase-noise-to-signal ratio, $\left(\dfrac{N}{S}\right)_T = \displaystyle\int_{f_L}^{f_H} P(f_m)\, df_m,$

$$(2\text{-}97)$$

where $P(f_m)=$ envelope function of the noise spectrum in 1 Hz bandwidth,

$f_L=$ low end of the frequency band (Hz),

$f_H=$ high end of the band (Hz).

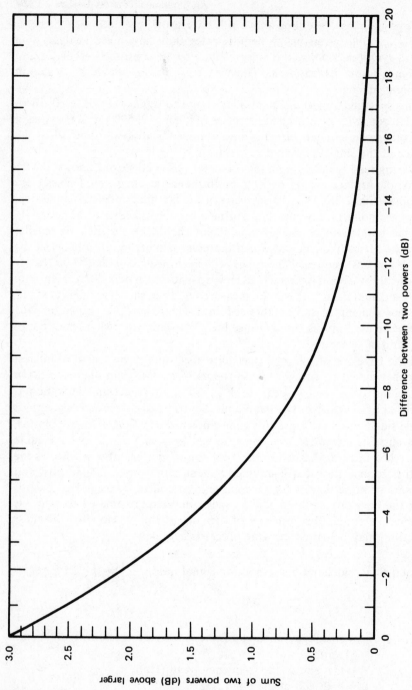

Figure 2-62. Sum of two powers expressed in decibels. Courtesy of ITT Defense Communications Division.

Difference between two powers (dB)

Sum of two powers (dB) above larger

154

The function $P(f_m)$ is determined from the knowledge of the system. For example, if the requirement specifies close-in noise, the synthesizer is designed so that the noise at the final synthesizer frequency is determined by the noise of the reference source. Figure 2-40, curve 1, is the phase-noise plot of a typical 1 MHz reference source. The slope of the curve is 6 dB/octave in the band of interest, hence the noise of the synthesizer can be approximated by the equation $P(f_m) = k/f_m^2$, where k is the slope of the line segment to be integrated, $f_L = 400$ Hz, and $f_H = 600$ Hz.

For $n \neq -1$ (Ref. 48),

$$\int_{f_1}^{f_2} k f^n \, df = k \left(\frac{f_2^{n+1}}{n+1} - \frac{f_1^{n+1}}{n+1} \right).$$

Hence

$$\int_{f_L}^{f_H} P(f_m) \, df_m = \int_{f_L}^{f_H} \frac{k}{f_m^2} \, df_m = k \left(\frac{1}{f_L} - \frac{1}{f_H} \right).$$

The specified total single sideband-phase-noise-to-signal ratio is -80 dB, so that

$$10 \log_{10} \left(\frac{N}{S} \right)_T = -80 \text{ dB}$$

or

$$\left(\frac{N}{S} \right)_T = 10^{-8}$$

and

$$-k = \frac{10^{-8}}{1/600 - 1/400},$$

$$k = 1.2 \times 10^{-5}.$$

The equation of the envelope is

$$f(f_m) = 10 \log_{10} \left(\frac{k}{f_m^2} \right) \text{ dB}.$$

Thus, at $f_m = f_L = 400$ Hz

$$f(f_L) = 10 \log_{10} \left(\frac{1.2 \times 10^{-5}}{1.6 \times 10^5} \right) \cong -101 \text{ dB}$$

and at $f_m = f_H = 600$ Hz

$$f(f_H) = 10\log_{10}\left(\frac{1.2 \times 10^{-5}}{3.6 \times 10^5}\right) \cong -105 \text{ dB}.$$

A straight line drawn through these two points (see Fig. 2-63), is the -80 dB requirement expressed in 1 Hz bandwidth.

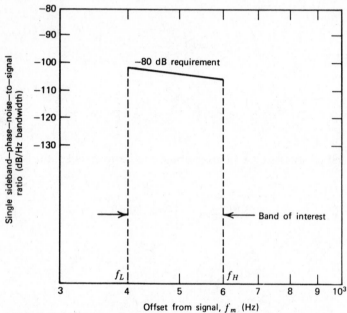

Figure 2-63.   A vhf phase-noise requirement, referred to 1 Hz bandwidth.

Figure 2-64 demonstrates that, if the single-loop synthesis approach of Fig. 2-1 were selected, the noise of the reference source multiplied to the final synthesizer frequency, curve 3, would exceed the -80 dB requirement by more than 50 dB. Table 2-4 will assist the designer in following the mathematical steps used in computing $f(f_m)$ and plotting the curves of Fig. 2-64. Since phase noise increases or decreases, as the case may be, by a ratio equal to $20\log_{10}n$ in decibels, where n is the frequency multiplication number or division ratio, respectively, the voltage ratio columns of Table 2-4 were used throughout the computations to determine by how many decibels a curve should rise or fall as the signal passes through a frequency multiplier or divider.

Notice that curve 1 does not fall by 60 dB, as the divider ratio of 1000

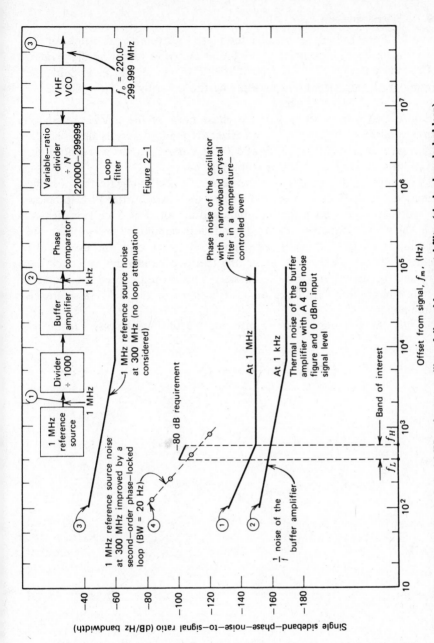

Figure 2-64. Phase-noise analysis, 1 MHz reference oscillator followed by a crystal filter (single phase-locked loop).

157

would lead one to believe, because the noise spectrum at the output of the divider is limited by the noise of the buffer amplifier following the divider.

The noise rise of 109 dB from curve 2 to curve 3 is caused by the phase-locked loop effectively multiplying the noise by N. Here the worst-case ratio of 299999 was chosen.

Figure 2-64 demonstrates that the phase noise of the 1 MHz reference source multiplied to the final synthesizer frequency exceeds the -80 dB requirement by more than 50 dB at 500 Hz, curve 3, unless a second-order PLL with a 20 Hz bandwidth is employed, curve 4.

The results of the analysis performed in Fig. 2-65 indicate that, for the estimated uhf VCO noise to be reduced by the second-order PLL to below the -80 dB requirement, the loop bandwidth should be 5 kHz or greater. As is shown in Chapter 5, the loop bandwidth cannot be greater than the sampling frequency, which in the example cited is 1 kHz, and even if one could reduce the VCO noise by making the loop bandwidth approach the sampling frequency, the total synthesizer noise would exceed the requirement by a prohibitively large amount because of the 1 MHz reference source noise.

Neither the reference source noise nor the uhf VCO noise can be re-

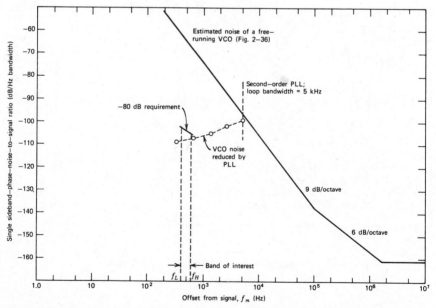

Figure 2-65.   Phase-noise analysis, vhf VCO phase locked to a reference source (second-order phase-locked loop).

duced to the required degree. The single-loop digital synthesizer, therefore, is not the right approach for this particular application.

To demonstrate a significant, but not adequate, improvement in phase-
noise performance an auxiliary phase-locked loop is added to the system as
shown in Fig. 2-2. For convenience, the two-loop synthesizer is redrawn in
Fig. 2-66 with the frequencies and division ratios indicated to point out the
multiplication factors along the main and auxiliary PLL paths of synthesis
that have to be considered in determining the synthesizer noise due to the
reference source.

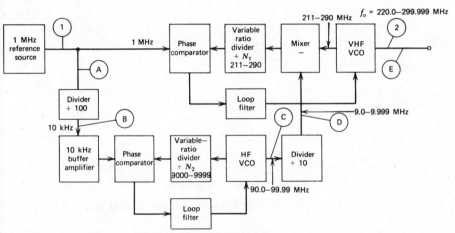

Figure 2-66. A double-loop digital frequency synthesizer.

A two-loop approach allows the bandwidth of the main loop to exceed 5
kHz so that the uhf VCO noise is neglected in further analysis. If a 5 dB
margin is assumed, the highest allowable multiplication factor along either
path of synthesis is either 40 dB, which is the difference between curve 1 in
Fig. 2-64 and the −80 dB requirement, or 50 dB, the difference between
curve 2 and the requirement, depending on whether the 1 MHz reference
signal is directly multiplied to the final frequency or first divided by a ratio
greater than 14 dB, which is the difference between curves 1 and 2 in Fig.
2-64.

Since the selected multiplication factor along the main-loop path in Fig.
2-66 exceeds the 40 dB limit by approximately 9 dB, one would expect the
phase-noise performance of this two-loop synthesizer to fall short also of
meeting the −80 dB requirement by the same amount. This is demonstrated in Fig. 2-67, which shows a step-by-step procedure for determining
the phase noise of the reference source multiplied to the final synthesizer

frequency. To assist the designer in adding two noise powers (or any number), the curve in Fig. 2-62 is provided. The use of this curve is self-explanatory.

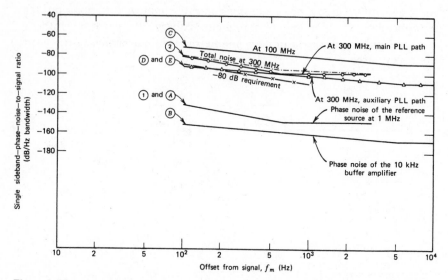

Figure 2-67. Phase-noise analysis, 1 MHz reference oscillator followed by a crystal filter (double-loop frequency synthesizer).

Finally, Fig. 2-68 is a summary of the uhf VCO noise, the total reference source noise at the final synthesizer frequency, and the sum of these two noise sources. As one would expect, the total system phase noise is approximately equal to the noise of the reference source. The contribution of the uhf VCO is less than 0.5 dB and could be made smaller by widening the main-loop bandwidth even further.

At the circuit design phase during which the uhf VCO gain constant is determined, the noise due to the uhf VCO reactance control circuit is estimated, using Eq. 2-72, and compared to the requirement. Further widening of the main-loop bandwidth may be necessary to reduce the VCO noise to below the level specified.

To keep this presentation simple, the phase noise of the hf VCO in the auxiliary PLL has been neglected. Also, no consideration has been given to spurious outputs that result from mixing two signals in the main PLL mixer. It is suggested that the interested student explore these areas on his own and determine (*a*) whether or not the approach shown in Fig. 2-66 is practical if the spurious-output requirement is -80 dB, (*b*) what problems

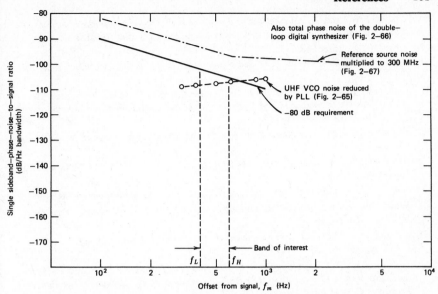

Figure 2-68. Estimated phase noise of the double-loop digital synthesizer shown in Fig. 2-66.

are associated with the choice of frequencies given in the figure, and (*c*) what can be done to make the system meet a -90 dB spurious-output requirement.

References

1. Mouw, R. B. and S. M. Fukuchi. "Broadband Double Balanced Mixer/Modulators," Part I. *The Microwave Journal*, March 1969, pp. 131–134.

2. Steiner, J. W. "An Analysis of Radio Frequency Interference due to Mixer Intermodulation Products," *IEEE Transactions on Electromagnetic Compatibility*, January 1964, pp. 62–68.

3. Pollack, H. W. and M. Engelson. "An Analysis of Spurious Response Levels in Microwave Receivers," *The Microwave Journal*, December 1962, pp. 72–78.

4. Markel, J. D. "Shrinking Intermodulation," *EDN*, August 1967, pp. 56–65.

5. Olson, W. R. and R. V. Salcedo. "Mixer Frequency Charts," *Frequency*, March/April 1966, pp. 24–25.

6. Shores, M. W. "Chart Pinpoints Receiver Interference Problems," *EDN*, January 15, 1969, pp. 43–46.

7. Pearl, B. "How to Determine Spur Frequencies," *EDN*, October 1965, pp. 128–129.

8. Westwood, D. H. "Rid Mixer of Spurious Signals," *Electronic Design*, August 16, 1966, pp. 210–216.

9. Fairley, D. O. "Noise Considerations for Solid State Microwave Sources in High Capacity FM Radio Systems," *Telecommunications*, April 1968, pp. 11–16.

10. George, S. F. and J. W. Wood. *Ideal Limiting*, Part 1. (Washington, D. C.: U. S. Naval Research Laboratory, AD 266069, October 2, 1961).

10a. Magnus, W. and F. Oberhettinger. *Formulas and Theorems for the Functions of Mathematical Physics* (New York: Chelsea Publishing Company, 1954), p. 1.

11. Cross, T. G. "Intermodulation Noise in FM Systems Due to Transmission Deviation and AM/PM Conversion," *The Bell System Technical Journal*, December 1966, pp. 1749–1773.

12. Chapman, R. C. and J. B. Millard. "Intelligible Crosstalk Between Frequency Modulated Carriers through AM-PM Conversion," *IEEE Transactions on Communication Systems* June 1964, pp. 160–166.

13. *ITT Reference Data for Radio Engineers*, 5th ed. (New York: Howard W. Sams & Company, 1968).

14. Goldman, S. *Frequency Analysis, Modulation and Noise* (New York: McGraw-Hill Book Company, 1948).

15. Bell Telephone Laboratories. *Transmission Systems for Communications.*, 1970.

16. Warren, W. B. "Suppression of Spurious Signals by Frequency Division," *Frequency*, October 1968, pp. 15–17.

17. Tykulsky, A. "Spectral Measurements of Oscillators," *Proceedings of the IEEE*, February 1966, p. 306.

18. Horn, C. H. "A Carrier Suppression Technique for Measuring S/N and Carrier/ Sideband Ratios Greater than 120 dB," *Proceedings of the Annual Symposium on Frequency Control*, May 1969, pp. 223–235.

19. Engelson, M. and R. Breaker. "Interpreting Incidental FM Specifications," *Frequency Technology*, February 1969, pp. 13–15.

20. Engelson, I. "Pinning Down 'Frequency Stability'," *EDN*, May 15, 1969, pp. 43–50.

21. Dimitrios, J. T. "Spurious Modulation in Phase-locked Oscillator Systems," *Frequency*, September/October 1965, pp. 28–30.

22. Shields, R. B. "Review of the Specification and Measurement of Short-Term Stability," *The Microwave Journal*, June 1969, pp. 49–55.

23. Nelson, J. N. and R. D. Frost. "AM and FM Noise in Low-noise TWT Amplifiers with Integral Power Supplies," *The Microwave Journal*, April 1971, pp. 45–50.

24. Schwartz, M. *Information Transmission, Modulation, and Noise* (New York: McGraw-Hill Book Company, 1959).

25. Bagdady, E. J., R. N. Lincoln, and B. D. Nelin. "Short-Term Frequency Stability: Characterization, Theory, and Measurement," *Proceedings of the IEEE*, July 1965, pp. 704–722.

26. Ondria, J. G. "A Microwave System for Measuring of AM and FM Noise Spectra," *IEEE Transactions on Microwave Theory and Techniques*, Vol. MTT-16, No. 9 (September 1968), pp. 767–781.

27. Meyer, D. G. *An Ultra Low Noise Direct Frequency Synthesizer*. John Fluke Mfg. Company, Seattle, Washington July 1970.

28. Shields, R. B. "Review of the Specification and Measurement of Short-Term Stability," *The Microwave Journal*, June 1969, pp. 49–55.

29. Hewlett Packard Company, Frequency and Time Standards, *Application Note 52*, Section III, pp. 3-1 through 3-7. November 1965.

30. Kelly, J. "Phase Jitter and Its Measurement," *Telecommunications*, July 1970, pp. 28–31.

31. Van Duzer, V. "Short-Term Stability Measurements," IV-87, *Interim Proceedings of the Symposium on the Definition of Short-Term Frequency Stability*, X-521-64-380 (Greenbelt, Md.: Goddard Space Flight Center, 1964).

32. Halford, D. Lectures at 1969 National Bureau of Standards Seminar on Frequency and Time Stability.

33. Cutler, L. S. "Some Aspects of the Theory and Measurements of Frequency Fluctuations in Frequency Standards." Proceedings IEEE. NASA Symposium on the Definition and Measurement of Short Term Frequency Stability, November 1964, pp. 89–100.

34. Barnes, J. A. and others. "Characterization of Frequency Stability," *National Bureau of Standards Technical Note* 394, issued October 1970.

35. Johnson, J. B. "Thermal Agitation of Electricity in Conductors," *Physical Review*, Vol. 32 (1928), p. 97.

36. Nyquist, H. "Thermal Agitation of Electric Charge in Conductors," *Physical Review*, Vol. 32 (1928), p. 110.

37. Grove, A. S. "Don't Just Fight Semiconductor Noise," *Electronic Design*, August 16, 1969, pp. 228–235.

38. Halford, D., A. E. Wainwright, and J. A. Barnes. "Flicker Noise of Phase in RF Amplifiers and Frequency Multipliers: Characterization, Cause, and Cure," *Proceedings of 20th Annual Symposium on Frequency Control*, April 1968, pp. 340–341.

39. Leeson, D. B. "A Simple Model of Feedback Oscillator Noise Spectrum" *Proceedings of the IEEE*, Vol. 54, No. 2 (February 1966).

40. Cutler, L. S. and C. L. Searle. "Some Aspects of the Theory and Measurement of Frequency Fluctuations in Frequency Standards," *Proceedings of the IEEE*, February 1966, pp. 136–154.

41. Leeson, D. B. "Short Term Stable Microwave Sources, "*The Microwave Journal*, June 1970, pp. 59–69.

42. Stewart, J. L. "Frequency Modulation Noise in Oscillators," *Proceedings of the IRE*, March 1956, pp. 372–376.

43. Hafner, E. "Stability of Crystal Oscillators," *Proceedings of the 14th Annual Symposium on Frequency Control*, 1960, pp. 192–199.

44. Halford, D. "Phase Noise in RF Amplifiers and Frequency Multipliers," U. S. Government Memorandum to J. A. Barnes, Chief. 253.00, National Bureau of Standards, October 25, 1967.

45. Halford, D. "Phase Noise in RF Amplifiers and Frequency Multipliers," U. S. Government Memorandum to J. A. Barnes, Chief. 253.00 National Bureau of Standards, October 30, 1967.

45a. Davenport, W. B. "Signal-to-Noise Ratios in Band-Pass Limiters," *Journal of Applied Physics*, Vol. 24, No. 6 (June 1953), pp. 720–727.

46. Morrison, R. *Grounding and Shielding Techniques in Instrumentation* (New York: John Wiley and Sons, 1967).

47. Zverev, A. I. *Handbook of Filter Synthesis* (New York: John Wiley and Sons, 1967).

48. Bigsbee, E. M. *Five-Place Mathematics Tables* (Ames, Iowa: Littlefield, Adams & Company, 1955), p. 151.

49. General Radio Company. Frequency Synthesizers Application Note 1, "Reviewing Switching-Speed Performance," 1975.

3 Shielding

The analysis of spurious outputs presented in Chapter 2 provides guidance in selecting a system that can be implemented. Unfortunately, a good system design achieved on paper is not sufficient for the successful development of a product satisfying the stringent requirements of electronic systems. Spurious outputs, in addition to being generated in frequency synthesis, result from the following:

1. A defect in circuit design, such as a lack of circuit decoupling, which allows spurious leakages through power supply lines.
2. An improper circuit layout technique, such as poor grounding.
3. Inadequate shielding practices.

The circuit design and layout techniques associated with spurious outputs are outside the scope of this book. The subject is treated extensively in Refs. 1 to 12. However, a few general rules applicable to the design of an ultralow-spurious-output synthesizer will be formulated.

1. Take action to suppress all extraneous signals. Even signals whose frequencies fall outside the band of the circuitry may be translated in band by a converter action taking place in amplifiers, multipliers, mixers, dividers, and oscillators.
2. If circuits operating at various frequencies are biased from the same dc power source, provide broadband isolation at the dc power input of each circuit. This rule stresses preference for resistor-capacitor (RC) over inductor-capacitor (LC) filtering.
3. When both analog and digital circuits are utilized in frequency synthesis, provide two dc power supplies. The use of separate power supplies prevents high-rise-time transients generated by the digital circuits from reaching the analog circuits.
4. Synthesize at frequencies above 3 kHz (preferably above 10 kHz). Problems associated with low impedance fields are most severe at low frequencies.

164

5. Place power supply transformers as far from the circuitry used in frequency synthesis as practicable to minimize pickup at the power line frequency.
6. A conducting strip on a PC (printed circuit) board or a wire carrying rf current is an antenna that can both transmit and receive. Make these conductors as short as possible.
7. Proceed with circuit layouts such that the paths of signals which are periodic pulses start and terminate within the same module, that is, are not transmitted from one module to another by way of cables. (A practical example of this approach is given at the end of this chapter. Also see the section on GI/ESD synthesizers, Chapter 7.) The extra cost for components necessary to achieve this effect is well justified in view of the cost of (a) engineering effort required to identify the problem otherwise incurred and (b) filters used to control levels of spurious signals originating from cable radiation.
8. To prevent generation of ground return currents in shields do not mount rf-current-carrying components on the shields.

Two approaches are widely used in the design of shielding enclosures for synthesizers. One approach is to disregard possible problems associated with shielding in the initial stage of design and to work on solving such problems as they arise during system testing. This approach is costly and time consuming and often does not lead to satisfactory results within the cost and time frames of the project because solutions to some basic shielding problems require extensive mechanical redesign of the equipment. The opposite of this is a brute-force approach that results in an overdesigned, oversized, and costly system.

In this chapter we attempt to provide guidance as to what physical phenomena are of frequent occurrence in the electromechanical design of synthesizers, to assist the reader in gaining a basic knowledge of phenomena that are detrimental to the design, to present several analytical aids, and to help the reader in acquiring some insight into radiation-and-pickup problems that do not easily lend themselves to analytical investigation by describing these problems, giving examples of practical solutions, demonstrating applications of these techniques in the construction of synthesizers presently in use (see also Chapter 7), and supplementing the material with numerous references, which the reader is urged to examine.

3-1   Electrostatic Fields

Electrostatic fields are produced by oscillating charges. The mechanism of electrostatic coupling is through mutual capacitance, whereby a voltage in one circuit results in a current in another circuit because of capacitive coupling between the two circuits.

Absence of a Shield

One of the most common configurations of capacitive coupling consists of two points (*A* and *B* in Fig. 3-1) above a ground plane. These points could be the turret ends of two ceramic standoff terminals used for mounting electrical components. Stray capacitances exist between points *A* and *B* and between each point and the ground, C_{AB}, C_{AG}, and C_{BG}, respectively. If a voltage V_{AG} is applied between *A* and the ground, there will develop a voltage V_{BG} between point *B* and the ground because of a stray capacitive coupling between the two points. The magnitude of this voltage is

$$V_{BG} = V_{AG} \left(\frac{C_{AB}}{C_{AB} + C_{BG}} \right). \tag{3-1}$$

In case of V_{BG} applied to point *B*,

$$V_{AG} = V_{BG} \left(\frac{C_{AB}}{C_{AB} + C_{AG}} \right). \tag{3-2}$$

To illustrate this case we assume that the circuit considered is a times-3 multiplier and a multipole band-pass filter. The filter selects the third harmonic of a 1 MHz signal and attenuates the spurious signals at the fundamental frequency and at all undesired harmonics of 1 MHz by 90 dB referred to the final output. Let the voltage at the multiplier input $V_{1\,\text{MHz}}$ = 500 mV rms, the output voltage $V_{3\,\text{MHz}} = 100$ mV rms, the load capacitance $C_l = 15$ pF, and the stray feedback capacitance of the circuit due to the layout and other factors $C_f = 0.001$ pF. Then the leakage of the input

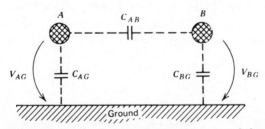

Figure 3-1.   Stray capacitance associated with two points above a ground plane.

signal to the output is

$$V_{1\,\text{MHz, spur}} = V_{1\,\text{MHz}}\left(\frac{C_f}{C_f + C_l}\right) \cong 33.3 \ \mu\text{V rms.}$$

This is only 69.6 dB below the desired output. If one compares the leakage levels of undesired harmonics of 1 MHz, similar results are obtained. In general, each frequency synthesis operation involves signals at two or more frequencies with a required suppression of spurious signals varying between 60 and 110 dB. Due consideration should be given, therefore, to stray capacitive coupling at the initial stage of circuit design and layout.

In the example given above, a 90 dB isolation may be attained by inserting an electrostatic shield between the multiplier and the filter. Depending on whether or not several paths of coupling exist, other steps may have to be taken as well. For example, if the multiplier dc power supply lead (in case of point-by-point wiring) were wired parallel and in close proximity to the leads connecting the input and output of the multiplier-filter configuration to the appropriate rf connectors, a spurious leakage exceeding 90 dB would probably occur even if a shield were used. In such cases coupling is reduced by increasing the distance between the leads or decreasing their length, as Eqs. 3-3 and 3-4, associated with Fig. 3-2, indicate. These equations describe the capacitance between two parallel circular conductors in terms of the length and radius of each conductor, L and b, respectively, and the separation between the conductors, D:

$$C = \frac{\pi\epsilon L}{\cosh^{-1}(D/2b)} \ \text{F} \qquad (3\text{-}3)$$

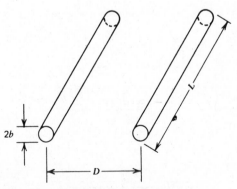

Figure 3-2.   Capacitance of two parallel circular conductors.

and when the radius of each conductor is small and the separation is large,

$$C = \frac{\pi \epsilon L}{\log(D/b)} \ \text{F,} \qquad (3\text{-}4)$$

where ϵ is a dielectric constant of the medium (in farads per meter) and L, b, and D are expressed in meters (Ref. 13, p. 119).

A better approach is to use the metal plate on which all electrical components of the circuit are mounted as an electrostatic shield by placing all rf-current-carrying components on one side of the plate and running the power supply lead on the other side. A similar technique can be employed in the case of a printed circuit board and sufficient electrostatic isolation obtained at no extra cost.

Although the problems associated with capacitive coupling often cannot be dealt with analytically within the time frame of a project because of the very large number of cases that require investigation, minimizing stray capacitance in circuit layout whenever possible should be a matter of routine for every designer.

For convenience, we include a few other expressions for stray capacitance commonly encountered in practice. A familiar expression for the capacitance between two parallel plates, Fig. 3-3, in terms of physical parameters is

$$C = \frac{\epsilon S}{d} \ \text{F,} \qquad (3\text{-}5)$$

where d = distance between two plates (m),
  S = area of each plate (m²),
  ϵ = dielectric constant of the medium between the plates (F/m).

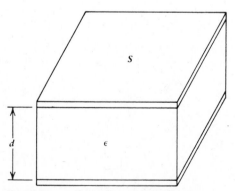

Figure 3-3.   Capacitance of two parallel plates.

The equation for a conducting cylinder of radius b and a ground plane at a distance h from the cylinder, Fig. 3-4, is (Ref. 13, p. 118)

$$C = \frac{2\pi\epsilon L}{\log\left(h + \sqrt{h^2 - b^2} \, / b\right)}, \qquad (3\text{-}6)$$

where L = length of conducting cylinder (m),
ϵ = dielectric constant of the medium between the cylinder and the ground plane (F/m).

Equations 3-3 to 3-6 are derived in Ref. 13, pp. 114–120.

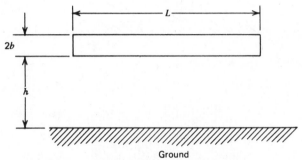

Figure 3-4.   Capacitance of a conducting cylinder above a ground plane.

Electrostatic Shields

When a metal shield is placed between two points in space and is electrically connected to a ground plate, as shown in Fig. 3-5, it assumes plate potential and nearly all lines of force extending from one point to another are diverted to the shield, which, therefore, represents an almost perfect electrostatic shield.

A mesh of wire in good contact with the ground at a point is sufficient to provide good electrostatic shielding because the screening effect of a metal shield is not affected by small holes or long, narrow slots. Such an arrangement of parallel wires or strips of tin foil, Fig. 3-6, is used whenever screening against low-frequency electrostatic fields is required, but eddy currents within the screen are undesirable. At high frequencies, however, the impedance of the screen or of the lead that connects the screen to the ground may become comparable with the impedance of the capacitance between the screen and the source of radiation. In such a case, the screen

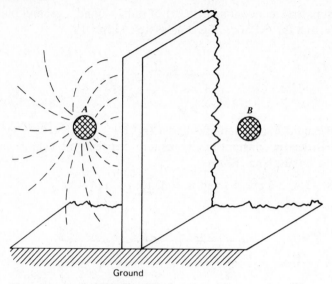

Ground

Figure 3-5.   An electrostatic shield.

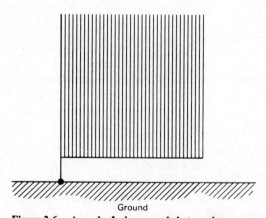

Ground

Figure 3-6.   A mesh of wire grounded at a point.

potential no longer is zero and the screen does not provide the required isolation. This is demonstrated as follows.

From Fig. 3-7

$$V_{SG} = V_{AG}\left(\frac{j\omega L_{SG}}{j\omega L_{SG} + 1/j\omega C_{AS}}\right) = V_{AG}\left(\frac{1}{1 - 1/\omega^2 L_{SG} C_{AS}}\right)$$

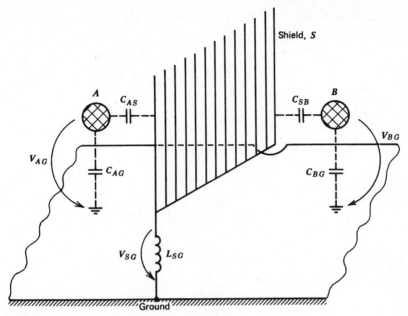

Figure 3-7.   A mesh of wire at high frequencies.

and

$$V_{BG} = V_{SG} \left(\frac{C_{SB}}{C_{SB} + C_{BG}} \right).$$

Therefore

$$V_{BG} = V_{AG} \left(\frac{1}{1 - 1/\omega^2 L_{SG} C_{AS}} \right) \left(\frac{C_{SB}}{C_{SB} + C_{BG}} \right). \qquad (3\text{-}7)$$

Equation 3-7 indicates that the voltage at point B due to a capacitive-inductive coupling to point A, V_{BG}, is dependent on frequency so that, as the frequency increases, V_{BG} approaches $V_{AG}[C_{SB}/(C_{SB} + C_{BG})]$ and the screening effect accordingly decreases.

Ungrounded or Improperly Grounded Shields

For reasons of economy the common practice has been to plate individual parts of a module with a poorly conductive or nonconductive finish such as anodizing and to assemble the module using screws, eyelets, or rivets to hold parts in place. Under these conditions electrical contact between the module and internal shields is poor and intermittent; therefore leakage of spurious signals is excessive at times, resulting in unreliable equipment.

This situation is described in Fig. 3-8 and is analyzed below. Again, as in the previous cases,

$$V_{SG} = V_{AG} \left(\frac{C_{AS}}{C_{AS} + C_{SG}} \right)$$

and

$$V_{BG} = V_{SG} \left(\frac{C_{SB}}{C_{SB} + C_{BG}} \right).$$

Hence

$$V_{BG} = V_{AG} \left(\frac{C_{AS}}{C_{AS} + C_{SG}} \right) \left(\frac{C_{SB}}{C_{SB} + C_{BG}} \right). \tag{3-8}$$

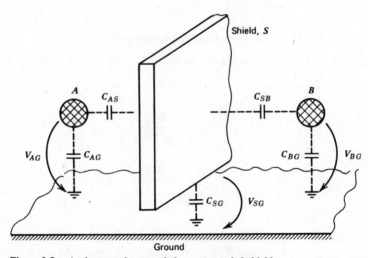

Figure 3-8.   An improperly grounded or ungrounded shield.

Hence the poorer the electrical contact between the ground and the shield (the smaller C_{SG}), the greater V_{BG} and the poorer the screening effect of the shield are.

Screening Provided by the Cover

Whenever spacing between a cover and a chassis can be made small as shown in Fig. 3-9, adequate electrostatic screening between two points in the chassis can be achieved without insertion of a shield because most of

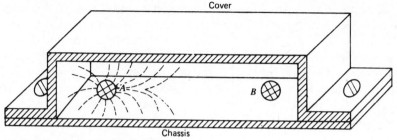

Figure 3-9. Screening provided by a cover.

the lines of force of the field generated at one point will be drawn toward the cover and the chassis.

Screening by Separation

Capacitive coupling may be reduced by spacing the points of interest (i.e., by reducing stray capacitance). This approach is used whenever space is not at a premium.

3-2 Electromagnetic Fields

Electric and magnetic components of the time-varying field, associated with the voltage and current, respectively, exist simultaneously in any simple ac circuit, although one of the field components can be stronger than the other. An energy field that contains both electric and magnetic components is called an electromagnetic field. The ratio of electric intensity to magnetic intensity in such a field is the impedance of the field. It is expressed as

$$Z = \frac{E \ (\text{V/m})}{H \ (\text{A/m})} . \tag{3-9}$$

The electromagnetic field whose impedance is equal to 376.7 Ω, the impedance of free space, is a plane-wave field. It is so named because a relatively small portion of the spherical wave front arriving at the shield is plane. Fields with impedance greater than 376.7 Ω are called electric or high-impedance fields; fields with impedance smaller than 376.7 Ω, magnetic or low-impedance fields.

The frequency of a field, f, is the rate at which the field polarity alternates. The wavelength of a field, λ, is the distance that electromagnetic energy travels in one cycle. Both electric and magnetic fields gradually become plane-wave fields. All fields, if allowed to propagate freely through space removed from any other electromagnetic source, will have an impedance of 376.7 Ω beyond approximately one wavelength from their source (Ref. 14, p. 7).

In synthesizer work spurious magnetic fields present far more difficult problems than electric fields, and in most practical cases measures taken to suppress magnetic fields are more than adequate to suppress electric fields also. For this reason due attention is given to studies of magnetic field sources and attenuation of magnetic radiation.

3-3 Magnetic Fields

Magnetic fields are generally produced by the motion of charged particles. Such fields exist around wires carrying current, rf coils, and transformers. The mechanism of magnetic energy transfer is through mutual inductance between two circuits: current flowing in a wire in one circuit produces flux that induces voltage in another circuit.

Shielding of equipment against magnetic fields is costly and problematic at low frequencies, between approximately 10 Hz and 3 kHz, when attenuation of the field depends primarily on the absorption loss of the shielding material. At high frequencies available shielding materials provide magnetic field attenuation that is more than adequate for most practical purposes, and problems arise mainly from leakage of spurious signals through discontinuities in shields, such as holes and slots, and interwiring pickup.

The most common source of magnetic radiation is a wire carrying rf current. The magnetic intensity of a current element in an infinite nonconducting medium is described by (Ref. 15, p. 87)

$$H_\phi = \frac{IL\sin\theta}{4\pi r^2}, \qquad (3\text{-}10)$$

where I = current in a wire (A),
L = length of the wire element (m),
r = distance from the element to a point in space (m),
and ϕ and θ are as defined in Fig. 3-10.

Another important source of magnetic radiation is a solenoid carrying rf current. A special case of practical importance is a very thin, closely wound solenoid. The magnetic field outside the solenoid is (Ref. 15, pp. 85, 86)

$$\mathbf{H} = \frac{\Phi L}{2\pi\mu r^3}(\cos\theta\mathbf{a}_r + \sin\theta\mathbf{a}_\theta), \qquad (3\text{-}11)$$

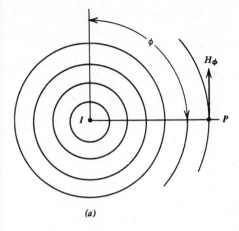

(a)

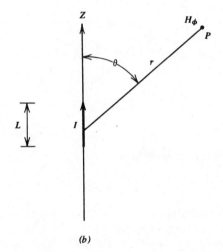

(b)

Figure 3-10. Magnetic intensity of a current element in an infinite nonconducting medium. (a) Top view; (b) side view.

where Φ = magnetic flux emerging from one end of the solenoid and converging to the other end,

L = length of the solenoid,

μ = permeability of the medium,

r = distance from the solenoid to a point P in space,

$\mathbf{a}_r$ and $\mathbf{a}_\theta$ = unit vectors in directions of r and θ, respectively,

and θ is as defined in Fig. 3-11.

In both cases cited above, the magnetic field intensity at a sensitive-to-pickup circuit depends on the current drawn by the source of spurious

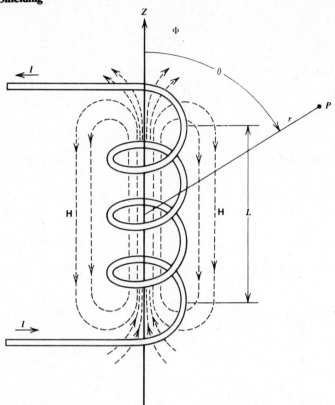

Figure 3-11.   Magnetic field of a solenoid carrying current.

radiation, the length of the radiating element, and the distance from the radiating source to a sensitive-to-pickup circuit. The reduction of the field with distance is 12 dB/octave for a current element and 18 dB/octave for a solenoid.

3-4   Electromagnetic Shields

In working on solutions of shielding problems, the nature of the field (electric or magnetic), the frequency of the field, the distance from the source of spurious radiation, and the isolation provided by spacing between each source of radiation and a sensitive-to-pickup circuit in the absence of any shield or with a known shield have to be considered.

For this purpose sources of radiation and sensitive-to-pickup circuits should be identified as soon as the synthesizer block diagram is prepared and operating rf power levels are established. The circuit designer may have to perform some preliminary circuitry testing in the initial phase of system development to determine the isolation provided by spacing.

With this information available the designer can proceed with the selection of a shielding material.

Shielding Effectiveness

Shielding effectiveness is a measure of how effective a conducting barrier is in preventing the propagation of electromagnetic energy. It is defined (Ref. 12, p. 2-39) as

$$SE = R + A + B \tag{3-12}$$

or

$$SE = 20 \log_{10} \frac{E_1}{E_2} = 20 \log_{10} \frac{H_1}{H_2}, \tag{3-13}$$

where SE = shielding effectiveness (dB),
R = reflection power loss of the first and second boundary (dB),
A = absorption power loss (dB),
B = B-factor (dB), which is neglected if A is greater than 10 dB,
E_1 = intensity of incident electric field,
E_2 = intensity of electric field passing the shield,
H_1 = intensity of incident magnetic field,
H_2 = intensity of magnetic field passing the shield.
Figure 3-12 describes the shielding process schematically.

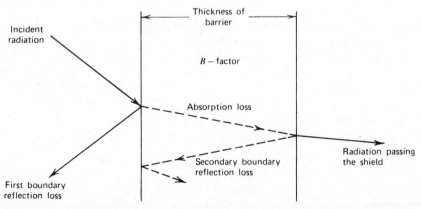

Figure 3-12. Shielding process associated with a metal barrier. Filtron Company, Inc., West-bury, N.Y., *Interference Reduction Guide for Design Engineers,* **Vol. I, August 1, 1964. Reprinted by permission.**

The shielding effectiveness of a metal barrier is a function of the frequency and impedance of the incident field, the distance from the source of radiation, and such parameters of the barrier as thickness, permeability, and conductivity. For clarity the discussion of shielding effectiveness is presented in two parts: (*a*) electric and magnetic waves (near fields) and (*b*) plane waves (far fields).

Shielding against Near Fields

As defined in Eq. 3-12, the shielding effectiveness of a conducting barrier consists of three parts, R, A, and B. In most practical problems associated with synthesizer design the contribution of B to shielding effectiveness is negligible and is, therefore, omitted from the following discussion.

If the effect of multiple reflections inside the shield is neglected, the total reflection loss from both surfaces is

$$R = 20 \log_{10} \left| \frac{(Z_s + Z_w)^2}{4 Z_s Z_w} \right|, \tag{3-14}$$

where R = reflection loss (dB),
$\quad Z_s$ = intrinsic impedance of conducting barrier (Ω),
$\quad Z_w$ = impedance of incident field (Ω).
Hence a large impedance mismatch results in a large reflection loss.

The intrinsic impedance of a conducting barrier is

$$Z_s = (1 + j) \sqrt{\frac{\mu f}{2G}} \; (3.69 \times 10^{-7}), \tag{3-15}$$

where μ = relative magnetic permeability of shield referred to free space,
$\quad f$ = frequency of incident field (Hz),
$\quad G$ = relative conductivity of shield referred to copper.
For copper μ and G are equal to 1 at frequencies between 60 and 10^8 Hz. For aluminum the corresponding values are 1 and 0.60, respectively. For iron the value of G is 0.17, and the values of μ versus frequency are given in Table 3-1.

For high-impedance fields the impedance of the incident field is

$$Z_{wE} = -j \left(\frac{0.71 \times 10^{12}}{fr} \right), \tag{3-16}$$

where Z_{wE} = impedance of electric field (Ω),
$\quad r$ = distance from the source of radiation to the shield (in.).

Table 3-1.   Relative Magnetic Permeability of Iron Referred to Free Space^a versus Frequency ^{b}

Frequency (Hz)	Relative permeability
60	1000
10^3	1000
10^4	1000
1.5×10^5	1000
10^6	700
3.0×10^6	600
10^7	500
1.5×10^7	400
10^8	100
10^9	50
1.5×10^9	10
10^{10}	1

^{a}Permeability of free space $= 1.26 \times 10^{-6}$ H/m.
^{b}Filtron Company, Inc., Westbury, N. Y., Inter-ference *Reduction Guide for Design Engineers*, Vol. 1, August 1, 1964. Reprinted by permission.

Equation 3-16 is a good approximation of the field impedance of a very short nonresonant dipole of length L when

$$L \ll \mu \ll \lambda.$$

For low-impedance fields the impedance of the incident field is

$$Z_{wH} = +j(0.2 \times 10^{-6}) fr, \qquad (3\text{-}17)$$

where $Z_{wH} =$ impedance of magnetic field (Ω), and f and r are defined above. Equation 3-17 is an accurate estimate of the field impedance of a very small loop with diameter D when

$$D \ll \mu \ll \lambda.$$

The absorption loss inside the shield is an I^2R loss and is dissipated in the form of heat. It is the same for both electric and magnetic fields and is expressed as

$$A = (3.338 \times 10^{-3})(t) \sqrt{Gf\mu} , \qquad (3\text{-}18)$$

where A = absorption loss of the shield (dB),
     t = thickness of the shield (mils),
and G, f, and μ are defined above.

The expressions for R and A, given in Ref. 12, pp. 2-44 to 2-47, are included in this chapter for instructive purposes only. Practical analysis is easier to perform with the aid of nomograms, Figs. 3-13, 3-14, 3-15, and 3-17 (Ref. 16).

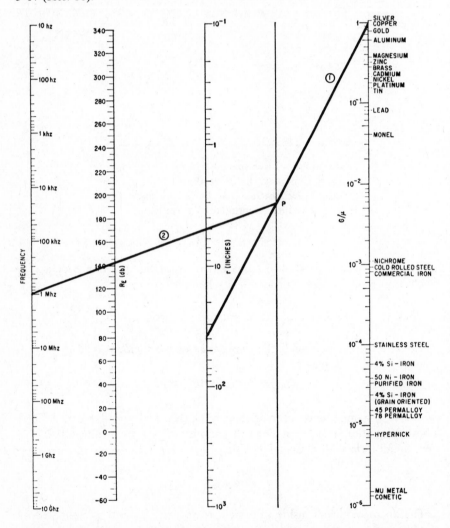

Figure 3-13.   Electric field reflection losses. *Electronics,* **April 17, 1967; copyright McGraw-Hill, Inc., 1967. Reprinted by permission.**

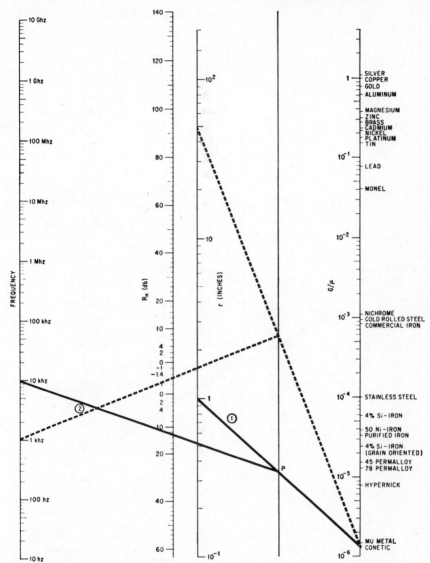

Figure 3-14.   Magnetic field reflection losses. *Electronics*, **April 17, 1967; copyright McGraw-Hill, Inc., 1967. Reprinted by permission.**

Electric field reflection loss, R_E, is determined from Fig. 3-13. In the example cited in the figure the distance from a source of radiation to the shield, r, is 40 in. and the material is copper. Line 1 is drawn through these points, intersecting the unmarked scale at point P. Line 2, drawn through

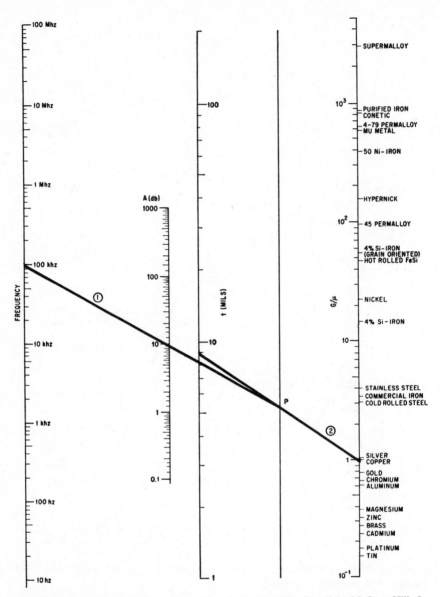

Figure 3-15. Absorption losses. *Electronics*, April 17, 1967; Copyright McGraw-Hill, Inc., 1967. Reprinted by permission.

182

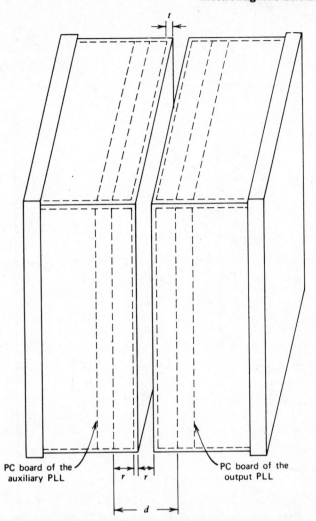

Figure 3-16.   Two modules of the synthesizer considered in the example on p. 185.

P and the frequency of the field (1 MHz), intersects R_E at 142 dB. Negative values of reflection loss indicate increased coupling, which can occur at certain combinations of parameters.

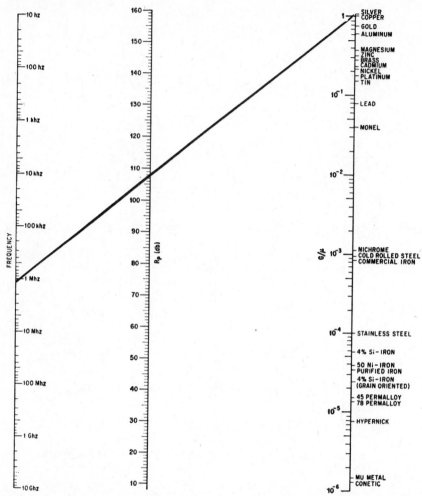

Figure 3-17.   Plane-wave reflection losses. *Electronics,* **April 17, 1967; copyright McGraw-Hill, Inc., 1967. Reprinted by permission.**

Magnetic field reflection loss, R_H, is determined from Fig. 3-14 in the same manner as electric field loss. The broken-line example indicates that coupling to the shield is enhanced at certain distances and frequencies.

The thickness of material required for a given absorption loss is obtained from Fig. 3-15. In the example considered, a 10 dB absorption loss is to be achieved with a copper shield at 100 kHz. Line 1 is drawn between $A = 10$ dB and $f = 100$ kHz, intersecting the unmarked scale at P. Point P

and the material (copper) fix line 2 and determine the minimum thickness of 9 mils.

A numerical example best illustrates a practical application of electromagnetic shielding theory. Consider a coherent indirect synthesizer consisting of an output and auxiliary digital phase-locked loops (Fig. 3-16). Assume the following design parameters.

1. Each phase-locked loop (PLL) is wired on a printed circuit board located in a rectangular metal box with a cover. The preferred wall thickness of both modules, t, is 62 mils. This wall thickness is not mandatory, however, and, if necessary, may be increased to 125 mils.
2. The two modules are mounted adjacently to each other as shown in Fig. 3-16. The distance between spurious sources and sensitive-to-pickup circuitry, d, is approximately $\frac{1}{2}$ in.
3. The maximum operating rf power level of circuits identified as spurious sources and located in the auxiliary PLL module is $+10$ dBm.
4. The circuit of highest sensitivity to pickup is the error-voltage varicap of the voltage-controlled oscillator located in the output PLL module. The allowable level of pickup at the VCO error voltage input is -120 dBm at any frequency between 500 Hz and 30 MHz.
5. The radiation-and-pickup efficiency of spurious sources and high-sensitivity circuits is unity, that is, the required isolation is 130 dB.

To keep the cost of the circuitry low it is desirable to design the synthesizer so that the lowest frequency of spurious sources is 1 kHz. Raising this frequency to 10 kHz would result in a more costly circuit design. The task of the designer is to determine whether or not a 130 dB isolation can be obtained at 1 kHz, using a low-cost material for module construction.

Let us evaluate aluminum as a possible choice for shielding. Aluminum is inexpensive and lightweight and lends itself to casting. Since an equal amount of isolation is provided by the walls of each module, the requirement for one shield is 65 dB, one half of the total requirement. The spacing from the source to the shield, r, is $\frac{1}{8}$ in. Therefore the design parameters are as follows: $t = 62$ mils, $SE = 65$ dB, and $r = 0.125$ in. We determine the shielding effectiveness of aluminum at frequencies of interest with the help of the nomographs in Figs. 3-13, 3-14, and 3-15.

Table 3-2 summarizes the results of the computations. Two conclusions can be reached from studying the data presented in the table: (1) aluminum is an excellent shield against electric fields at frequencies between 1 kHz and 30 MHz; and (2) the magnetic field energy attenuation provided by aluminum at low frequencies under the conditions specified above is far from adequate.

Table 3-2.   Shielding Effectiveness of Aluminum at $r = \frac{1}{8}$ in. and $t = 62$ mils

Parameter	Frequency			
(dB)	1 kHz	10 kHz	1 MHz	30 MHz
R_E	282	251	190	145
R_H	−1	4	22	37
A	5	16	170	900
SE				
E-field	287	267	360	Greater than 1000
H-field	4	20	192	937
Total attenuation due to				
two shields, H-field	8	40	384	Greater than 1000

By doubling the distance from the source to the shield and the aluminum thickness we increase the reflection and absorption losses; see Table 3-3. This improvement by itself is not sufficient at both 1 and 10 kHz. However, for purposes of simplifying the analysis it was assumed that there was no power loss in energy transfer from the spurious sources of radiation to the sensitive-to-pickup circuits. As the magnetic wave propagates in space, there is an energy loss with distance even if no shield is used. The isolation thus achieved depends on the frequency of the field, the type of magnetic field ($1/r^2$ or $1/r^3$), the circuit separation, and the layout. Although methods of estimating this type of loss are known in some simple configurations, such as two parallel wires considered later in this chapter, measurements conducted at an early stage of the system development constitute a more practical approach to determining what the loss is. We shall assume in this example that the measured isolation at the initial $\frac{1}{2}$-in. spacing between the output and auxiliary PLLs ($d = \frac{1}{2}$ in.) is 40 dB at 1 and 10 kHz. Moreover, at least 12 dB of isolation is gained because of doubling the separation. Hence the total isolation provided by 1 in. spacing and two aluminum shields 125 mils thick placed $\frac{1}{4}$ in. away from the source of radiation is 76 and 132 dB at 1 and 10 kHz, respectively. It is clear that aluminum does not, by far, meet the isolation requirement at 1 kHz. Of course, further separation of the output and auxiliary PLLs would eventually result in a 130 dB isolation at 1 kHz, but in most practical cases synthesizer size is of prime importance and means other than spacing are employed to attain a specified system performance.

Let us evaluate cold-rolled steel as a possible shielding material at $r = \frac{1}{8}$

Table 3-3. Magnetic Field Shielding Effectiveness of Aluminum at
$r = \frac{1}{4}$ **in. and** $t = 125$ **mils**

Parameter (dB)	Frequency (kHz)	
	1	10
R_H	2	10
A	10	30
SE, H-field	12	40
Total isolation due to two shields	24	80
Isolation due to the initial spacing between the two PC boards ($d = \frac{1}{2}$ in.)	40	40
Isolation due to doubling of the spacing	12	12
Total isolation, H-field (two shields)	76	132

in. ($d = \frac{1}{2}$ in.) and with two different thicknesses, $t_1 = 62$ mils and $t_2 = 125$ mils. Table 3-4 summarizes the results of the computations. Since the measured isolation for $\frac{1}{2}$ in. spacing is 40 dB, the total isolation provided by the spacing and two cold-rolled steel shields 62 mils thick placed $\frac{1}{8}$ in. away from the source of radiation is 92 and 120 dB at 1 and 10 kHz, respectively. Doubling the shield thickness, we arrive at 110 and 182 dB.

From these data we conclude that neither aluminum nor steel provides a good shield against a magnetic field at 1 kHz, and other materials, such as Mu-metal, which is costly and shock sensitive, must be employed to obtain the required 130 dB attenuation of magnetic field at 1 kHz. At 10 kHz aluminum provides adequate attenuation with proper material thickness and spacing that do not exceed practical limits, even though a high degree of isolation is required. However, if the synthesizer size were a governing requirement, choice of the cold-rolled steel would result in a smaller volume at the cost of an increase in weight.

Table 3-4. Magnetic Field Shielding Effectiveness of Cold-Rolled Steel at $r = \frac{1}{8}$ in., $t_1 = 62$ mils, and $t_2 = 125$ mils

Parameter (dB)	Frequency (kHz)			
	1		10	
R_H	14		4	
A				
$\quad t_1$	12		36	
$\quad t_2$		21		67
SE, H-field				
$\quad t_1$	26		40	
$\quad t_2$		35		71
Total attenuation due to two shields (H-field)				
$\quad t_1$	52		80	
$\quad t_2$		70		142
Isolation due to $\frac{1}{2}$ in. spacing	40		40	
Total isolation				
$\quad t_1$	92		120	
$\quad t_2$		110		182

Shielding against Far Fields

Usually, if a shield performs adequately with respect to electric and magnetic fields, it displays satisfactory performance with respect to plane-wave fields also. The equations for reflection loss in the case of a plane wave are the same as Eqs. 3-14 and 3-15, but the field impedance is equal to the impedance of free space, or 376.6 Ω. Hence a simplified expression for reflection loss is (Ref. 12, p. 2-47)

$$R_p = 108.2 + 10\log_{10}\left(\frac{G \times 10^6}{\mu f}\right) \text{ dB.} \qquad (3\text{-}19)$$

The expression for absorption loss is the same for plane waves as it is for electric or magnetic waves. It is given by Eq. 3-18.

The shielding effectiveness of a metal for plane waves can be easily determined from Fig. 3-15, the absorption loss nomogram, and from Fig. 3-17, the reflection loss nomogram. Plane-wave reflection loss, R_p, is calculated by drawing a line connecting the frequency with the appropriate value of G/μ.

Shielding Discontinuities

Leakage of spurious signals through discontinuities in shields is a serious problem for the designer of rf equipment. Provisions have to be made for intercircuit connections; modules must be designed with covers for easy access to circuitry; ventilation of equipment is needed to remove excess heat. These requirements necessitate the presence of apertures through which undesirable radiation takes place.

A simplified description of the mechanism involved is given in Ref. 17, p. 7. When the lines of magnetic field incident on a conducting barrier with a slot are parallel to the slot, a current I_i is induced by the magnetic field in the upper layer of the barrier exposed to the field (see Fig. 3-18a). The slot in the barrier interferes with the flow of current, so that a voltage E_i is developed across it. If the barrier is thin, there is also a voltage E_o, which nearly equals E_i, on the opposite side of the barrier (see Fig. 3-18b). This voltage is a source of spurious field generation on the side of the barrier opposite the incident field and constitutes a degradation of the barrier shielding effectiveness.

By changing the slot into a row of small holes one reduces the interference presented by the slot to the flow of induced current (and hence reduces E_i and E_o); this is the principle of rf gasketing, considered further in this chapter. There are, however, some situations wherein proper positioning of electronic components eliminates the necessity for gasketing. Consider, for example, a solenoid mounted on two Teflon standoffs in compartment 1 of a two-compartment module with its axis parallel to a slot between the cover and a partition, as shown in Fig. 3-19a. Under these conditions the lines of the magnetic field that is produced by the solenoid are also parallel; hence the current induced in the shield is perpendicular to the slot, resulting in maximum leakage through the slot. If the solenoid is positioned as shown in Fig. 3-19b, the direction of the current is parallel to the slot, and the resistance presented by the slot to the current flow is significantly smaller than in the first case described above and so is the magnitude of the induced voltage, that is, the magnitude of the magnetic field leaking through the slot to compartment 2.

Often leakage can be minimized not only by proper positioning of radiation sources but also by changing the nature of discontinuities. It was demonstrated that radiation is smaller through a hole than through a slot.

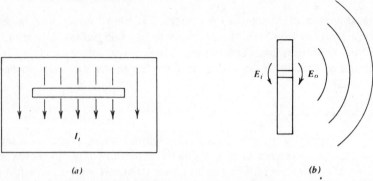

(a)   (b)

Figure 3-18. Radiation through discontinuity in a shield. (*a*) Current induced in a conductor by magnetic field; (*b*) spurious signal radiation due to a discontinuity.

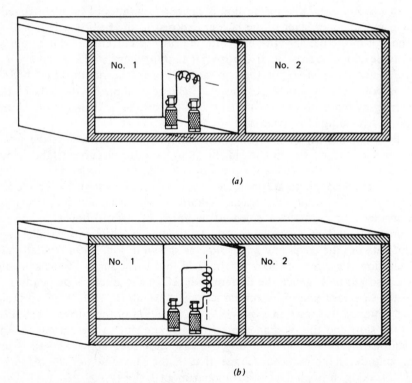

(a)

(b)

Figure 3-19. Magnetic radiation through a slot. (*a*) Axis of solenoid parallel to the slot; (*b*) axis of solenoid perpendicular to the slot.

Hence, whenever mechanical design requirements permit, a row of small apertures is preferred to a long rectangular opening. Another important consideration is that the field intensity at a given distance from a circular aperture be proportional to the cube of the radius of the hole, assuming negligible thickness of the hole (Ref. 11c, p. 18). Hence, spurious radiation can be significantly reduced by minimizing the hole dimension. Further reduction of radiation is achieved by converting the aperture into a waveguide operating below cutoff. Waveguides below cutoff operate as high-pass filters attenuating all signals whose frequencies are lower than the cutoff frequency of the waveguide.

For the circular waveguide attenuator shown in Fig. 3-20a (Ref. 12, p. 2-128)

$$A_c = 31.95 \left(\frac{L_c}{D} \right) \sqrt{1 - \left(\frac{Df}{6920} \right)^2} \qquad (3\text{-}20)$$

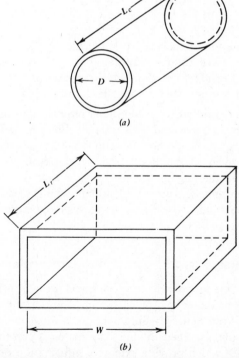

(a)

(b)

Figure 3-20.   Waveguides below cutoff. (a) Circular waveguide; (b) rectangular waveguide.

and

$$f_{co-c} = \frac{6920}{D},$$ (3-21)

where A_c = attenuation of circular waveguide (dB),
     L_c = depth of cylinder (in.),
     D = inside diameter of cylinder (in.),
     f = frequency (MHz),
    f_{co-c} = cutoff frequency of cylinder (MHz).
   For the rectangular waveguide attenuator shown in Fig. 3-20b (Ref. 12, p. 2-125)

$$A_r = 27.3 \left(\frac{L_r}{W} \right) \sqrt{1 - \left(\frac{Wf}{5910} \right)^2}$$ (3-22)

and

$$f_{co-r} = \frac{5910}{W}$$ (3-23)

where A_r = attenuation of rectangular waveguide (dB),
     L_r = depth of rectangular waveguide (in.),
    W = largest inside cross-sectional dimension (in.),
    f_{co-r} = cutoff frequency of rectangular waveguide (MHz).

   When mechanical design considerations require that large apertures be used, the apertures are covered with such shielding material as perforated metal sheets, wire mesh, or honeycomb. The design equations for such shields are not given in this book because electrical test data for these shields, as well as the shields themselves, are available on request from various manufacturers. The interested student who is performing his or her own design of these shields will find the requisite equations in Ref. 12, pp. 2-92 to 2-116.

3-5   Design Considerations

A thorough knowledge of electromechanical design principles, among which packaging has always been of foremost importance, is essential for low-cost, trouble-free, reliable equipment. A designer who is not familiar with these principles is urged to study the material presented in Refs. 11, 12, 18, 19, and 20, which provide extensive information on the grounding, bonding, cabling, and packaging techniques, fabrication processes, and so forth preferred by users and manufacturers of electronic equipment. In this section only those electromechanical considerations that are integral parts of good shielding practices are discussed.

Packaging

Good rf packaging provides the following:

1. Shielding of internal circuitry from compartmental rf energy leakage and prevention of the energy leakage to the outside environment.
2. Isolation of various circuit stages to prevent undesirable feedback and coupling.
3. Power line filtering to attenuate propagation of rf signals within an assembly module and among modules.
4. Low-impedance rf grounds.

These four features have to be achieved with a small-volume, light-weight, and low-cost design. As is true of almost everything concerning synthesizers, there is no unique approach that satisfies all these requirements. Many new techniques have been developed and tried out in practice. However, a number of approaches of modular design have found wide application and enjoyed long life.

It has been common practice to use a flat-plate, L-, U-, or H-chassis whenever low-degree electrostatic shielding is required. Figures 3-21 to 3-24 describe the conceptual design of these chassis. A flat-plate chassis is used when easy access to components is to be provided and space limitations do not exist. An L-chassis is employed when vertical mounting of indicators and manual controls is required. A U-chassis is more effective in space utilization than the flat-plate and L-chassis when some of the electrical components are disproportionately large. An H-chassis is formed by putting two U-chassis back to back. Of these four configurations, the H-chassis gives the highest degree of electrostatic shielding and is the most efficient in space utilization. Reference 18, Chapter 3, provides design and layout information for these chassis.

The metal used in manufacturing the chassis is aluminum because aluminum is inexpensive and lightweight. It also exhibits excellent shielding characteristics against electrostatic fields.

Whenever greater attenuation of spurious fields between circuits or between the chassis and the outside environment is needed, a box chassis, or modular construction, is utilized. A module may consist of a single compartment, Fig. 3-25, if isolation from the outside environment is the only shielding requirement, or it may have the multicompartment, egg-crate structure shown in Fig. 3-26.

Modular construction is commonly used in shielding circuitry against electrostatic, magnetic, and electromagnetic fields. Modules are manufac-

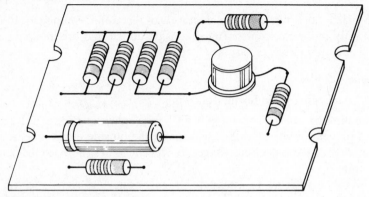

Figure 3-21.   Flat plate chassis.

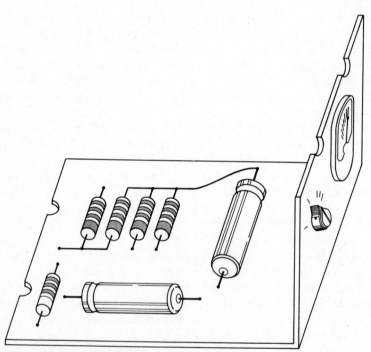

Figure 3-22.   L-shaped chassis.

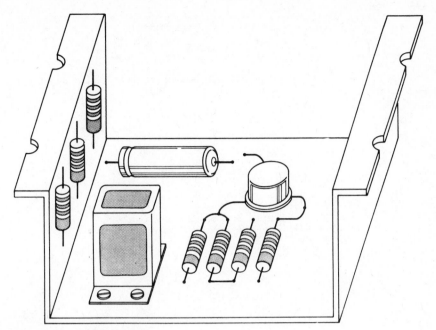

Figure 3-23.   U-shaped chassis.

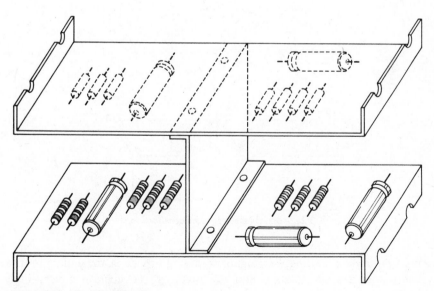

Figure 3-24.   H-shaped chassis.

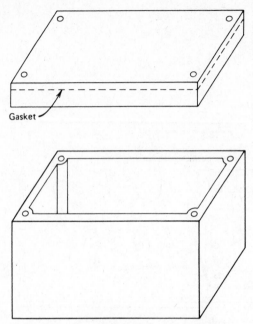

Gasket

Figure 3-25.   Box chassis, single compartment.

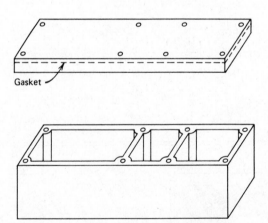

Gasket

Figure 3-26.   Box chassis, egg crate construction.

tured by milling, casting, brazing, or soldering processes, depending on the required degree of isolation and the quantity. To achieve a very high degree of isolation, milling is used in the construction of modules. However, because of the high manufacturing costs associated with milling operations, the use of this process is limited to prototype work. Whenever large quantities are involved, casting is used instead.

The brazing process is usually associated with aluminum chasis (aluminum brazing). A less frequently used copper brazing applies to low-carbon steel that can be heated in a furnace, alloy steel such as nickel, and silicon steel (Ref. 18, pp. 79–82).

Soldering is the cheapest of all processes if small quantities are involved, but also provides the least mechanical strength. Problems associated with the maximum thicknesses of material that can be soldered together impose limitations on the degree of isolation achieved with soldered modules.

Metals used in modular construction are aluminum, steel, brass, and high-permeability alloys.

Figure 3-27 is a photograph of a module manufactured by the milling

Figure 3-27. Box chassis, egg crate construction: milling process of manufacture. Courtesy of ITT Defense Communications Division.

technique. To simplify component assembly and maintenance, electrical components are mounted on removable gold-plated metal boards, which provide an excellent rf ground. The boards are attached to the floor of a plug-in aluminum module. Radio-frequency bypass capacitors, ferrite beads, and RC filters furnish power supply isolation. In this type of construction, even in prototype work, rf isolation is more than 100 dB at 300 MHz.

Soldering techniques of modular construction are demonstrated in Fig. 3-28. A brass module is divided into a large number of compartments isolating rf circuits, which operate at various frequencies, from each other. Power supply isolation is achieved by mounting all dc components and running power supply lines on the outside of the module and by utilizing feedthrough capacitive filters to connect dc to rf circuitry. In this type of construction isolation is more than 80 dB at 300 MHz.

In both cases described above, stage-to-stage radiation is reduced by having the inner side of the covers (omitted in Figs. 3-27 and 3-28) lined

(a)

(b)

Figure 3-28. Box chassis, egg crate construction: soldering process of manufacture. (a) Top view; (b) bottom view. Courtesy of Polarad Electronics Instruments, Division of Polarad Electronics Corporation.

with rf gaskets. Pola Sheet II, manufactured by Metex Corporation, is the gasket material used; however, other shielding materials or methods of mounting gaskets may be considered.

Gaskets

All practical electronic housings must have some means of access. This results in shield discontinuities in the form of long, narrow slots. To reduce rf leakage through these slots, a device making many points of contact along each discontinuity should be employed.

A row of closely spaced screws may provide adequate leakage reduction. Figure 3-29 demonstrates the shielding effectiveness of a $\frac{1}{2}$-in.-wide metal to a metal joint fastened with screws. With a 2-in. spacing between screws, shielding effectiveness of over 80 dB is achieved at 200 MHz.

In many applications the use of a large number of screws is not desirable. In such cases conductive gaskets are utilized. A gasket used for the purpose of rf shielding should exhibit the following:

1. Minimum thickness that will allow for the expected surface discontinuities of the joint.
2. Correct height and resistance to pressure.
3. Adequate resiliency to allow for the frequency of opening and closing the joint.

Before chosing a gasket, the frequency range of interfering signals, the nature of the expected field (electrostatic, magnetic, or electromagnetic), the degree of attenuation, and the available contact area should be estimated. This may require that some preliminary testing be performed. However, only after the shielding requirements are established, can a gasket satisfying these requirements be chosen.

Successful efforts have been made by manufacturers to improve the design and extend the frequency range of gaskets. It would be of little use, therefore, to classify gaskets with respect to their shielding performances at a particular moment in time. Instead, various kinds of gaskets and ways of mounting them in electronic equipment will be briefly discussed.

The least expensive kind of rf gasket is a spring strip (rf fingerstock) of beryllium-copper (or phosphor-bronze), shown in Fig. 3-30. The strip is usually welded to one of the surfaces to be joined. The most widely used type is the flat gasket. A flat gasket is made of a knitted wire mesh or short segments of wire embedded in a nonconductive material. Two ways of mounting these gaskets are shown in Fig. 3-31. Other types of gaskets are described in Fig. 3-32. The choice of gasket is governed by considerations of shielding effectiveness, cost, and space, and by the form of discontinuities. The mounting methods employed with round knitted mesh strip are

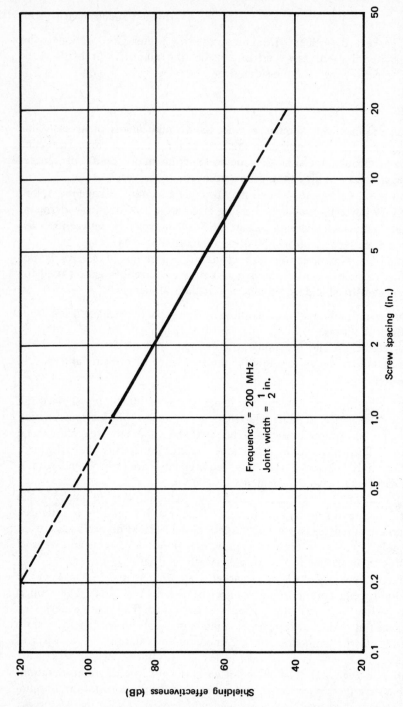

Figure 3-29. Shielding effectiveness of a metal-to-metal joint versus screw spacing. Designer's Guide on Electromagnetic Compatibility, Bulletin No. 4, "System Design of Electronic Equipment," Electronic Industries Association, April 1965. Reprinted by permission.

Screw spacing (in.)

Shielding effectiveness (dB)

Frequency = 200 MHz
Joint width = $\frac{1}{2}$ in.

shown in Fig. 3-33. In this case a rectangular knitted mesh strip can be used instead of the round strip with equal success. Single round mesh strip and double round mesh with attachment fin have the advantage of ease of attachment to the chassis. The two gaskets in Fig. 3-32e and f can be used in various ways, one of which is described in Fig. 3-33c and in Fig. 3-34a, respectively. In the event that both moisture and RFI seals are required, a combination of rubber and mesh gaskets is utilized. This is shown in Fig. 3-34b and c.

The recommended pressure of 20 psi should be uniformly applied to a metal gasket to achieve optimum performance. Pressures in excess of 30 psi tend to destroy the elastic properties of gaskets (Ref. 21, p. 107).

The use of gaskets is expensive, as the description of mounting methods demonstrates. It is of great importance, therefore, to establish shielding requirements and to conduct an investigation of other techniques of

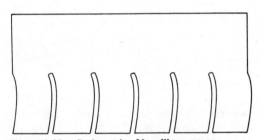

Figure 3-30.   Spring strip of beryllium-copper.

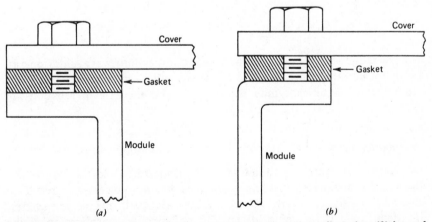

Figure 3-31.   Mounting of flat conductive gaskets. (a) External lip configuration; (b) internal lip configuration.

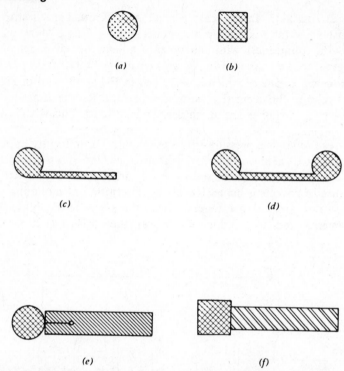

Figure 3-32.   Radio-frequency gaskets. (*a*) **Round knitted mesh strip;** (*b*) **rectangular knitted mesh strip;** (*c*) **single round mesh strip;** (*d*) **double round mesh with attachment fin;** (*e*) **aluminum extrusion with mesh strip permanently crimped on;** (*f*) **combination fluid and rf gasketing (mesh strip bonded to rubber member). Filtron Company, Inc., Westbury, N.Y.,** *Interference Reduction Guide for Design Engineers,* **Vol. I, August 1, 1964. Reprinted by permission.**

leakage reduction (e.g., converting discontinuities into waveguides operating below cutoff) before deciding to use gaskets. In many cases such effort can result in significant reduction of cost and increase in reliability of equipment.

Degradation of Shielding Effectiveness Due to Corrosion and Plating

The successful control of radiation through apertures and the choice of a metal with adequate shielding properties do not necessarily result in long-term satisfactory performance. The original performance will eventually deteriorate if the initial design did not include consideration of the prevention of corrosion failures.

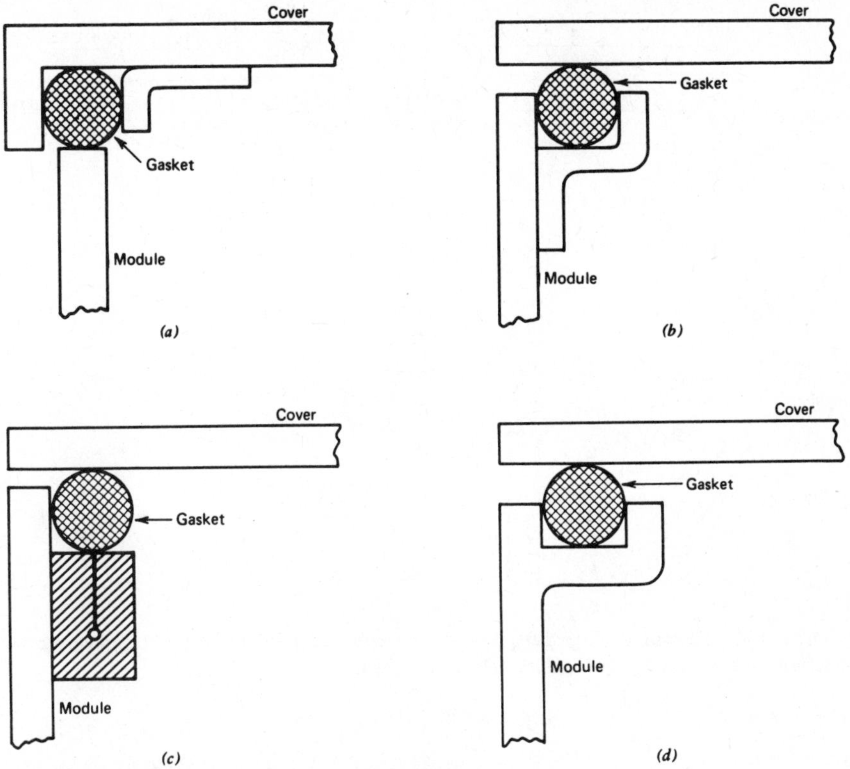

Figure 3-33. Mounting of round conductive gaskets.

Corrosion is a very complex form of material deterioration. It may appear as corrosion fatigue, stress corrosion, stress corrosion cracking, crevice corrosion, pitting, dealloying, fretting, or galvanic corrosion. Reference 22 discusses various types of corrosion. This text considers only galvanic corrosion because it governs both the choice of shielding material and gaskets and the choice of the plating used to prevent corrosion.

Galvanic corrosion is an electrochemical phenomenon analogous to a voltaic cell. For corrosion to take place, the following must exist (Ref. 22, p. 17):

1. The anode, where the corrosion takes place.
2. The metallic path, which conducts the electrons from one site to another.
3. The cathode, where the reduction action occurs.
4. The electrolyte, which presents an electrolytic path from the anode area to the cathode area.

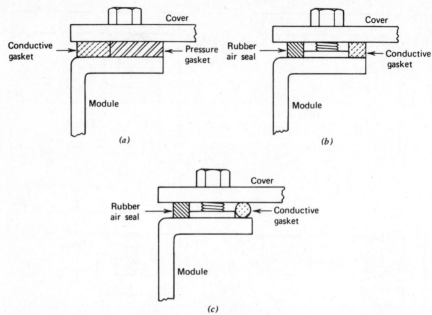

Figure 3-34. Mounting of gaskets. (a) Pressurized-conductive gasket; (b) gasket against moisture and RFI; (c) gasket against moisture and RFI.

Galvanic corrosion occurs when dissimilar metals are joined by a common electrolytic solution, such as an aqueous solution of salts, acids, or alkalis, which may originate as a thin film or condensed moisture. This solution subsequently dissolves ionizable substances from the atmosphere or from the metal surfaces themselves. Ions present in the electrolytic solution provide a path for current to travel through the solution between metals of different electrode potentials. Under these conditions the more active metal (the anode) corrodes, while the more noble metal (the cathode) remains unaffected (Ref. 23, p. 32). Table 3-5 lists the most commonly used metals in relation to their electrode potentials and indicates permissible combinations of metals that do not produce galvanic corrosion to any significant (for all practical purposes) degree. This table should be consulted before the chassis material, plating, and gasket are chosen.

Preventing moisture from reaching contact surfaces and gaskets is one method of corrosion control. It is expensive, however, when surfaces have to be disassembled for maintenance or repair. Plating the surfaces is an inexpensive way of corrosion control, but plating affects the shielding effectiveness of metals and should be carefully chosen. Figure 3-35 indicates the effect of four protective coatings on the shielding effectiveness of

Table 3-5. Galvanic Series of Metals[a]

	Group	Metallurgical category	EMF (V)	Permissible couples[b, c]
↑	1	Gold, solid and plated; gold-platinum alloys; wrought platinum	+0.15	
Protected End (Cathodic or more noble)	2	Rhodium; graphite	+0.05	
	3	Silver, solid or plated; high silver alloys	0	
	4	Nickel, solid or plated; monel; high nickel-copper alloys; titanium	−0.15	
	5	Copper, solid or plated; low brasses or bronzes; silver solder; German silver; high copper-nickel alloys; nickel-chrome alloys; austenitic stainless steels (301, 302, 304, 309, 316, 321, 347)	−0.20	
	6	Commercial yellow brasses and bronzes	−0.25	
	7	High brasses and bronzes; naval brass; muntz metal	−0.30	
	8	18% Chromium-type corrosion-resistant steels (440-430, 431, 446 17-7PH, 17-4PH)	−0.35	
	9	Chromium, plated; tin, plated; 12% chromium-type corrosion-resistant steel (410, 416, 420)	−0.45	
	10	Tinplate, terneplate; tin-lead solders	−0.50	
	11	Lead, solid or plated; high lead alloys	−0.55	
	12	Aluminum, wrought alloys of the duralumin type (2014, 2024, 2017)	−0.60	
	13	Iron, wrought, gray, or malleable; plain carbon and low alloy steels; armco iron	−0.70	
	14	Aluminum wrought alloys other than duralumin; type 6061, 7075, 5052, 5056, 1100, 3003; cast alloys of the silicon type (355, 356)	−0.75	
Corroded End (Anodic or least noble)	15	Aluminum, cast alloys other than silicon type; cadmium, plated and chromated	−0.80	
	16	Hot-dip zincplate; galvanized steel	−1.05	
↓	17	Zinc, wrought; zinc-base die-cast alloys; zinc, plated	−1.10	
	18	Magnesium and magnesium-base alloys, cast or wrought	−1.60	

[a]Reprinted from MIL-STD-1250 (MI).
[b]members of groups connected by lines are considered as permissible couples; however, this should not be construed as being devoid of galvanic action. Permissible couples represent a low galvanic effect.
[c]○ indicates the most cathodic member of the series; ● represents an anodic member; the arrows indicate the anodic direction.

205

aluminum against an electric field. The degradation in shielding effectiveness is a function of the conductivity of applied coating; materials with lower conductivity degrade shielding effectiveness. more than materials with high conductivity. Anodizing is nonconductive and should be avoided if a high degree of shielding is required. Instead, iridite No. 14 should be used. Iridite No. 14 is of a complex chromium-chromate nature and is generated by a reaction that occurs when an aluminum part is immersed in the iridite solution. The film becomes an integral part of the metal itself rather than being superimposed. Iridite No. 14 provides excellent corrosion resistance and has very little effect on the electrical characteristics of aluminum over a wide frequency range (Ref. 12, p. 2-86).

Shielding Materials

Electrostatic shields are made of nonmagnetic high-conductivity metals such as aluminum, brass, copper, and zinc. Table 3-6 compares the conductivities of various metals to the conductivity of copper. These data, in conjunction with the expressions for shielding effectiveness, Eqs. 3-12 to 3-19, explain why nonmagnetic metals are better electrostatic shields than magnetic metals; in addition to providing high absorption loss, they are characterized of high reflection loss.

Purely magnetic fields require metals based on ferromagnetic elements, which have high permeability. Unfortunately, these materials have poor conductivities (low reflection loss); hence the magnetic shielding that they provide depends primarily on absorption losses. A magnetic shield is essentially a low-reluctance path in which the magnetic field is contained. Metals such as permalloy, Hypernick, and Mu-metal are commonly used as magnetic shields. Iron and steel have also been employed when weight and size requirements are not critical. High shielding effectiveness is obtained with these metals by compensating with a much greater thickness.

A word of caution is in order about high-permeability metals. They exhibit strain sensitivities in proportion to their permeabilities. To achieve optimum permeability these metals are annealed. The process of annealing takes place after the fabrication of shields. In their final state the materials are soft and should be handled with care if their shielding effectiveness is to be maintained.

Radio-Frequency Wire Interference

Spurious signals may be transferred from one circuit to another by interconnecting wires. The case of coupling between unshielded wires is easy to understand; it needs no explanation. For a good understanding of the problem associated with shielded wires, consider a shielded wire over a

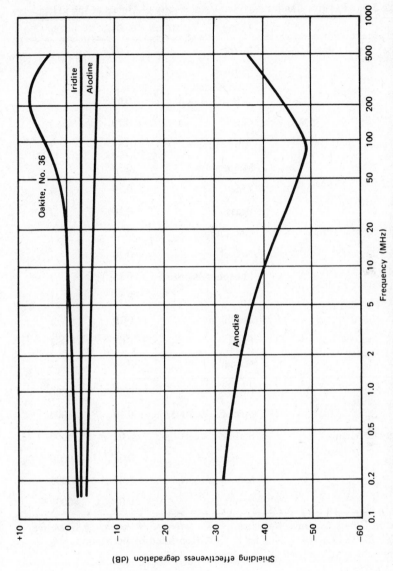

Figure 3-35. Shielding effectiveness degradation caused by finishes of aluminum. *Designer's Guide on Electromagnetic Compatibility*, Bulletin No. 4, "System Design of Electronic Equipment," Electronic Industries Association, April 1965. Reprinted by permission.

207

Table 3-6.   Relative Conductivity and Permeability of Metals at 150 kHz[a]

	Metal	G	μ
	Silver	1.05	1
	Copper		
	Annealed	1.00	1[b]
	Hard-drawn	0.97	1
	Gold	0.70	1
	Aluminum	0.61	1
	Magnesium	0.38	1
	Zinc	0.29	1
	Brass	0.26	1
Nonmagnetic Metals	Cadmium	0.23	1
	Nickel	0.20	1
	Phosphor-bronze	0.18	1
	Tin	0.15	1
	Beryllium	0.10	1
	Lead	0.08	1
	Monel	0.04	1
	Iron	0.17	1000
	Steel, SAW 1045	0.10	1000
Magnetic Metals	Stainless steel	0.02	1000
	Hypernick	0.06	80000
	Mu-metal	0.03	80000
	Permalloy	0.03	80000

[a]Filtron Company, Inc., Westbury, N. Y., *Interference Reduction Guide for Design Engineers*, Vol. 1, August 1, 1964. Reprinted by permission.
[b]Conductivity of copper $= 1.72 \ \mu\Omega/cm^3$.

ground plane, Fig. 3-36. The leakage portion of the return current, i_e, is conducted by the ground plane. If the leakage current were zero, the magnetic field due to the current in the center conductor would be identically balanced by the field produced by the current in the shield, and the shield would be 100% effective. The larger the leakage current, the more ineffective the shield is. Leakage current increases as the resistance of

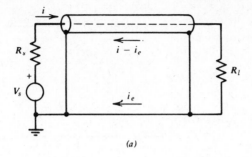

(a)

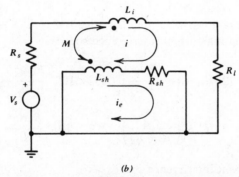

(b)

Figure 3-36.   Shielded wire over a ground plane. (a) Schematic circuit; (b) equivalent circuit. *IEEE Transactions on Electromagnetic Compatibility*, Vol. EMC-9, No. 2 (September 1967). Reprinted by permission.

the shield becomes larger than the reactance of the shield (Ref. 24, p. 39). At frequencies below 3 kHz the magnetic fields of shielded and unshielded wires are approximately equal because the return currents of both wires flow through the ground plane (Ref. 25, p. 37).

Figure 3-37 describes an open-wire-line equivalent of a shielded wire. The magnetic field radiated by the equivalent line is the same as the field radiated by the shielded wire. The parameter M in Fig. 3-36b is the mutual inductance between the center conductor and the shield. The inductances of the center conductor and the shield are L_i and L_{sh}, respectively. The resistance of the shield is R_{sh}.

Four cases of electromagnetic interference for parallel shielded and unshielded wires that are of practical importance (Ref. 25) are summarized below.

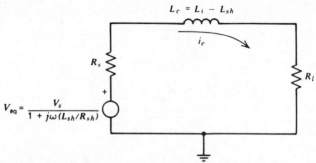

Figure 3-37. **Open-wire-line equivalent of a shielded wire line.** *IEEE Transactions on Electromagnetic Compatibility*, Vol. EMC-9, No. 2 (September 1967). Reprinted by permission.

1. Open-wire-to-open-wire coupling, Fig. 3-38*a*
   (*a*) Alternating current interference source

$$e_c = (2\pi f L_m li)\left(\frac{R_c}{R_c + R_d}\right)\left|1.032 \times 10^{-6}\frac{R_b R_d}{L_1 L_2} - 1\right|, \qquad (3\text{-}24)$$

$$e_d = (2\pi f L_m li)\left(\frac{R_d}{R_c + R_d}\right)\left(1.032 \times 10^{-6}\frac{R_b R_c}{L_1 L_2} + 1\right). \qquad (3\text{-}25)$$

   (*b*) Transient interference source

$$V_c = \left(\frac{L_m lI}{\tau}\right)\left(\frac{R_c}{R_c + R_d}\right)\left(1.032 \times 10^{-6}\frac{R_b R_d}{L_1 L_2} - 1\right), \qquad (3\text{-}26)$$

$$V_d = \left(\frac{L_m lI}{\tau}\right)\left(\frac{R_d}{R_c + R_d}\right)\left(1.032 \times 10^{-6}\frac{R_b R_c}{L_1 L_2} + 1\right), \qquad (3\text{-}27)$$

where

$$\tau = \frac{t_r}{2.2}. \qquad (3\text{-}28)$$

2. Shielded-wire-to-open-wire coupling, Fig. 3-38*b*
   (*a*) Alternating current interference source

$$e_c = e_i\left(\frac{R_c}{R_c + R_d}\right), \qquad (3\text{-}29)$$

$$e_d = e_i\left(\frac{R_d}{R_c + R_d}\right), \qquad (3\text{-}30)$$

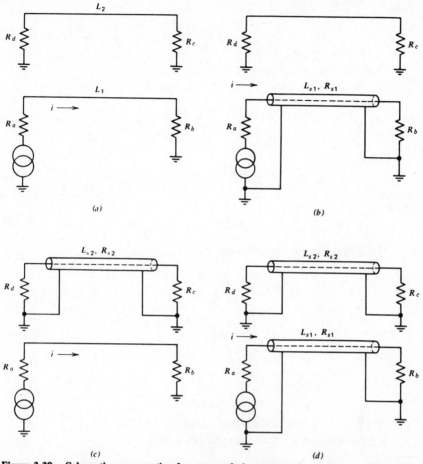

Figure 3-38. Schematics representing four cases of electromagnetic coupling. (*a*) Open-wire-to-open-wire coupling; (*b*) shielded-wire-to-open-wire coupling; (*c*) open-wire-to-shielded-wire coupling; (*d*) shielded-wire-to-shielded-wire coupling. Reprinted by permission. *EDN Magazine*, July 1, 1969.

where

$$e_i = 2\pi f L_m l i \alpha_{s1}. \tag{3-31}$$

(*b*) Transient interference source

$$V_c = V_i \left(\frac{R_c}{R_c + R_d} \right), \tag{3-32}$$

$$V_d = - V_i \left(\frac{R_d}{R_c + R_d} \right), \tag{3-33}$$

where

$$V_i = -\frac{L_m lI}{\tau} A_{s1}.$$ (3-34)

3. Open-wire-to-shielded-wire coupling, Fig. 3-38c
   (a) Alternating current interference source

$$e_c = e_i \left(\frac{R_c}{R_c + R_d} \right),$$ (3-35)

$$e_d = e_i \left(\frac{R_d}{R_c + R_d} \right),$$ (3-36)

where

$$e_i = 2\pi f L_m li\alpha_{s2}.$$ (3-37)

   (b) Transient interference source

$$V_c = V_i \left(\frac{R_c}{R_c + R_d} \right),$$ (3-38)

$$V_d = -V_i \left(\frac{R_d}{R_c + R_d} \right),$$ (3-39)

where

$$V_i = -\frac{L_m lI}{\tau} A_{s2}.$$ (3-40)

4. Shielded-wire-to-shielded-wire coupling, Fig. 3-38d
   (a) Alternating current interference source

$$e_c = e_i \left(\frac{R_c}{R_c + R_d} \right),$$ (3-41)

$$e_d = e_i \left(\frac{R_d}{R_c + R_d} \right),$$ (3-42)

where

$$e_i = 2\pi f L_m li\alpha_{s1}\alpha_{s2}.$$ (3-43)

   (b) Transient interference source

$$V_c = V_i \left(\frac{R_c}{R_c + R_d} \right),$$ (3-44)

$$V_d = -V_i \left(\frac{R_d}{R_c + R_d} \right),$$ (3-45)

where

$$V_i = -\frac{L_m lI}{\tau} A_{ss}. \qquad (3\text{-}46)$$

The parameters used in Eqs. 3-24 to 3-46 and in Figs. 3-38 to 3-44 are as follows:

Subscript 1 = source line parameters.

Subscript 2 = victim line parameters.

α_s = ac shield attenuation factor, a function of fL_s/R_s, obtained from Fig. 3-42.

A_{ss}, A_s = transient shield attenuation factors, functions of $(L_s/R_s)/\tau$, obtained from Fig. 3-43.

C = line capacitance ($\mu F/ft$).

d = diameter of unshielded wire (in.).

d_i = outer diameter of inner conductor of shielded line (in.), obtained from Table 3-7 for several standard coaxial cables.

Table 3-7. Coaxial Cable and Shielded Wire Parameters[a]

Type RG #/U	Braid[b]	Z_0 (Ω)	d_s (in.)	d_i (in.)	$L_s{}^c$ ($\mu H/ft$)	R_s ($m\Omega/ft$)	L_s/R_s (μsec)
9B	D	50.0	0.280	0.085	0.20	0.80	250
6A	D	75.0	0.185	0.028	0.23	1.15	200
5B	D	50.0	0.181	0.053	0.23	1.20	190
13A	D	75.0	0.280	0.043	0.20	0.95	210
8A	S	50.0	0.285	0.086	0.20	1.35	150
55A	D	50.0	0.116	0.035	0.26	2.55	100
29	S	53.5	0.116	0.032	0.26	4.75	55
58C	S	50.0	0.116	0.035	0.26	4.70	55
59A	S	75.0	0.146	0.022	0.24	3.70	65
141	S	50.0	0.116	0.036	0.26	4.70	55
142	D	50.0	0.116	0.039	0.26	2.20	118
122	S	50.0	0.096	0.029	0.27	5.70	47
223	D	50.0	0.116	0.035	0.26	2.55	100
188	S	50.0	0.060	0.019	0.30	7.60	40
180	S	95.0	0.103	0.011	0.26	6.00	43
A[d]	S		0.057	0.023	0.30	16.00	19
B[d]	S		0.056	0.030	0.30	6.20	48

[a]*EDN Magazine*, July 1, 1969. Reprinted by permission.

[b]D = double; S = single.

[c]Valid for $b = 2$ in. For $b \neq 2$ in., use Fig. 3-41 to compute inductance.

[d]A and B are #22 and #20 shielded hookup wires, respectively.

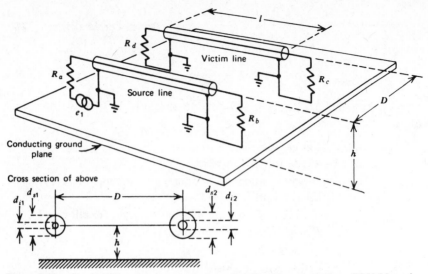

Figure 3-39. Coupled lines over a ground plane. Reprinted by permission. *EDN Magazine*, July 1, 1969.

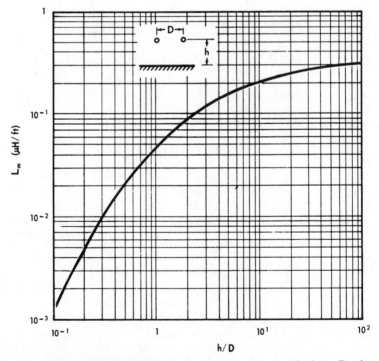

Figure 3-40. Mutual inductance between two wires over a ground plane. Reprinted by permission. *EDN Magazine*, July 1, 1969.

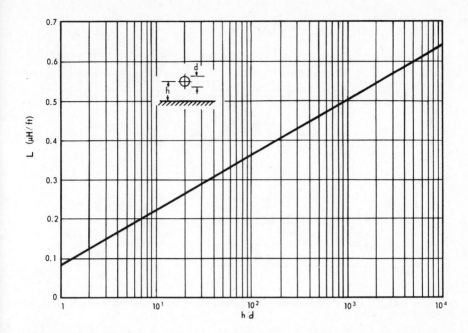

Figure 3-41. Inductance of a wire over a ground plane. Reprinted by permission. *EDN Magazine*, July 1, 1969.

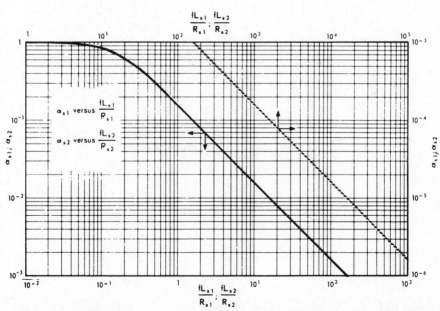

Figure 3-42. Alternating current shield attenuation factors. Reprinted by permission. *EDN Magazine*, July 1, 1969.

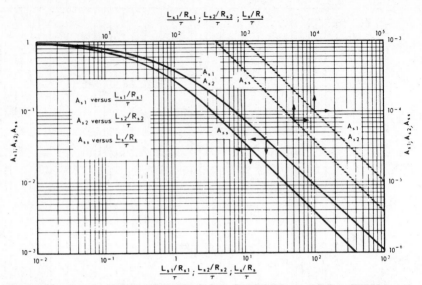

Figure 3-43. Transient shield attenuation factors. Reprinted by permission. *EDN Magazine,* July 1, 1969.

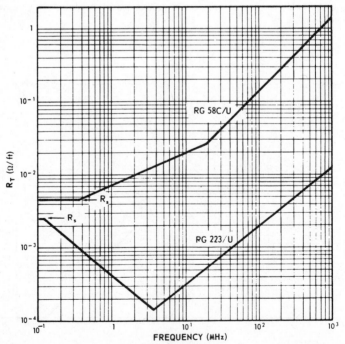

Figure 3-44. Transfer impedance versus frequency. Reprinted by permission. *EDN Magazine,* July 1, 1969.

d_s = inner diameter of outer conductor of shielded line (in.), obtained from Table 3-7 for several standard coaxial cables.

D = separation of coupled wires (in.).

e_i = induced ac voltage in victim line (volts rms).

e_c, e_d = coupled ac voltage at R_c, R_d (volts rms).

f = frequency (MHz).

h = height of coupled lines over the ground plane (in.).

i = source ac current (A).

I = peak source transient current (A), with transient current of the form

$$i = I\left[1 - \epsilon^{-(t/\tau)} \right]. \tag{3-47}$$

l = coupled line length (ft).

L = inductance of wire (μH/ft), obtained from Fig. 3-41 as a function of h/d.

L_s = inductance of shield of a shielded wire (μH/ft), obtained from Fig. 3-41 as a function of h/d_s. Table 3-7 gives L_s for several standard coaxial cables at $h = 2$ in.

L_m = mutual inductance between coupled lines (μH/ft), obtained from Fig. 3-40 as a function of h/D.

R_a, R_b, R_c, R_d = terminating resistances on source and victim lines (Ω).

R_s = dc and low-frequency resistance of shield on shielded lines (Ω/ft). Table 3-7 gives R_s for several common coaxial lines. Use R_T instead of R_s at frequencies above 200 kHz.

R_T = transfer resistance of shield on shielded lines (Ω/ft). It is a function of frequency but is nearly constant and is equal to R_s below about 200 kHz for typical braided cables. See Fig. 3-44 for variation of R_T with frequency for two typical coaxial cables.

τ = time constant of interfering transient (μsec).

t_r = 10 to 90% rise time of interfering transient (μsec).

V_c, V_d = peak-coupled transient voltage at R_c, R_d (V).

V_i = peak transient induced voltage in victim line (V).

Z_0 = characteristic impedance of cable (Ω).

Figure 3-39 defines the physical parameters of the coupled lines over a ground plane. This model represents the bulk of system wiring where the source and load are referenced to the structure, which is simulated in Fig. 3-39 by a conducting ground plane. Figures 3-40 to 3-44 and Table 3-7 are used as required in solving Eqs. 3-24 to 3-48.

Equations 3-24 to 3-46 are accurate provided that the series inductive reactance and shunt capacitive susceptance of the victim lines are negligible compared with the terminating impedances, that is, when

$$R_c + R_d \gg 2\pi f l L_2$$

and

$$\frac{R_c R_d}{R_c + R_d} \ll \frac{1}{2\pi f l C_2} = \frac{10^6 L_2}{1.032(2\pi f l)},$$

where L_2 = series inductance of victim line (μH/ft),
C_2 = shunt capacitance of victim line (μF/ft).

Open-wire-to-open-wire coupling equations, Eqs. 3-24 to 3-28, include both electric and magnetic components. Mutual capacitance, C_m, is contained implicitly in the interrelationship among mutual capacitance, mutual inductance, and self-inductance as

$$C_m = \frac{1.032 L_m}{L_1 L_2} \qquad \text{pF/ft.} \tag{3-48}$$

If one or both lines are shielded, Eqs. 3-29 to 3-46, and their lengths are electrically short, the coupling will be predominantly magnetic. As was mentioned above, the external magnetic fields of unshielded and shielded wires are approximately the same at frequencies below 3 kHz, so that cases 2, 3, and 4 are reduced to the case of a pair of unshielded wires at those frequencies.

In transient analysis the shield resistance, R_s, is conveniently used instead of the transfer resistance, R_T, at frequencies up to 200 kHz. It has been found to be adequate even when dealing with the fastest transients. Above 200 kHz R_s is replaced by R_T (Ref. 25, p. 37).

A numerical example of ac interference between two open-wire parallel lines is given next to illustrate the analysis of the coupling between wires. Consider two unshielded parallel wires 0.05 ft long, $1\frac{3}{4}$ in. above a ground plane, separated 7/8 in. from each other. Let the wire size be No. 20 AWG ($d_1 = d_2 = 0.032$ in.). Let the terminating resistances be as follows: $R_a = 50$ Ω, $R_b = 150$ Ω, $R_c = 2$ kΩ, and $R_d = 150$ Ω. Let the frequency of the interfering signal be 0.15 MHz, and the source open-circuit voltage, e, be equal to 10 V rms.

From Fig. 3-40 with $h/D = 1.75/0.875 = 2$,

$$L_m = 0.086 \ \mu\text{H}/\text{ft}.$$

From Fig. 3-41 with $h/d_1 = h/d_2 = 1.75/0.032 = 54.7$,

$$L_1 = L_2 = 0.33 \ \mu\text{H}/\text{ft}.$$

If the small impedance due to L_1 and the shunting effect of C_1 are neglected,

$$i = \frac{e}{R_a + R_b} = \frac{10}{200} = 0.05 \ \text{A}.$$

From Eq. 3-24,

$$e_c = 6.28 \times 0.15 \times 0.086 \times 0.05 \times 0.05 \times \frac{2000}{2150} \times \left| \frac{1.032 \times 10^{-6} \times 150 \times 150}{0.33 \times 0.33} - 1 \right|$$

$$= 1.5 \times 10^{-4} \ \text{V}.$$

From Eq. 3-25,

$$e_d = 6.28 \times 0.15 \times 0.086 \times 0.05 \times 0.05 \left(\frac{150}{2150} \right) \left(\frac{1.032 \times 10^{-6} \times 150 \times 2000}{0.33 \times 0.33} + 1 \right)$$

$$= 5.4 \times 10^{-5} \ \text{V}$$

If the desired signal across R_c, e_c, were 1 V rms and the requirement for spurious-signal suppression were 90 dB below the signal, the spurious interference caused by the coupling between the wires would be

$$20 \log_{10} \left(\frac{1}{5.4 \times 10^{-5}} \right) \approx 76 \ \text{dB}$$

below the signal, exceeding the requirement by 14 dB.

It is important in the design of high-purity signal generators that the cable diagrams of the equipment be prepared, and estimated levels and waveforms of all ac signals carried by various cables be indicated. Cables that carry high-level signals, particularly pulses, should not be harnessed together with cables used to transmit low-level high-purity signals. If this cannot be avoided, mathematical analysis and tests should be performed and levels of interference determined. The results of the analysis and testing should govern the choice of an approach. For example, packaging may be conducted in such a manner as to eliminate a need for cabling between circuits generating high-purity signals, or the power of such signals transmitted through cables may be increased to make cable interference negligible. Cable interference, if not taken care of, presents a costly problem of spurious-signal reduction in synthesizers.

Pulse Interference

There are two basic types of interference: narrowband and broadband. Narrowband interference is described by a signal whose power distribution is confined to a narrow frequency range, Fig. 3-45a. A spurious intermodulation product generated in adding two signals is a kind of narrowband interference. Another example of such interference is a single-frequency spurious signal electrically or magnetically coupled into a circuit. Broadband interference is described by a signal whose power is distributed over a wide frequency range, Fig. 3-45b. Any steep, sharp-angled, or short-

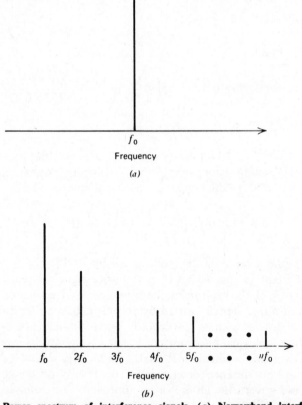

f_0

Frequency

(a)

f_0 $2f_0$ $3f_0$ $4f_0$ $5f_0$ • • • nf_0

Frequency

(b)

Figure 3-45. Power spectrum of interference signals. (a) Narrowband interference; (b) broadband interference.

duration nonsinusoidal current or voltage waveform is classified as. a broadband interference (Ref. 12, Vol. I, pp. 1–6 to 1–20). The most commonly encountered broadband interference is produced by a repetitive rectangular pulse. Such a highly nonsinusoidal signal may be either generated from a sine wave, to achieve a frequency multiplication effect, or produced by digital circuits used in frequency synthesis. In any event, a pulse train generates strong, varying magnetic fields that easily transfer to other circuits. This problem arises from the fact that a pulse-train waveform has high harmonic content. Figure 3-46 and Eqs. 3-49 and 3-50 describe an ideal repetitive pulse (Ref. 26, p. 42-12):

$$C_n = 2A\left(\frac{T_0}{T}\right)\left[\frac{\sin(n\pi T_0/T)}{n\pi T_0/T}\right], \tag{3-49}$$

where C_n = peak amplitude of the nth harmonic (nth coefficient of Fourier series),

$n = 1, 2, 3, \ldots$,

and

$$\text{pulse repetition rate} = \frac{1}{T}. \tag{3-50}$$

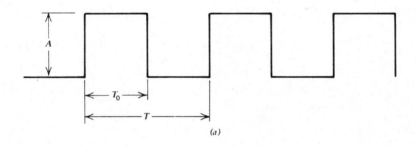

(a)

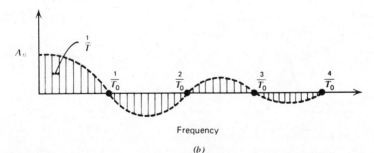

Frequency

(b)

Figure 3-46.   An ideal pulse train. (a) Waveshape; (b) harmonic spectrum.

In practice, pulses have finite rise and decay times and resemble trapezoids. Figure 3-47 and Eqs. 3-51 and 3-52 describe a trapezoidal pulse train (Ref. 26, p. 42-13):

$$C_n = 2A\left(\frac{T_0 + t_1}{T}\right)\left[\frac{\sin(n\pi t_1/T)}{n\pi t_1/T}\right]\left\{\frac{\sin[n\pi(T_0 + t_1)/T]}{n\pi(T_0 + t_1)/T}\right\} \quad (3\text{-}51)$$

and

$$t_1 = t_r = t_d \quad (3\text{-}52)$$

where t_r = pulse rise time.
$\quad t_d$ = pulse decay time.

Under these conditions, the shorter the rise time the more harmonics the waveform contains, the greater is the intensity of the interference field, and the greater is the voltage induced in the neighboring conductors.

Another form of pulse interference is a single pulse. Circuits such as rf switches and relays give rise to transients. These transients are electrically or magnetically coupled to analog and digital circuits and often cause them to malfunction. This kind of interference is described next.

Consider two capacitively coupled circuits as shown in Fig. 3-48a and b (Ref. 12, pp. 2-206 to 2-209). Assume that $v_s(t)$ is a step voltage of magnitude V_p initiated at a time $t = 0$, where $v_s(t)$ is defined as follows:

$$v_s(t) = v_{s1}(t)\left(\frac{R_{l1}}{R_{l1} + R_{s1}}\right). \quad (3\text{-}53)$$

To simplify the expression for the output voltage, $v_o(t)$, assume that $C_{1g} = C_{2g}$ and $R_s = R_l$, where

$$R_s = \frac{R_{s1}R_{l1}}{R_{s1} + R_{l1}} \quad (3\text{-}54)$$

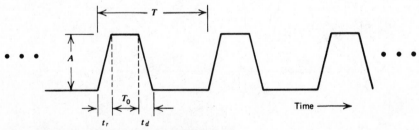

Figure 3-47.   Waveshape of a trapezoidal pulse train.

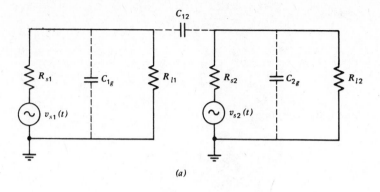

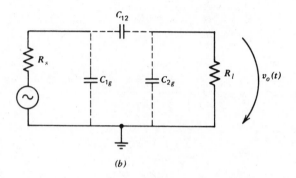

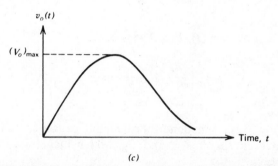

Figure 3-48. Transient interference. (*a*) Two capacitively coupled circuits; (*b*) equivalent circuit; (*c*) transient response. Filtron Company, Inc., Westbury, N.Y., *Interference Reduction Guide for Design Engineers*, Vol. I, August 1, 1964. Reprinted by permission.

and

$$R_l = \frac{R_{s2}R_{l2}}{R_{s2} + R_{l2}} \tag{3-55}$$

Then the output voltage, $v_o(t)$, is

$$v_o(t) = \frac{V_p R_l}{2R_{s1}} \left[e^{-(\gamma t/\tau)} - e^{-(t/\tau)} \right], \tag{3-56}$$

where

$$\tau = R_l C_{1g} \tag{3-57}$$

and

$$\gamma = \frac{\left[1 + (C_{1g}/C_{12}) \right] - 1}{\left[1 + (C_{1g}/C_{12}) \right] + 1}. \tag{3-58}$$

Let

$$\alpha = 1 + \frac{C_{1g}}{C_{12}}. \tag{3-59}$$

Then the maximum value of $v_o(t)$ (Fig. 3-48c) is

$$(V_o)_{max} = V_p \left(\frac{R_l}{R_{s1}} \right) \left[\frac{\gamma \alpha}{2\sqrt{\alpha^2 - 1}} \right]. \tag{3-60}$$

The parameter $(V_o)_{max}$ is a function of the coupling capacitance, C_{12} and the capacitances in each circuit, C_{1g} and C_{2g}, which include any circuit capacitance plus strays.

A thorough understanding of pulse interference permits the designer of electronic equipment to prevent it by providing either adequate broadband filtering or sufficient isolation, or both.

The following example demonstrates an application of the wire and pulse interference theory discussed above.

Consider an shf synthesizer consisting of a single digital phase-locked loop and a times-10 frequency multiplier. The uhf PLL operates over a 300 to 400 MHz frequency range in 10 kHz increments. The final synthesizer frequency is 3 to 4 GHz variable in 100 kHz steps.

Figure 3-49 is a functional block diagram of the synthesizer. The basic parts of the PLL are a uhf voltage-controlled oscillator, programmable divider, and phase comparator. The uhf VCO is locked to a 10 kHz reference signal derived by frequency division from a 1 MHz crystal-controlled oscillator. A two-stage isolation amplifier, loop filter, digital-to-analog (D/A) converter, and data processor comprise the rest of the synthesizer.

The isolation amplifier attenuates harmonics of 10 kHz which fall within the VCO frequency range. These spurious signals, appearing at the input of the programmable divider at a level of − 80 dBm, result from an electrostatic coupling between the input and the output of the divider. The divider output signal is a 4V fast-rise-time pulse with a 10 kHz repetition rate.

The loop filter consists of a Twin-T and low-pass filters designed to attenuate spurious signals at the fundamental and harmonics of the 10 kHz reference signal leaking to the output of the phase comparator.

The D/A converter provides a coarse tuning voltage to the uhf VCO.

Upon receipt of a frequency command word, the programmable divider is set to divide by the required division ratio and the uhf VCO is tuned to the one tenth of the desired output frequency close enough for the PLL acquisition to take place. Frequency control data conversion and distribution are accomplished by the data processor, which selects the ratio N (30,000 to 40,000), VCO band (1, 2, or 3), and value of the VCO coarse tuning voltage.

In the initial paper design stage of the project the synthesizer circuitry is separated into six groups, or modules, for the best packaging arrangement with respect to module size. This modular breakdown is shown in Fig. 3-49.

The five modules $A2$ to $A6$ are aluminum rectangular boxes with individual covers. The modules are plugged into a mounting bracket, which constitutes part of the synthesizer main frame.

The 1 MHz reference oscillator, module $A1$, is a nonuniform-size module attached directly to the main frame.

The purpose of the analysis is to identify the most important sensitive-to-pickup circuits and sources of radiation and to evaluate the proposed module breakdown with respect to spurious signals. The analysis is limited to examining radiation and pickup problems associated with the module harness. The spurious FM requirement for the synthesizer is − 80 dB with a 6 dB margin at the shf output. Hence the requirement is − 106 dB at the uhf VCO output.

A visual examination of Fig. 3-49 leads to the conclusion that the most sensitive-to-pickup circuit is the uhf VCO with its two high-sensitivity

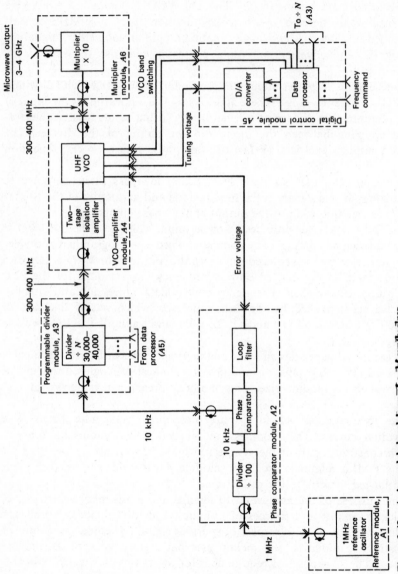

Figure 3-49.   A six-module breakdown of a shf synthesizer.

226

inputs associated with error and tuning voltages, and that the sources generating high-level spurious signals are the phase comparator and programmable divider. The phase comparator is completely enclosed in module $A2$. Hence, except for power supply leakage and radiation through apertures not considered in this exercise, it cannot contaminate the synthesizer output. (It is assumed that the loop filter provides adequate attenuation of the ripple voltage at the fundamental and harmonics of 10 kHz leaking to the phase comparator output.) The programmable divider, however, may generate spurious FM by way of (*a*) coupling of the 10 kHz output signal to the error voltage line, location 1 in Fig. 3-50; (*b*) coupling that takes place between the divider control lines and tuning voltage line, location 2; and (*c*) coupling taking place between the divider control lines and error voltage line, location 3.

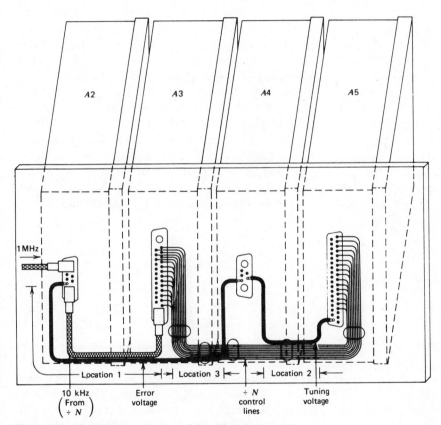

Figure 3-50.   Partial harness drawing of the uhf synthesis section.

Figure 3-50 shows the details of interconnections between modules $A2$ to $A5$ which play an important role in this analysis. For clarity, such interconnections as dc power supply leads are omitted from Fig. 3-50 and so is module $A6$.

In location 1 a coaxial shielded cable carrying a pulsed signal at 10 kHz from the programmable divider to the phase comparator is placed adjacent to an insulated wire used to supply the error voltage to the VCO. The length of the cable and wire in this location is approximately 3 in. outside and 2 in. inside module $A2$, including the connector, before the pair is separated. The cable is type RG-188/U and the insulated wire is No. 22 with an outside diameter (including the insulation) of approximately 50 mils.

Figure 3-51 is a schematic diagram and an enlarged cross-sectional view of the arrangement. The expression for the voltage that is coupled to the insulated wire and appears at the VCO error voltage input is given by Eqs. 3-30 and 3-31. In this particular case, $f = 10^{-2}$ MHz and $l = 0.415$ ft. Resistor R_d is the input impedance of the error voltage varicap circuit at 10 kHz, and R_c is the phase comparator output impedance. Let $R_d = 150$ kΩ and $R_c = 2$ kΩ. The separation of coupled wires D and the height of the coupled lines over the ground plane h are 0.075 and 0.05 in., respectively. Hence the ratio h/D is 0.666. From Fig. 3-40 we obtain $L_m = 3 \times 10^{-2}$ μH/ft. Similarly, Fig. 3-41 gives us the value of $L_s = 0.075$ μH/ft for $h/d_s = 0.833$, where $d_s = 0.06$ in. for RG-188/U cable (see Table 3-7). Having determined L_s, we find from Fig. 3-42 that $\alpha_{s1} = 0.85$ for $fL_s/R_s = 9.88 \times 10^{-2}$, where the low-frequency resistance of the shield, R_s, is given in Table 3-7. (For RG-188/U cable $R_s = 7.6 \times 10^{-3}$ Ω.) Substituting these numbers in place of the corresponding unknowns in Eqs. 3-30 and 3-31, we obtain

$$e_d = (6.28 \times 10^{-2} \times 3 \times 10^{-2} \times 0.415 \times 0.85)i \left(\frac{1.5 \times 10^5}{1.502 \times 10^5} \right)$$

$$\cong (6.65 \times 10^4)i,$$

where i is the ac current of the source line. This current is computed in the following manner.

Let the programmable divider load R_b equal to 10 kΩ. The pulse amplitude A is 4 V, and pulse duration T_0 is 20 μsec. The rise and fall times are 0.1 μsec each; hence $t_1 = 0.1$ μsec. The period of the pulsed wave T is 0.1 msec. Using Eq. 3-51 for $n = 1$, we compute the peak amplitude of

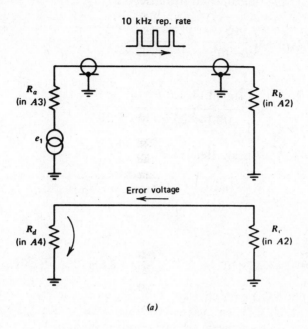

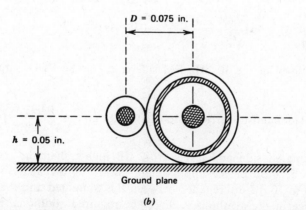

Figure 3-51. An example of ac interference between a coaxial shielded cable and an insulated wire. (a) Schematic diagram; (b) enlarged view of the cable and wire configuration (location 1 in Fig. 3-50).

the voltage at the fundamental frequency as

$$A_{10\,kHz} = 8\left(\frac{20 \times 10^{-6}}{10^4}\right)\left\{\frac{\sin\left[(3.14 \times 0.1 \times 10^{-6})/10^{-4}\right]}{(3.14 \times 0.1 \times 10^{-6})/10^{-4}}\right\}$$

$$\times\left\{\frac{\sin\left[(3.14 \times 20.1 \times 10^{-6})/10^{-4}\right]}{(3.14 \times 20.1 \times 10^{-6})/10^{-4}}\right\}$$

or $A_{10\,kHz} = 1.275$ V peak. Hence

$$i_{10\,kHz} = \frac{A_{10\,kHz}}{R_b} = \frac{1.275}{10^4} = 1.275 \times 10^{-4} \text{ A peak}$$

and

$$e_d = 6.65 \times 10^{-4} \times 1.275 \times 10^{-4} = 8.5 \times 10^{-8} \text{ V peak.}$$

Let the uhf VCO gain constant with respect to the error voltage input $\alpha_{VCO} = 1.5$ MHz/V. Then the frequency deviation of the VCO signal due to a 10 kHz ripple voltage at the error voltage input is

$$\Delta f_{peak} = \alpha_{VCO}e_d = 1.5 \times 10^6 \times 8.5 \times 10^{-8} = 0.1275 \text{ Hz,}$$

$$\Delta f_{rms} = \frac{\Delta f_{peak}}{\sqrt{2}} = \frac{0.1275}{1.414} \cong 0.09 \text{ Hz.}$$

Hence the level of the 10 kHz FM sidebands of the VCO output is (Eq. 2-19)

$$\text{single sideband-to-carrier ratio (dB)} = 20\log_{10}\left(\frac{0.09}{1.414 \times 10^4}\right) = 104.$$

The estimated sideband levels are 2 dB higher than the requirement. Moreover, because of the approximate nature of the analysis, one can expect a 6 to 10 dB discrepancy between the computed and the measured performance of the synthesizer. Since the required additional isolation is small, the recommended approach is to use a coaxial shielded cable for the error voltage signal. It is shown in the analysis of coupling taking place in location 2 that such cable as RG-188/U provides ample isolation.

Similar computations can be performed at several harmonics of 10 kHz. The task is left to the interested student.

Let us estimate levels of pickup in location 2 next. Here coupling takes place between the insulated wire carrying the uhf VCO tuning voltage signal and every control line from the data processor to the programmable divider. The control signals are nonperiodic pulses. The transitions take place only when the synthesizer frequency is changed. Therefore these signals do not introduce any interference. However, in the breadboard stage of circuit development the presence of a 10 mV rms ripple at 10 kHz was detected on all control lines. This spurious signal resulted from poor isolation between outputs and control inputs of the individual counters utilized in the programmable divider. Each control line is bypassed with a 0.01 μF capacitor. More effective filtering cannot be used because of a fast tuning time requirement imposed on the synthesizer performance.

The analysis follows the same steps as those taken in regard to location 1. To save time, only the worst-case coupling, that is, the coupling between a pair of insulated wires adjacent to each other, is investigated. The wires are No. 22 with an outside diameter, including the insulation, of 50 mils. Hence h and D are 25 and 50 mils, respectively. The length of the wires in location 2 is 1 in., or $l = 0.0835$ ft. The frequency of the interfering signal is 10^{-2} MHz. The tuning voltage input impedance, R_d, is 150 kΩ, and the D/A converter output impedance R_c is 10 kΩ at 10 kHz. The ratio $h/D = 0.5$; hence $L_m = 2 \times 10^{-2}$ μH/ft.

The open-wire-to-open-wire expression, Eq. 3-25, is used to compute the spurious ripple voltage coupled to the tuning voltage line. Since the inductances L_1 and L_2 in Eq. 3-25 are the same for both wires, $L = L_1 = L_2$. The diameter of No. 22 conductor is 25.35 mils, so that $h/d = 0.99$ and $L \cong 0.085$ μH/ft. The impedance of the control lines, R_b, is approximately 1 kΩ at 10 kHz. Substituting these values in place of the corresponding unknowns in Eq. 3-25, we obtain

$$e_d = (6.28 \times 10^{-2} \times 2 \times 10^{-2} \times 8.35 \times 10^{-2})i \left(\frac{150}{160} \right)$$

$$\times \left(\frac{1.032 \times 10^{-6} \times 10^3 \times 1.5 \times 10^5}{0.085} - 1 \right)$$

$$= (0.18)i.$$

The ripple voltage in the source line, V_r, is 10 mV rms. The load, R_b, is 1 kΩ at 10 kHz. Hence the ripple current in the source line is

$$i = \frac{V_r}{R_b} = \frac{10^{-2}}{10^3} = 10^{-5} \text{ A rms}$$

and

$$e_d = 0.18 \times 10^{-5} = 1.8 \times 10^{-6} \text{ V rms.}$$

The maximum VCO gain constant associated with the tuning voltage input is 5 MHz/V, so that

$$\Delta f_{rms} = 5 \times 10^6 \times 1.8 \times 10^{-6} = 9 \text{ Hz rms.}$$

Using Eq. 2-19, we obtain

$$\text{single sideband-to-carrier ratio (dB)} = 20 \log_{10} \left(\frac{9}{1.414 \times 10^4} \right)$$

$$= -64,$$

which is 42 dB above the requirement.

We shall try to reduce the coupling by replacing the unshielded wire used for the tuning voltage with the coaxial cable, type RG-188/U. The analysis follows the steps associated with the investigation of pickup in location 1. Equations 3-30 and 3-31 are used again, and the parameters of interest are as follows:

$$R_d = 150 \text{ k}\Omega,$$

$$R_c = 10 \text{ k}\Omega,$$

$$f = 10^{-2} \text{ MHz,}$$

$$L_m = 3 \times 10^{-2} \text{ }\mu\text{H/ft,}$$

$$l = 0.0835 \text{ ft,}$$

$$\alpha_{s1} = 0.85,$$

$$i = 10^{-5} \text{ A rms.}$$

Hence

$$e_d = 6.28 \times 10^{-2} \times 3 \times 10^{-2} \times 8.35 \times 10^{-2} \times 10^{-5} \times 0.85 \left(\frac{1.5 \times 10^5}{1.51 \times 10^5} \right)$$

or

$$e_d = 1.33 \times 10^{-9} \text{ V rms.}$$

For a VCO gain constant of 5 MHz/V, the frequency deviation due to e_d is

$$\Delta f_{rms} = 5 \times 10^6 \times 1.33 \times 10^{-9} = 6.65 \times 10^{-3} \text{ Hz rms}$$

and

$$\text{single sideband-to-carrier ratio (dB)} = 20\log_{10}\left(\frac{6.65\times 10^{-3}}{1.414\times 10^4}\right)$$

$$= -126.5.$$

By using a shielded cable for the tuning voltage line we improved the isolation by more than 60 dB at 10 kHz. (This exercise also verifies our assumption that by using a shielded cable for the error voltage line we eliminate the problem of 10 kHz coupling in location 1.) The margin of 20 dB thus achieved would not be exceeded even if the 10 kHz spurious ripple on all control lines which is coupled to the tuning voltage line were adding in phase. The analysis of coupling in location 3 is left as an exercise to the interested student.

When VCO band switching is not mandatory, it may be more economical to tune the VCO continuously over the specified band. This approach usually results in an increase in VCO sensitivity to pickup by way of the tuning voltage input because of the relatively high VCO gain constants associated with broadband tuning.

For example, if the uhf VCO considered above were tuned from 300 to 400 MHz without band switching, the VCO gain constant referred to the tuning voltage input would probably be higher than 10 MHz/V at 300 MHz, a greater than 6 dB increase in VCO sensitivity to pickup. Under these conditions satisfactory performance would be difficult to guarantee even if a coaxial shielded cable were used for the VCO tuning voltage. In such cases a more appropriate step is to subdivide the synthesizer circuitry so that the cables which transmit high-level nonsinusoidal signals or lead to high-sensitivity circuits are removed from the harness.

Figure 3-52 is an example of circuit breakdown of the shf synthesizer which achieves this effect. By placing the programmable divider in the phase comparator module and the D/A converter in the VCO module, the cables carrying the 10 kHz periodic pulse signal and tuning voltage are removed from the synthesizer harness.

Placing the loop filter in the VCO module seems to be advantageous also; however, the effect of this change on the synthesizer spurious-output performance should be determined by analyzing the phase-locked loop transfer function. The subject of PLLs is treated in detail in Chapters 4 and 5.

By working out this example, we seek to point out the fact, among others, that an intimate knowledge of the circuitry involved is mandatory for analysis of most shielding problems. Mistakes made in estimating the values of such circuit parameters as the terminal resistances, VCO gain

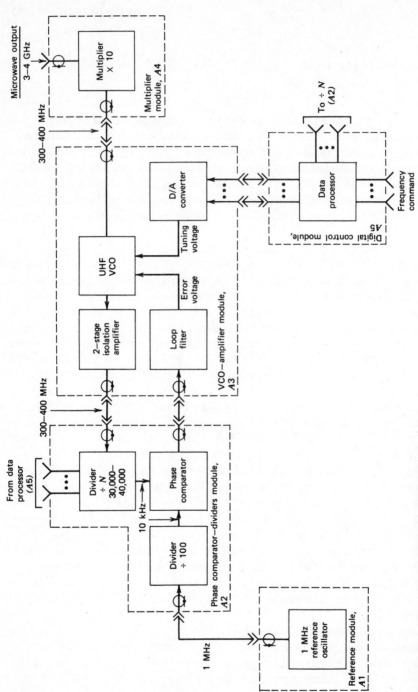

Figure 3-52. A five-module breakdown of the shf synthesizer.

constants, and waveshapes of interfering signals, as well as lack of aware-
ness of the presence of spurious signals carried by lines that theoretically
should be free of any interference, modify, if not invalidate, the results of
the analysis and may lead to either an underestimation of interference
levels or costly solutions.

References

1. Morrison, R. *Grounding and Shielding Techniques in Instrumentation* (New York: John Wiley and Sons, 1967).

2. Darbie, A. M. "Avoid the Pitfalls of Power-Supply Connections," *Electronic Design*, February 15, 1970, pp. D10–D20.

3. A Staff Report. "RFI/EMC and the Power Supply," *Frequency Technology*, February 1970, pp. 23–25.

4. Brown, H. C. "Get Rid of Ground-loop Noise," *Electronic Design*, No. 15, July 19, 1969, pp. 84–87.

5. Skopal, T. E. "Stop Noise Problems Before They Start," *Electronic Design*, No. 1, January 4, 1969, pp. 90–94.

6. Caso, L. F. "Solve Interference Problems Painlessly," *Electronic Design*, No. 25, December 6, 1970, pp. 78–83.

7. Cipperly, W. L. "Optimize Grounding System Design with Planar Circuits," *Frequency Technology*, November 1969, pp. 23–26.

8. Buchman, A. S. "Noise Control in Low Level Data Systems," *Electromechanical Design*, September 1962, pp. 64–81.

9. Widlar, R. J. and J. J. Kubinec. "Transmitting Data with Digital IC's," *The Electronic Engineer*, May 1969, pp. 58–61.

10. Heniford, W. "Muffling Noise in TTL," *The Electronic Engineer*, July 1969, pp. 63–69.

11. Electronic Industries Association, Engineering Department, 11 West 42d Street, New York. *Designer's Guide on Electromagnetic Compatibility*;
    (a) Bulletin No. 4, System Design of Electronic Equipment, April 1965;
    (b) Bulletin No. 5, Bonding of Electronic Equipment, February 1964;
    (c) Bulletin No. 7, Enclosures, Electronic Equipment, October 1966;
    (d) Bulletin No. 8, Cabling of Electronic Equipment, March 1965.

12. Filtron Company, New York. *Interference Reduction Guide for Design Engineers*, Vols. I and II (Washington, D. C.: Clearinghouse for Federal Scientific and Technical Information, Department of Commerce, AD 619666, August 1, 1964).

13. Hayt, W. H., Jr. *Engineering Electromagnetics* (New York: McGraw-Hill Book Company, 1958).

14. METEX Corporation, Edison, N. J. *Application Note* ME-31.

15. Schelkunoff, S. A. *Electromagnatic Fields* (New York: Blaisdell Publishing Company, 1963).

16. Cowdell, R. B. "Nomograms Solve Tough Problems of Shielding," *Electronics*, April 17, 1967, pp. 92–99.

17. Schreiber, O. P. "Some Useful Analogies for RF Shielding and Gasketing." Presented at the Tutorial Section of the Third National Symposium on Radio-Frequency Interference, Washington, D. C., June 12, 1961.

18. Caroll, J. M. *Mechanical Design for Electronics Production* (New York: McGraw-Hill Book Company, 1956).

19. TRIDENT Laboratories, Inc. *Navy Systems Design Guidelines Manual, Electronic Packaging* (Washington, D.C.: Superintendent of Documents, US Government Printing Office, NAVMAT P3940, May 1967).

20. Bureau of Naval Weapons. *Workmanship and Design Practices for Electronic Equipment* (Washington, D. C.: Superintendent of Documents, U. S. Government Printing Office, OP 2230, December 1962).

21. Bunk, D. S. and T. J. Donovan. "Electromagnetic Shielding," *Machine Design*, July 6, 1967, pp. 102–117.

22. Rossler, G. D. "Corrosion and the EMI/RFI Knitted Wire Mesh Gaskets," *Frequency Technology*, March 1969, pp. 15–24.

23. Rothenberg, R. A. "Corrosion Considerations Pertinent to the Use of Electromagnetic Shielding Gaskets," *Frequency Technology*, December 1968, pp. 32–35.

24. Mohr, R. J. "Coupling Between Open and Shielded Wire Lines over a Ground Plane," *IEEE Transactions on Electromagnetic Compatibility*, September 1967.

25. Mohr, R. J. "Interference Coupling—Attack It Early," *Electronic Design News*, July 1, 1969, pp. 33–41.

26. *ITT Reference Data for Radio Engineers*, 5th ed. (New York: Howard W. Sams & Company, 1968).

4 Analog Phase-Locked Loops

It was demonstrated in Chapter 1 that a phase-locked loop (PLL) is one of the building blocks commonly used in frequency synthesis. The complexity of this circuit requires that it be considered in much greater depth than other circuits, which are discussed in some detail in Chapter 6. This chapter and the next one are devoted to a description and evaluation of the parameters and performance of analog and digital PLLs, presented from a point of view important to frequency synthesis, that is, generation of frequency increments, optimization of phase noise, reduction of spurious outputs, and improvement in switching speed.

4-1 Basic Principles of Feedback Systems

This section presents a brief summary of the concepts and terminology of feedback control systems, which play an important role in the analysis of PLLs. Figure 4-1 is a block diagram of a feedback control system. The forward gain is $KG(s)$; $KG(s)H(s)$ is the open-loop gain; K is a constant

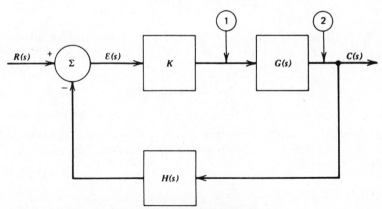

Figure 4-1. Equivalent block diagram of a feedback system.

with respect to the Laplace variable, s; and $G(s)$, as well as $H(s)$, is a frequency-dependent quantity. The system transfer function, that is, the closed-loop response of the system to variations in s of the reference signal, $R(s)$, is

$$\frac{C(s)}{R(s)} = \frac{\text{forward gain}}{1 + (\text{open-loop gain})} \qquad (4\text{-}1)$$

or

$$\frac{C(s)}{R(s)} = \frac{KG(s)}{1 + KG(s)H(s)}. \qquad (4\text{-}2)$$

The transfer function from any point in the system to the output is simply the forward gain from that point to the output divided by $[1 + (\text{open-loop gain})]$, or

$$\text{transfer function from point 1 (see Fig. 4-1) to output} = \frac{G(s)}{1 + KG(s)H(s)}$$

and

$$\text{transfer function from point 2 to output} = \frac{1}{1 + KG(s)H(s)}.$$

Of significant importance in synthesizer design are the system noise bandwidth, B_n, which is defined as

$$B_n = \frac{1}{2\pi} \int_0^\infty \left| \frac{C(j\omega)}{R(j\omega)} \right|^2 d\omega \text{ Hz}, \qquad (4\text{-}3)$$

and the 3 dB bandwidth, B_{3dB}, in hertz, which is obtained by substituting $j\omega_{3dB}$ for s in Eq. 4-2, where $\omega_{3dB} = 2\pi B_{3dB}$, and solving for B_{3dB} so that

$$\left| \frac{C(j\omega_{3dB})}{R(j\omega_{3dB})} \right|^2 = (0.707)^2. \qquad (4\text{-}4)$$

One measure of feedback control system performance is the steady-state error, that is, the error remaining after all transients have died out. The equation for the error is

$$\mathcal{E}(s) = \frac{R(s)}{1 + KG(s)H(s)}. \qquad (4\text{-}5)$$

The steady-state error, ε_{ss}, is computed by applying the final value theorem of Laplace transforms to the error equation for various inputs. Three types of input driving functions are considered in this book: a step, a ramp, and a parabolic function. These are presented analytically in Table 4-1. The final value theorem states that

$$\lim_{t \to \infty} \epsilon(t) = \lim_{s \to 0} s\mathcal{E}(s), \qquad (4\text{-}6)$$

provided that $\mathcal{E}(s)$ is the Laplace transform of $\epsilon(t)$ and $\epsilon(t)$ is stable, that is, all poles of $s\mathcal{E}(s)$ lie in the left half of the s-plane. Hence, in a stable feedback control system, the steady-state error is

$$\varepsilon_{ss} = \lim_{s \to 0} s\mathcal{E}(s). \qquad (4\text{-}7)$$

For a step input $R(s) = A/s$. Using Eqs. 4-5 and 4-7, one solves for the steady-state error:

$$\varepsilon_{ss} = \lim_{s \to 0} \frac{A}{1 + KG(s)H(s)}$$

or

$$\varepsilon_{ss} = \frac{A}{1 + \lim_{s \to 0} KG(s)H(s)}. \qquad (4\text{-}8)$$

For a ramp input $R(s) = v/s^2$, and the steady-state error is

$$\varepsilon_{ss} = \lim_{s \to 0} \frac{v}{s + sKG(s)H(s)}$$

Table 4-1. Laplace Transforms of a Step, a Ramp, and a Parabolic Function.

Function	Mathematical expression (time domain)	Laplace transform (frequency domain)
Step	A	$\dfrac{A}{s}$
Ramp	vt	$\dfrac{v}{s^2}$
Parabolic	$\frac{1}{2}at^2$	$\dfrac{a}{s^3}$

or

$$\varepsilon_{ss} = \frac{v}{\lim_{s \to 0} sKG(s)H(s)} . \qquad (4-9)$$

For a parabolic input $R(s) = a/s^3$, and the steady-state error is

$$\varepsilon_{ss} = \lim_{s \to 0} \frac{a}{s^2 + s^2 KG(s)H(s)}$$

$$= \frac{a}{\lim_{s \to 0} s^2 KG(s)H(s)} . \qquad (4-10)$$

For unity feedback, $H(s) = 1$, the open-loop gain can be expressed as a ratio of factored polynomials:

$$KG(s) = \frac{k'(s+a)(s+b)\dots}{s^n(s+\alpha)(s+\beta)\dots} , \qquad (4-11)$$

where k' is equal to K times some constant generated in the process of factoring the $KG(s)$ expression. For convenience, feedback control systems are classified in terms of Eq. 4-11 as type 0, 1, 2,..., etc., systems for $n = 0$, 1, 2,..., respectively. For example, a feedback control system that is characterized by the equation

$$KG(s) = \frac{k'}{s+\alpha}$$

is a type 0 system, and one that is characterized by

$$KG(s) = \frac{k'}{s(s+\alpha)}$$

is a type 1 system. We use these expressions to solve Eqs. 4-8, 4-9, and 4-10 for the three most commonly used types of feedback control systems as follows. For $n = 0$, Eq. 4-8 becomes

$$\varepsilon_{ss} = \frac{1}{1 + (k'ab\dots/\alpha\beta\dots)} = \text{constant},$$

Eq. 4-9 is

$$\varepsilon_{ss} = \infty,$$

and Eq. 4-10 is

$$\varepsilon_{ss} = \infty.$$

For $n = 1$, Eq 4-8 is

$$\varepsilon_{ss} = 0,$$

Eq. 4-9 is

$$\varepsilon_{ss} = \frac{\alpha\beta\ldots}{k'ab\ldots} = \text{constant},$$

and Eq. 4-10 is

$$\varepsilon_{ss} = \infty.$$

One computes the steady-state error for $n=2$ in a similar manner. Table 4-2 summarizes the results of the computations for $n=0$, 1, and 2 and for a step, a ramp and a parabolic input, and Figs. 4-2, 4-3, and 4-4 are sketches

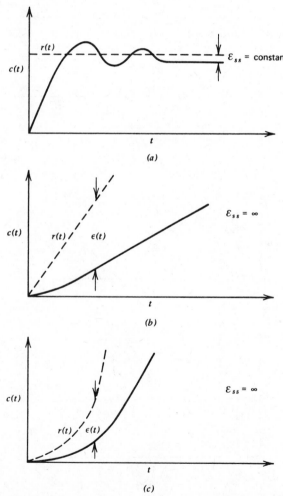

Figure 4-2. Steady-state error of type zero system. (*a*) Step input; (*b*) ramp input; (*c*) parabolic input.

of the output of an underdamped system under the same conditions drawn in the time domain.

Reduction of the steady-state error is limited by system stability considerations. It is the task of the designer to find the system performance parameters that result in a compromise between the magnitude of the steady-state error and the degree of stability.

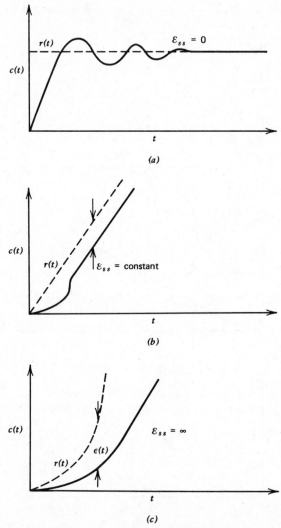

Figure 4-3.   Steady-state error of type 1 system. (a) Step input; (b) ramp input; (c) parabolic input.

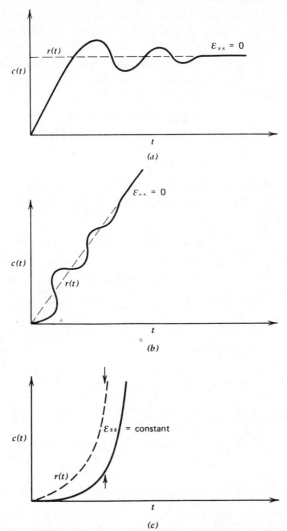

Figure 4-4.   **Steady-state error of type 2 system. (a) Step input; (b) ramp input; (c) parabolic input.**

A system is stable if any bounded input results in a bounded output. The stability of a feedback control system can be analyzed in a number of ways, which are described in Ref. 1, Chapters 4 and 5. In the present book the stability is defined in terms of the Bode diagram, that is, a plot of the open-loop gain in decibels and the open-loop phase shift in degrees versus frequency. The Bode plot analysis is based on the fact that a feedback control system will become unstable if the open-loop gain of the system exceeds unity at the frequency for which the open-loop phase shift is equal to ±180 degrees.

The stability conditions can be stated as follows. Assume that the phase angle of the open-loop gain, $KG(j\omega)H(j\omega)$, is ±180 degrees at some frequency ω_1. Then a feedback control system is

(a)   stable if $20\log_{10}|KG(j\omega_1)H(j\omega_1)|$ is negative;
(b)   marginally stable if $20\log_{10}|KG(j\omega_1)H(j\omega_1)|$ is zero;
(c)   unstable if $20\log_{10}|KG(j\omega_1)H(j\omega_1)|$ is positive.

Table 4-2.   Performance of Feedback Control Systems with Respect to a Step, a Ramp, and a Parabolic Input

Type of system	ε_{ss} for step input	ε_{ss} for ramp input	ε_{ss} for parabolic input
0	Constant	∞	∞
1	0	Constant	∞
2	0	0	Constant

In evaluating the stability of a feedback control system, one speaks of the gain margin, A_m, defined as

$$A_m \overset{\Delta}{=} -20\log_{10}|KG(j\omega_1)H(j\omega_1)| \text{ dB}, \qquad (4\text{-}12)$$

and the phase margin, Θ_m, defined as

$$\Theta_m \overset{\Delta}{=} +180 + \Theta(j\omega_c) \text{ degrees}, \qquad (4\text{-}13)$$

where

ω_c = frequency at which the open-loop gain is unity,

$\Theta(j\omega)$ = open-loop phase shift.

In a stable system $A_m \geqslant 10$ dB and $\Theta_m \geqslant 30$ degrees. Figure 4-5 is the Bode diagram of a typical feedback control system. To simplify the analysis a linear approximation of the gain and phase plots is used. With proper gain and phase margins designed into the system, this approximation is quite adequate for all practical purposes.

The Bode diagrams are very useful in ac compensation of unstable feedback control systems, as well as in phase-noise and spurious-output analysis of such systems, as is demonstrated in the corresponding sections of this chapter.

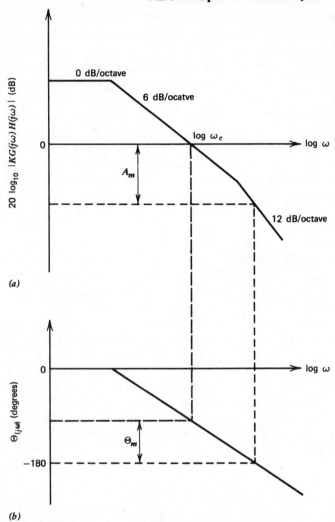

Figure 4-5. Linearized Bode plot of a feedback system. (a) Open-loop gain in decibels; (b) open-loop phase in degrees.

4-2 Performance of a Phase-Locked Loop

Most of the published material on this subject evaluates the performance of a phase-locked loop in the presence of a modulated low-level reference signal at a constant or variable frequency (see Refs. 2 to 9). Under these conditions the signal-to-noise ratio at the reference input is low, and, therefore, tracking information carried by the reference signal and extracting it are of prime importance; the PLL is used as a narrowband FM receiver or tracking filter, and the reduction of noise bandwidth plays a governing role in PLL design.

Typical synthesizer requirements are quite different. First of all, the level of the reference signal is chosen by the designer to be any convenient value. Second, the signal-to-noise ratio at the reference input is quite high (of the order of 100 dB or higher). Third, the functions that an effectively designed PLL is to perform are generation of frequency increments, frequency up-or downconversion, frequency multiplication, frequency division, or a combination of these. These functions are to be accomplished under a condition of optimum utilization of the reference and VCO low-phase-noise characteristics, which often results in a wide loop bandwidth contrary to the requirement encountered in the design of tracking systems. In general, a provision for adjusting loop bandwidth to any practical value dictated by phase-noise requirement is an important consideration in PLL design.

Before proceeding with the presentation of an optimum PLL design, basic principles of PLL operation and loop parameters, such as hold-in and capture ranges, are briefly discussed.

Figure 4-6 is a block diagram of a basic PLL. The loop consists of a voltage-controlled oscillator, a phase detector and a loop filter. The VCO gain constant is K_{VCO} rad/V-sec; the phase-detector gain constant is K_ϕ V/rad: $F(s)$ is the Laplace transform of the loop filter transfer function; and $E_m \sin \phi$ is the beat-note voltage. The maximum value of the beat-note voltage, when the loop is locked, is

$$E_m = \left[\text{half of the phase-detector range (rad)} \right] K_\phi \text{ V}, \qquad (4\text{-}14)$$

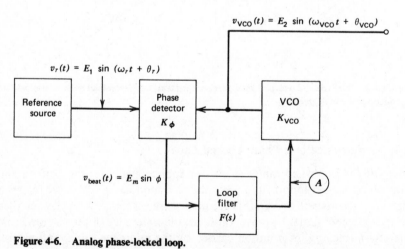

Figure 4-6.   Analog phase-locked loop.

and

$$\phi = \theta_r - \theta_{\text{VCO}} \text{ rad.} \tag{4-15}$$

If a sinusoidal phase detector with a ± 1 rad operating range is assumed, the maximum value of the beat-note voltage is

$$E_m = K_\phi \text{ V.} \tag{4-16}$$

Let the loop initially be open at point A shown in Fig. 4-6; that is, let $f_r - f_0' = \Delta f$, where f_0' is the free-running VCO frequency. Under this condition a difference beat note at Δf is generated at the output of the phase detector, which operates in a mixer mode. The sum and higher-order intermodulation products are also present at the phase-detector output, but it is assumed that their frequencies are significantly higher than the loop filter bandwidth, so that these products do not appear at point A. Assume that the loop is closed and VCO frequency is manually or electronically tuned so as to reduce Δf. The difference beat note is still present at the output of the phase detector, but its nature can no longer be simply defined because it phase-modulates the VCO. (See Refs. 6 and 8 for a detailed discussion of the locking phenomenon.) When Δf is made to be equal to or smaller than the capture range of the PLL, the ac beat note disappears and the phase detector generates a dc voltage that tunes the VCO to the locked frequency, f_0, by sensing the phase difference between the reference and VCO signals. Once the VCO is locked, any drift in VCO frequency that is within the hold-in range of the PLL is instantaneously compensated for by the phase-detector error voltage, which provides a reactance change in the VCO tuned circuit equal and opposite to the change caused by external factors such as ambient temperature. (For a detailed description of VCO circuits see Chapter 6.)

Transfer Functions

It has been customary to use the feedback control system theory in analysis of phase-locked loops. Figure 4-7 is a linearized equivalent block diagram of the basic PLL. It is assumed that a sinusoidal phase detector linearly operates over a ± 1 rad phase range so that $E_m = K_\phi$. According to the feedback control systems analogy, the forward gain is $K_\phi K_{\text{VCO}} F(s)/s$, and so is the open-loop gain. The system transfer function is

$$\frac{\Theta_o(s)}{\Theta_r(s)} = \frac{K_\phi K_{\text{VCO}} F(s)}{s + K_\phi K_{\text{VCO}} F(s)}. \tag{4-17}$$

Similarly, the transfer functions from points 1, 2, and 3 to the output are as

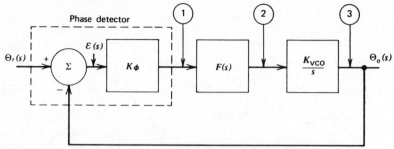

Figure 4-7.   Linearized block diagram of an analog phase-locked loop.

follows:

$$\text{transfer function from point 1 to output} = \frac{K_{\text{VCO}}F(s)}{s + K_{\phi}K_{\text{VCO}}F(s)}, \qquad (4\text{-}18)$$

$$\text{transfer function from point 2 to output} = \frac{K_{\text{VCO}}}{s + K_{\phi}K_{\text{VCO}}F(s)}, \qquad (4\text{-}19)$$

$$\text{transfer function from point 3 to output} = \frac{1}{1 + \left[K_{\phi}K_{\text{VCO}}F(s)/s \right]}. \qquad (4\text{-}20)$$

The phase error function is

$$\mathcal{E}(s) = \Theta_r(s) - \Theta_o(s) = \frac{s\Theta_r(s)}{s + K_{\phi}K_{\text{VCO}}F(s)}. \qquad (4\text{-}21)$$

The importance of Eqs. 4-17 to 4-21 in PLL design will become evident as the material of the following sections is presented.

First-Order Phase-Locked Loop

A phase-locked loop utilizing no loop filter, that is, when $F(s) = 1$, is called a "first-order PLL" because the highest power of s in the denominator of the system transfer function is 1. The open-loop gain of a first-order PLL, in the absence of a feedback function, $H(s)$, is equal to the forward gain, which is $K_{\phi}K_{\text{VCO}}/s$. The dc loop gain is $K_{\phi}K_{\text{VCO}}$. This is a type 1 system because $KG(s) = K_{\phi}K_{\text{VCO}}/s$ or $n = 1$. The transfer function of a first-order PLL is of the form

$$\left[\frac{\Theta_o(s)}{\Theta_r(s)} \right]_{\text{first order}} = \frac{K_{\phi}K_{\text{VCO}}/s}{1 + (K_{\phi}K_{\text{VCO}}/s)}$$

$$= \frac{1}{1 + s(1/K_{\phi}K_{\text{VCO}})}, \qquad (4\text{-}22)$$

and the error function is

$$\mathcal{E}(s) = \frac{s\Theta_r(s)}{s + K_\phi K_{\text{VCO}}}.$$ (4-23)

The steady-state phase error resulting from a step change of input phase of magnitude $\Delta\theta_r$, that is, for $\Theta_r(s) = \Delta\theta_r/s$, is

$$\varepsilon_{ss} = \lim_{s \to 0} \frac{s\Delta\theta_r}{s + K_\phi K_{\text{VCO}}} = 0,$$ (4-24)

and the steady-state error resulting from a ramp input phase, or from a step change in reference frequency of magnitude $\Delta\omega$, that is, for $\Theta_r(s) = \Delta\omega/s^2$, is a constant:

$$\varepsilon_{ss} = \lim_{s \to 0} \frac{\Delta\omega}{s + K_\phi K_{\text{VCO}}} = \frac{\Delta\omega}{K_\phi K_{\text{VCO}}}.$$ (4-25)

Equations 4-24 and 4-25 indicate that a first-order PLL will eventually track out any step change in input phase that is within the system hold-in range and will follow a step frequency change with a phase error that is proportional to the magnitude of the frequency step and inversely proportional to the dc loop gain. The loop will behave in the same manner with respect to a phase or frequency change of the VCO signal.

Of importance to the designer is the first-order PLL response to a ramp change in frequency, that is, when the reference frequency is linearly changing with time at a rate of $d\Delta\omega/dt$ rad sec^2 and $\Theta_r(s) = d\Delta\omega/dt/s^3$. A practical example of such a case can be found in shipboard-to-satellite communications, where the reference frequency is Doppler-corrected with Doppler frequency shift changing at a specified rate. Under these conditions the steady-state phase error is

$$\varepsilon_{ss} = \lim_{s \to 0} \frac{d\Delta\omega/dt}{s^2 + sK_\phi K_{\text{VCO}}} = \infty$$ (4-26)

because in a practically realizable system $K_\phi K_{\text{VCO}} \neq \infty$. This means that above some critical value of rate of change of reference frequency the loop will no longer stay locked. The inverse is also true: if, in an attempt to achieve locking, VCO frequency is linearly swept at a rate above a critical value, locking will not take place.

The time response of a first-order PLL to a step change in reference phase or frequency does not have an overshoot. For example, the time constant associated with the exponential rise of the output phase due to a step change in the input phase is $K_\phi K_{VCO}$, that is,

$$\theta_o(t) = |\theta_i(t)| [1 - e^{-K_\phi K_{VCO} t}].$$

A high dc loop gain results in a fast loop response. The sketches drawn in Fig. 4-8a, b, and c show in the time domain the response of a first-order PLL to the three types of input function previously considered in the frequency domain.

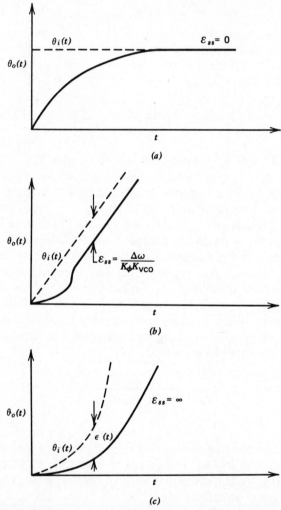

Figure 4-8. Steady-state error of a first-order phase-locked loop. (a) Step input; (b) ramp input; (c) parabolic input.

Of great importance to phase-noise analysis is the system 3 dB bandwidth, B_{3dB}, which describes PLL behavior as a phase-noise filter. This is the frequency, sometimes called the crossover frequency, f_c, at which the magnitude of the open-loop gain equals unity. The system behaves as a low-pass filter with respect to the phase noise of the reference signal; that is, components of the reference noise with rates below f_c appear at the output of the PLL unattenuated, whereas components with rates above f_c are attenuated by the loop at a 6 dB/octave rate. Conversely, a PLL behaves as a high-pass filter with respect to VCO noise.

Solving Eq. 4-4 for the first-order loop, one obtains

$$\omega_{3dB} = K_\phi K_{VCO} \text{ rad/sec}$$

or

$$B_{3dB} = \frac{K_\phi K_{VCO}}{2\pi} \text{ Hz.} \qquad (4\text{-}27)$$

A similar expression can be obtained from

$$\frac{K_\phi K_{VCO}}{\omega_c} = 1.$$

Hence

$$\omega_c = K_\phi K_{VCO} \text{ rad/sec}$$

or

$$f_c = \frac{K_\phi K_{VCO}}{2\pi} \text{ Hz.} \qquad (4\text{-}28)$$

The loop noise bandwidth, B_n, defined in Eq. 4-3 was derived for the first-order loop in Ref. 2 as

$$B_n = \frac{K_\phi K_{VCO}}{4} \text{ Hz.} \qquad (4\text{-}29)$$

The concept of an equivalent noise bandwidth, B_n, is more useful in receiver design, when one deals with the total system noise, than in analysis of synthesizer systems. In this book phase-noise, analysis is performed working with the spectral distribution of noise in the frequency domain, and it is B_{3dB} that is used as the measure of PLL performance. An example of such analysis, Chapter 2, demonstrates the use of B_{3dB}. More is said on this subject later in Chapter 4.

The derivation of Eqs. 4-22 to 4-29 is based on the assumption that either the phase detector is linear over the total phase range of operation or the steady-state error is small. When a sinusoidal phase detector is used and the error is large, a correction factor has to be introduced in all of the above equations. T. J. Rey (Ref. 6) derived a model of the PLL that includes the nonlinearities of the phase detector. His equivalent block diagram of a basic PLL is shown in Fig. 4-9. Using this model, Rey derived

$$\sin \varepsilon_{ss} = \frac{\Delta \omega}{K_\phi K_{VCO}}, \qquad (4\text{-}30)$$

$$\frac{\Theta_o(s)}{\Theta_r(s)} = \frac{K_\phi K_{VCO} F(s)}{s + K_\phi K_{VCO} F(s) \cos \varepsilon_{ss}}, \qquad (4\text{-}31)$$

where

$$\cos \varepsilon_{ss} = \sqrt{1 - \left(\frac{\Delta \omega}{K_\phi K_{VCO}} \right)^2} \qquad (4\text{-}32)$$

and for a first-order loop

$$\left[\frac{\Theta_o(s)}{\Theta_r(s)} \right]_{\text{first order}} = \frac{K_\phi K_{VCO}}{s + K_\phi K_{VCO} \cos \varepsilon_{ss}}, \qquad (4\text{-}33)$$

$$B_n = \frac{K_\phi K_{VCO}}{4 \cos \varepsilon_{ss}} \text{ Hz.} \qquad (4\text{-}34)$$

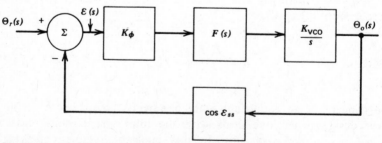

Figure 4-9. Equivalent block diagram of an analog phase-locked loop.

It can be shown that

$$B_{3\,dB} = \frac{K_\phi K_{VCO}}{2\pi} \sqrt{1 + \left(\frac{\Delta\omega}{K_\phi K_{VCO}}\right)^2} \text{ Hz.} \qquad (4\text{-}35)$$

Equations 4-34 and 4-35 indicate that the noise bandwidth and the 3 dB bandwidth of a first-order PLL utilizing a sinusoidal phase detector depend on the VCO mistuning, $\Delta\omega$. This is a highly undesirable feature in low-noise synthesizers because it prevents one from maintaining an optimum phase-noise performance over a specified temperature range and operating frequency band. Fortunately, the variation in $B_{3\,dB}$ is not large because the minimum value of $B_{3\,dB}$ occurs when $\Delta\omega = 0$ or

$$(B_{3\,dB})_{min} = \frac{K_\phi K_{VCO}}{2\pi},$$

and the maximum value of $B_{3\,dB}$ occurs when $\Delta\omega = K_\phi K_{VCO}$ or

$$(B_{3\,dB})_{max} = \frac{K_\phi K_{VCO}}{2\pi} \sqrt{2} = \sqrt{2} \, (B_{3\,dB})_{min}.$$

Phase locking, which is discussed next, is a nonlinear phenomenon and linear approximations cannot be used in the analysis. Consider a PLL utilizing a sinusoidal phase detector that operates over ± 1 rad. The output of the phase detector is $E_m \sin\phi$, where E_m and ϕ are defined on pp. 246 and 247 respectively. It was shown that the steady-state phase error resulting from a step change in reference frequency is, Eq. 4-30,

$$\varepsilon_{ss} = \sin^{-1}\left(\frac{\Delta\omega}{K_\phi K_{VCO}}\right),$$

where $\Delta\omega$ is the magnitude of the step reference frequency change or the VCO frequency error, that is, the difference between the free-running and the locked VCO frequency. Equation 4-30 implies that the maximum VCO frequency error that can be compensated for by a first-order PLL is equal to the dc loop gain, or

$$\Delta\omega_{hold\text{-}in} = K_\phi K_{VCO} \text{ rad/sec} \qquad (4\text{-}36)$$

and

$$\Delta f_{hold\text{-}in} = \frac{K_\phi K_{VCO}}{2\pi} \text{ Hz.} \qquad (4\text{-}37)$$

It is demonstrated in Ref. 6 that the capture range, that is, the maximum difference between the VCO free-running and the locked frequency at which locking is possible without skipping cycles, is also equal to the dc loop gain, or

$$\Delta\omega_{\text{capture}} = \Delta\omega_{\text{hold-in}} = K_\phi K_{\text{VCO}} \text{ rad/sec} \tag{4-38}$$

and

$$\Delta f_{\text{capture}} = \Delta f_{\text{hold-in}} = \frac{K_\phi K_{\text{VCO}}}{2\pi} \text{ Hz.} \tag{4-39}$$

If it is important to keep the steady-state phase error small, one can increase the dc loop gain. However, this eventually leads to an unstable or noisy PLL. Even if these two considerations were not a limiting factor, $K_\phi K_{\text{VCO}}$ could not be made arbitrarily large in a practically realizable system for the following reasons. The parameter K_ϕ depends on the range over which the phase detector operates and the amplitude of the phase-detector input rf voltages. If, for example, the phase detector is of a sinusoidal type, its range cannot be greater than ± 1 rad. The rf voltages, on the other hand, are limited by the dc power supply voltage used in the system. Introduction of a dc amplifier at the output of the phase detector increases K_ϕ only if the phase detector is a low-power device; otherwise, the dc amplifier suffers from the same limitations as the phase detector. To achieve a small error, a special kind of second-order PLL, described in the next section, is used.

If the dc loop gain is limited by phase-noise considerations (see Chapter 2, "Phase Noise in Oscillators"), to achieve a hold-in range adequate to compensate for an expected maximum VCO mistuning the VCO is designed to be as temperature-stable as possible and a tuning mechanism is provided with the best tuning accuracy that can be attained in a practical circuit. A VCO temperature drift of $\pm 0.1\%$ over -40 to $+80°C$ and a tuning accuracy of $\pm 0.5\%$ are within the practical limits of a temperature-compensated VCO performance. Phase noise generated in the VCO can be reduced, keeping $K_\phi K_{\text{VCO}}$ constant, by making K_ϕ as large, and K_{VCO} as small, as possible.

Another important PLL parameter is the phase acquisition time or the time to acquire phase lock when $\Delta\omega$ is made smaller than the capture range. Reference 10 gives the acquisition time for the first-order PLL approximately as

$$t_{\text{acq, phase}} \cong \frac{2}{K_\phi K_{\text{VCO}} \cos \varepsilon_{ss}} \log_e\!\left(\frac{2}{\gamma_{\text{lock}}}\right) \tag{4-40}$$

for small γ_{lock}, where $\gamma_{\text{lock}}=$ specified deviation from steady-state error at the time equal to $t_{\text{acq, phase}}$ (rad). For example, if a PLL is designed so that $K_\phi K_{\text{VCO}}=3.14\times10^3$ rad/sec and the expected $\Delta\omega=10^3$ rad/sec, the steady-state phase error is $\varepsilon_{ss}=0.322$ rad. On the assumption that the required $\gamma_{\text{lock}}=0.1$ rad the acquisition time is

$$t_{\text{acq, phase}}=\frac{2}{3.14\times10^3\times0.9483}\log_e\left(\frac{2}{0.1}\right)=2 \text{ m sec.}$$

Under the condition of $\Delta\omega<K_\phi K_{\text{VCO}}$ the loop locks in frequency without skipping cycles.

When the stability condition described on p. 244 is violated, self-maintained oscillations occur in the loop and phase-modulate the VCO. F. M. Gardner (Ref. 8) shows that the basic first-order loop is unconditionally stable for all values of dc loop gain. However, this simplified form of PLL is seldom used in frequency synthesis. The block diagram of a typical analog PLL used in an indirect synthesizer is shown in Fig. 4-10. Additional circuits are a mixer, a band-pass filter, and a low-pass filter. Sometimes a frequency multiplier or divider follows the band-pass filter. The function of such a PLL is to combine two varying-in-frequency signals at f_r and f_i, and for that purpose a mixer is added to the loop. The band-pass filter attenuates unwanted low-order intermodulation products generated in the process of mixing, and the low-pass filter prevents the reference signal leaking to the output of the phase detector from modulating the VCO.

A. J. Viterbi (Ref. 9) showed that a band-pass filter preceding a phase detector is equivalent to a low-pass filter following the phase detector, as shown in Fig. 4-11. Therefore a three-pole band-pass filter with a 3 dB bandwidth, B_1, and a low-pass filter with a 3 dB cutoff frequency, f_{c1}, can easily introduce enough additional phase shift to make the system unstable. This example is treated in the section of this chapter on stability (p. 275).

Second-Order Phase-Locked Loop

A phase-locked loop with a passive phase-lag filter that is described by the transfer function

$$F(s)=\frac{1+\tau_2 s}{1+\tau_1 s},\tag{4-41}$$

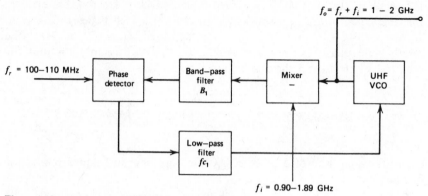

Figure 4-10. Analog phase-locked loop: an example.

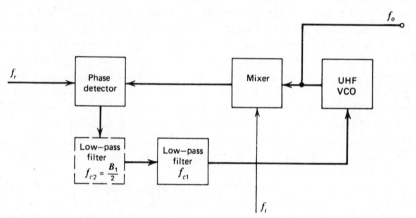

Figure 4-11. An equivalent to the system shown in Fig. 4-10.

where τ_1 and τ_2 are specified by the equations related to Fig. 4-12, is known as a second-order PLL because the highest power of Laplace variable, s, in the denominator of the PLL transfer function is 2.

The open-loop gain and the forward gain of a second-order PLL are

$$\frac{K_\phi K_{VCO}(\tau_2/\tau_1)\left[s+(1/\tau_2)\right]}{s\left[s+(1/\tau_1)\right]}.$$

This is still a type 1 system exhibiting the same steady-state performance as the performance of a first-order loop. The transfer function of a second-

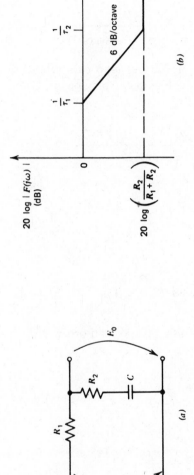

(a)

$$20 \log |F(j\omega)|$$
$$(\text{dB})$$

6 dB/octave

$$\frac{1}{\tau_1} \qquad \frac{1}{\tau_2}$$

$$\log \omega$$

$$20 \log \left(\frac{R_2}{R_1 + R_2} \right)$$

(b)

$$F(s) = \frac{1 + \tau_2 s}{1 + \tau_1 s}.$$

$$\tau_1 = (R_1 + R_2)C.$$

$$\tau_2 = R_2 C.$$

$$F(j\omega) = \frac{1 + j\omega\tau_2}{1 + j\omega\tau_1} = |F(j\omega)| \underline{/\Theta_F(\omega)}.$$

$$|F(j\omega)| = \sqrt{\frac{1 + (\omega R_2 C)^2}{1 + [\omega C (R_1 + R_2)]^2}}$$

$$\Theta_F(\omega) = \tan^{-1}(\omega\tau_2) - \tan^{-1}(\omega\tau_1) \text{ degrees.}$$

(d)

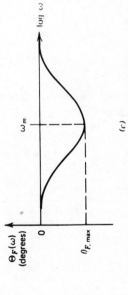

$$\Theta_F(\omega)$$
$$(\text{degrees})$$

$$\omega_m$$

$$\theta_{F, \max}$$

$$\log \omega$$

(c)

Figure 4-12. Phase-lag filter. (a) Schematic diagram; (b) amplitude response; (c) phase response; (d) related equations.

257

order PLL with a passive phase-lag filter is

$$\left[\frac{\Theta_o(s)}{\Theta_r(s)}\right]_{\text{second order}} = \frac{K_\phi K_{\text{VCO}}(1+\tau_2 s/1+\tau_1 s)}{s + K_\phi K_{\text{VCO}}(1+\tau_2 s/1+\tau_1 s)}$$

or

$$\left[\frac{\Theta_o(s)}{\Theta_r(s)}\right]_{\text{second order}} = \frac{K_\phi K_{\text{VCO}}(1/\tau_1)(1+\tau_2 s)}{s^2 + (1/\tau_1)(1+K_\phi K_{\text{VCO}}\tau_2)s + (K_\phi K_{\text{VCO}}/\tau_1)}. \tag{4-42}$$

For convenience, as will become obvious as various loop parameters are presented, Eq. 4-42 is expressed in terms of the loop damping factor, ζ, and natural frequency, ω_n, as

$$\left[\frac{\Theta_o(s)}{\Theta_r(s)}\right]_{\text{second order}} = \frac{s\omega_n\left[2\zeta - (\omega_n/K_\phi K_{\text{VCO}})\right] + \omega_n^2}{s^2 + 2\zeta\omega_n s + \omega_n^2},$$

where

$$\omega_n = \left(\frac{K_\phi K_{\text{VCO}}}{\tau_1}\right)^{1/2} \text{rad/sec} \tag{4-43}$$

and

$$\zeta = \frac{1}{2}\left(\frac{1}{\tau_1 K_\phi K_{\text{VCO}}}\right)^{1/2}(1 + \tau_2 K_\phi K_{\text{VCO}}). \tag{4-44}$$

It can be shown that the 3 dB bandwidth of a second-order PLL in terms of ζ and ω_n is

$$B_{3\text{dB}} = \frac{\omega_n}{2\pi}\left(b + \sqrt{b^2 + 1}\right)^{1/2} \text{Hz}, \tag{4-45}$$

where

$$b = \left[2\zeta^2 + 1 - \frac{\omega_n}{K_\phi K_{\text{VCO}}}\left(4\zeta - \frac{\omega_n}{K_\phi K_{\text{VCO}}}\right)\right].$$

F. M. Gardner (Ref. 8) derived the second-order PLL noise bandwidth as

$$B_n = \frac{\omega_n}{2}\left(\zeta + \frac{1}{4\zeta}\right) \text{Hz}. \tag{4-46}$$

The phase error function is, from Eq. 4-21,

$$\mathcal{E}(s) = \frac{s(1 + \tau_1 s)\Theta_r(s)}{\tau_1 s^2 + (1 + K_\phi K_{VCO} \tau_2)s + K_\phi K_{VCO}}.$$

(4-47)

For a step phase, a ramp phase, and a ramp frequency input the steady-state error is zero, $\Delta\omega / K_\phi K_{VCO}$, and infinite, respectively. Such behavior was expected because all type 1 systems display the same steady-state performance.

The transient behavior of a second-order PLL depends on loop parameters. It is described in terms of the loop natural frequency and damping factor by a set of equations derived for a high loop gain, that is, for $K_\phi K_{VCO} \gg \omega_n$, and summarized in Table 4-3. Plots of transient error versus time are given for various damping factors in Figs. 4-13, 4-14, and 4-15.

The critical value of rate of change of reference frequency above which the loop will no longer stay locked has been derived by A. J. Viterbi (Ref. 11) as

$$\left(\frac{d\Delta\omega}{dt}\right)_{f_r} = \omega_n^2.$$

(4-48)

The maximum rate at which VCO frequency can be swept in an attempt to achieve locking is less than $\omega_n^2/2$ for locking to be guaranteed, that is,

$$\left(\frac{d\Delta\omega}{dt}\right)_{f_{VCO}} < \frac{\omega_n^2}{2}.$$

(4-49)

If $d\Delta\omega/dt$ becomes larger than $\omega_n^2/2$, there is a possibility that the VCO frequency will be swept right through the capture range without locking. The maximum rate at which VCO control voltage may be swept is

$$\left|\frac{dE}{dt}\right|_{max} < \frac{\omega_n^2}{2K_{VCO}}$$

(4-50)

since

$$\frac{d\Delta\omega}{dt} = K_{VCO} \left|\frac{dE}{dt}\right|.$$

As in the case of a first-order PLL, the steady-state error for a ramp phase input, including the effect of phase-detector nonlinearities, is

$$\varepsilon_{ss} = \sin^{-1}\left(\frac{\Delta\omega}{K_\phi K_{VCO}}\right),$$

Table 4-3. Transient Phase Error of Second-Order PLL, $\epsilon(t)$, in Radians (High Loop Gain; $K_\phi K_{VCO} \gg \omega_n$)[a,b]

Damping factor	$\epsilon(t)$		
	for $\theta_r(t) = \Delta\theta,\ t > 0$	for $\theta_r(t) = \Delta\omega t$	for $\theta_r(t) = \tfrac{1}{2}\Delta\dot\omega t^2$
$\zeta < 1$	$\Delta\theta\left(\cos\sqrt{1-\zeta^2}\ \omega_n t - \dfrac{\zeta}{\sqrt{1-\zeta^2}}\sin\sqrt{1-\zeta^2}\ \omega_n t\right)e^{-\zeta\omega_n t}$	$\dfrac{\Delta\omega}{\omega_n}\left(\dfrac{1}{\sqrt{1-\zeta^2}}\sin\sqrt{1-\zeta^2}\ \omega_n t\right)e^{-\zeta\omega_n t}$	$\dfrac{\Delta\dot\omega t}{K_v} + \dfrac{\Delta\dot\omega}{\omega_n^2} - \dfrac{\Delta\dot\omega}{\omega_n^2}\Big(\cos\sqrt{1-\zeta^2}\ \omega_n t\big)e^{-\zeta\omega_n t}$ $+ \dfrac{\zeta}{\sqrt{1-\zeta^2}}\sin\sqrt{1-\zeta^2}\ \omega_n t\big)e^{-\omega_n t}$
$\zeta = 1$	$\Delta\theta(1-\omega_n t)e^{-\omega_n t}$	$\dfrac{\Delta\omega}{\omega_n}(\omega_n t)e^{-\omega_n t}$	$\dfrac{\Delta\dot\omega t}{K_v} + \dfrac{\Delta\dot\omega}{\omega_n^2} - \dfrac{\Delta\dot\omega}{\omega_n^2}(1+\omega_n t)e^{-\omega_n t}$
$\zeta > 1$	$\Delta\theta\left(\cosh\sqrt{\zeta^2-1}\ \omega_n t - \dfrac{\zeta}{\sqrt{\zeta^2-1}}\sinh\sqrt{\zeta^2-1}\ \omega_n t\right)e^{-\zeta\omega_n t}$	$\dfrac{\Delta\omega}{\omega_n}\left(\dfrac{1}{\sqrt{\zeta^2-1}}\sinh\sqrt{\zeta^2-1}\ \omega_n t\right)e^{-\zeta\omega_n t}$	$\dfrac{\Delta\dot\omega t}{K_v} + \dfrac{\Delta\dot\omega}{\omega_n^2} - \dfrac{\Delta\dot\omega}{\omega_n^2}\Big(\cosh\sqrt{\zeta^2-1}\ \omega_n t\big)e^{-\zeta\omega_n t}$ $+ \dfrac{\zeta}{\sqrt{\zeta^2-1}}\sinh\sqrt{\zeta^2-1}\ \omega_n t\big)e^{-\zeta\omega_n t}$
Steady-state error ϵ_{ss}	0	$\dfrac{\Delta\omega}{K_v}$ (not included above)	$\dfrac{\Delta\dot\omega t}{K_v} + \dfrac{\Delta\dot\omega}{\omega_n^2}$ (included above)

[a] From F. M. Gardner, *Phaselock Techniques*, John Wiley and Sons, New York, 1966.
[b] $K_v = K_\phi K_{VCO} F(0)$; $\Delta\dot\omega = d\Delta\omega/dt$.

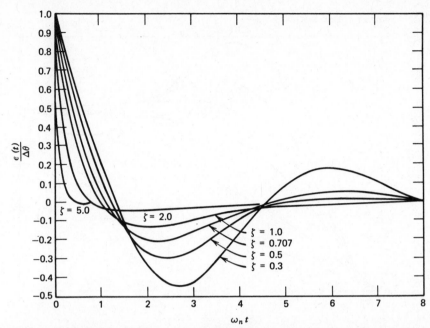

Figure 4-13. Transient phase error, $\epsilon(t)$, due to a step in phase, $\Delta\theta$. From L. A. Hoffman, *Receiver Design and the Phase-Locked Loop*, Aerospace Corporation, El Segundo, May 1963 (booklet prepared for Electronics and Space Exploration Lecture Series. Sponsored by Los Angeles IEEE). Reprinted by permission from F. M. Gardner, *Phaselock Techniques*, John Wiley and Sons, New York, 1966.

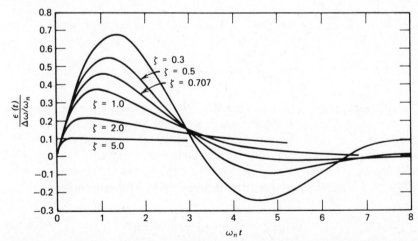

Figure 4-14. Transient phase error, $\epsilon(t)$, due to a step in frequency, $\Delta\omega$. (Steady-state velocity error, $\Delta\omega/K_v$, neglected.) From L. A. Hoffman, *Receiver Design and the Phase-Locked Loop*, Aerospace Corporation, El Segundo, May 1963 (booklet prepared for Electronics and Space Exploration Lecture Series. Sponsored by Los Angeles IEEE.). Reprinted by permission from F. M. Gardner, *Phaselock Techniques*, John Wiley and Sons, New York, 1966.

261

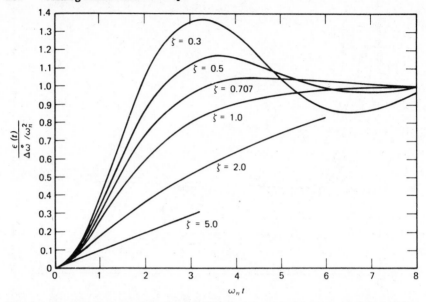

Figure 4-15. Transient phase error, $\epsilon(t)$, due to a ramp in frequency, $\Delta\dot\omega$. (Steady-state acceleration error, $\Delta\dot\omega/\omega_n^2$, included. Velocity error, $\Delta\dot\omega t/K_v$, neglected.) From L. A. Hoffman *Receiver Design and the Phase-Locked Loop*, Aerospace Corporation, El Segundo, May 1963 (booklet prepared for Electronics and Space Exploration Lecture Series. Sponsored by Los Angeles IEEE.). Reprinted by permission from F. M. Gardner, *Phaselock Techniques*, John Wiley and Sons, New York, 1966.

from which it can be deduced that the hold-in range of a second-order PLL is

$$\Delta\omega_{\text{hold-in}} = K_\phi K_{\text{VCO}} \text{ rad/sec.} \tag{4-51}$$

The capture range of a second-order PLL is always smaller than the hold-in range. This is a disadvantage because, having satisfied all other requirements of PLL performance, one would like to have the widest possible capture range in order to keep the cost and power consumption of VCO tuning circuitry at a minimum. The capture range of a second-order PLL with a passive phase-lag filter is given by F. M. Gardner as

$$\Delta\omega_{\text{capture}} = K_\phi K_{\text{VCO}} \left(\frac{\tau_2}{\tau_1} \right) \text{ rad/sec.} \tag{4-52}$$

When $\Delta\omega \leqslant \Delta\omega_{\text{capture}}$, the time to acquire phase lock is probably that described by Eq. 4-40 for the first-order loop. When $\Delta\omega > \Delta\omega_{\text{capture}}$, locking is still possible with some cycles of VCO frequency skipped as long as

$\Delta\omega < \Delta\omega_{\text{pull-in}}$, where, according to D. Richman (Ref. 12),

$$\Delta\omega_{\text{pull-in}} \cong \sqrt{2} \left[2\zeta\omega_n K_\phi K_{\text{VCO}} F(0) - \omega_n^2 \right]^{1/2} \text{ rad/sec}, \qquad (4\text{-}53)$$

where $F(0)$ is $F(s)$ at $s = 0$. This formula is valid for $\omega_n/K_\phi K_{\text{VCO}} F(0) < 0.4$ (moderate or high dc loop gains) but is very poor for low gains or for $\omega_n/K_\phi K_{\text{VCO}} F(0) > 0.5$. Richman (Ref. 3) derived an approximate value of the time required for a second-order loop to frequency-lock as

$$t_{\text{acq, freq}} \cong \frac{4(\Delta f)^2}{B_n^3} \text{ sec}, \qquad (4\text{-}54)$$

where both Δf and B_n are in hertz, and $\Delta f = \Delta\omega/2\pi$. The total time to acquire both frequency and phase lock, assuming that $\Delta\omega_{\text{capture}} < \Delta\omega < \Delta\omega_{\text{pull-in}}$, is

$$t_{\text{acq, total}} \cong t_{\text{acq, freq}} + t_{\text{acq, phase}}. \qquad (4\text{-}55)$$

Second-Order Phase-Locked Loop with a Perfect Integrator

It was demonstrated in the preceding sections of this chapter that the phase of the output signal of either a first- or a second-order PLL with a passive phase-lag filter depends on the VCO mistuning and varies, for example, with variations in ambient temperature, though the phase of the reference signal remains constant. Hence neither of these two types of PLL is in true phase lock. It would be more appropriate, therefore, to call this type of feedback control system a "phase control" system because of the utilization of the phase relationship between VCO and reference signals in achieving frequency locking.

A true PPL is the system whose loop filter transfer function is

$$F(s) = \frac{s+a}{s}. \qquad (4\text{-}56)$$

It is difficult to implement this function with a practical circuit, but there are a number of circuits with transfer functions approximating Eq. 4-56. For example, if $R_1 \gg R_2$ in the phase-lag filter shown in Fig. 4-12, the transfer function of this circuit is

$$F(s) \cong \frac{1 + \tau_2 s}{\tau_1 s} = \left(\frac{\tau_2}{\tau_1} \right) \left[\frac{s + (1/\tau_2)}{s} \right]. \qquad (4\text{-}57)$$

The quantity $\tau_2/\tau_1 = R_2/(R_1 + R_2)$ is very small, so that a high-gain amplifier is required to compensate for the reduction of loop gain introduced by this filter. Figure 4-16 is a schematic diagram of such a filter-amplifier configuration. A high-gain amplifier in the forward path is undesirable, however, because it increases the susceptibility of the system to internally generated or picked-up noise and spurious signals by a factor equal to $(R_1 + R_2)/R_2$, making A in Fig. 4-16 an ultrasensitive point in the loop. Placing the amplifier before the filter reduces the sensitivity of the system to pickup but introduces a new problem—the design of a dc-coupled amplifier capable of delivering a very high output voltage with a relatively small distortion.

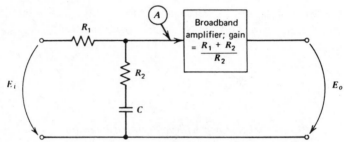

Figure 4-16.   An integrator, a phase-lag filter followed by a high-gain amplifier.

A preferred filter configuration is shown in Fig. 4-17.
The transfer function of this circuit is (Ref. 8)

$$F(s) = \frac{A(sCR_2 + 1)}{sCR_2 + 1 + (1 - A)sCR_1}. \qquad (4-58)$$

For the amplifier gain, A, large and for $R_1 \gg R_2$ this equation converts to

$$F(s) \cong \left(\frac{R_2}{R_1}\right)\left[\frac{s + (1/\tau_2)}{s}\right], \qquad (4-59)$$

where $\tau_2 = R_2C$. This filter is used in the analysis presented below.
The open-loop gain as well as the forward gain of a PLL with a perfect integrator is

$$\frac{K_\phi K_{\text{VCO}}(R_2/R_1)[s + (1/\tau_2)]}{s^2}.$$

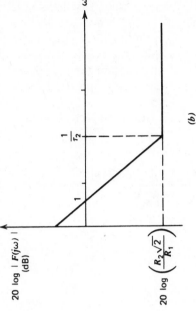

(a)

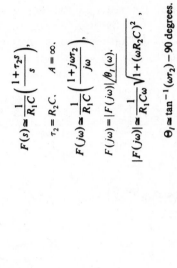

20 log | $F(j\omega)$ |
(dB)

$$\frac{1}{\tau_2}$$

1

$$20 \log\left(\frac{R_2\sqrt{2}}{R_1}\right)$$

(b)

$$F(s) \approx \frac{1}{R_1 C}\left(\frac{1+\tau_2 s}{s}\right),$$

$$\tau_2 = R_2 C. \qquad A = \infty.$$

$$F(j\omega) \approx \frac{1}{R_1 C}\left(\frac{1+j\omega\tau_2}{j\omega}\right),$$

$$F(j\omega) = |F(j\omega)|\underline{/\theta_1(\omega)}.$$

$$|F(j\omega)| \approx \frac{1}{R_1 C \omega}\sqrt{1+(\omega R_2 C)^2},$$

$$\Theta_1 \approx \tan^{-1}(\omega\tau_2) - 90 \text{ degrees.}$$

(d)

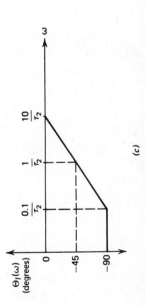

$\Theta_1(\omega)$
(degrees)

0

−45

−90

$$\frac{0.1}{\tau_2} \qquad \frac{1}{\tau_2} \qquad \frac{10}{\tau_2}$$

(c)

Figure 4-17. An integrator, a high-gain feedback amplifier. (a) Schematic diagram; (b) amplitude response; (c) phase response; (d) related equations.

265

It is a type 2 feedback control system used whenever true phase locking is required or when one expects the frequency of the reference signal to be changing at a known rate and wishes the loop to follow these changes. A type 2 PLL is also used in applications that permit large acquisition times because, theoretically, the loop exhibits an infinite capture range and, by utilizing this characteristic, one can save hardware and dc power dissipation by eliminating VCO tuning circuitry.

The transfer function of a second-order PLL with a perfect integrator, according to Eqs. 4-31 and 4-59, is

$$\left[\frac{\Theta_o(s)}{\Theta_r(s)} \right]_{\substack{\text{perfect} \\ \text{integrator}}} = \frac{K_\phi K_{\text{VCO}}(R_2/R_1)[s+(1/\tau_2)]}{s^2 + K_\phi K_{\text{VCO}}(R_2/R_1)s + (K_\phi K_{\text{VCO}}/\tau_2)(R_2/R_1)}.$$

$$(4\text{-}60)$$

Expressing Eq. 4-60 in terms of ζ and ω_n gives

$$\left[\frac{\Theta_o(s)}{\Theta_r(s)} \right]_{\substack{\text{perfect} \\ \text{integrator}}} = \frac{2\zeta\omega_n s + \omega_n^2}{s^2 + 2\zeta\omega_n s + \omega_n^2}, \qquad (4\text{-}61)$$

where

$$\omega_n = \left[\frac{K_\phi K_{\text{VCO}}}{\tau_2} \left(\frac{R_2}{R_1} \right) \right]^{1/2} \text{rad/sec} \qquad (4\text{-}62)$$

and

$$\zeta = \tfrac{1}{2} \left[K_\phi K_{\text{VCO}} \tau_2 \left(\frac{R_2}{R_1} \right) \right]^{1/2} \qquad (4\text{-}63)$$

It can be shown that the 3 dB bandwidth of a second-order PLL with a perfect integrator is

$$B_{3\text{dB}} = \frac{\omega_n}{2\pi} \left[2\zeta^2 + 1 + \sqrt{(2\zeta^2+1)^2 + 1} \right]^{1/2} \text{Hz} \qquad (4\text{-}64)$$

and the noise bandwidth (Ref. 9) is

$$B_n = \frac{K_\phi K_{\text{VCO}}(R_2/R_1) + 1/\tau_2}{4} \text{Hz.} \qquad (4\text{-}65)$$

The error function, according to Eqs. 4-21 and 4-59, is

$$\mathcal{E}(s) = \frac{s\Theta_r(s)}{s + K_\phi K_{VCO}(R_2/R_1)\{[s + (1/\tau_2)]/s\}}$$

or

$$\mathcal{E}(s) = \frac{s^2\Theta_r(s)}{s^2 + K_\phi K_{VCO}(R_2/R_1)s + (K_\phi K_{VCO}/\tau_2)(R_2/R_1)} . \qquad (4\text{-}66)$$

The steady-state error resulting from a step change of input phase of magnitude $\Delta\theta_r$, that is, for $\Theta_r(s) = \Delta\theta_r/s$, is

$$\varepsilon_{ss} = \lim_{s\to 0} s\mathcal{E}(s) = 0,$$

and for a step change of input frequency, $\Theta_r(s) = \Delta\omega/s^2$,

$$\varepsilon_{ss} = 0.$$

For a ramp change of input frequency, $\Theta_r(s) = (d\Delta\omega/dt)/s^3$,

$$\varepsilon_{ss} = \frac{d\Delta\omega/dt}{(K_\phi K_{VCO}/\tau_2)(R_2/R_1)}$$

or

$$\varepsilon_{ss} = \left(\frac{R_1}{R_2}\right)\frac{\tau_2(d\Delta\omega/dt)}{K_\phi K_{VCO}} . \qquad (4\text{-}67)$$

When the nonlinearities of the phase detector are taken into account, the steady-state error becomes

$$\varepsilon_{ss} = \sin^{-1}\left[\left(\frac{R_1}{R_2}\right)\frac{\tau_2(d\Delta\omega/dt)}{K_\phi K_{VCO}}\right]. \qquad (4\text{-}68)$$

Indeed, locking cannot take place at all and the loop will immediately fall out of lock when an excitation is applied at rates

$$\frac{d\Delta\omega}{dt} > \left(\frac{R_2}{R_1}\right)\frac{K_\phi K_{VCO}}{\tau_2} . \qquad (4\text{-}69)$$

The maximum rate at which VCO frequency may be swept in order for locking to take place is significantly lower than the value in Eq. 4-69. For acquisition to be certain, the maximum sweep rate is (Ref. 9)

$$\left(\frac{d\Delta\omega}{dt}\right)_{\text{VCO,max}} = \frac{1}{2\tau_2}\left(4B_n - \frac{1}{\tau_2}\right) \text{rad/sec}^2, \qquad (4\text{-}70)$$

where B_n is the noise bandwidth given by Eq. 4-63. In terms of the maximum rate at which VCO control voltage may be swept, this becomes

$$\left.\frac{dE}{dt}\right|_{\text{max}} = \frac{1}{2K_{\text{VCO}}\tau_2}\left(4B_n - \frac{1}{\tau_2}\right) \text{V/sec} \qquad (4\text{-}71)$$

since

$$\frac{d\Delta\omega}{dt} = K_{\text{VCO}}\frac{dE}{dt}.$$

It is been demonstrated by A. J. Viterbi (Ref. 9) that the pull-in range of a second-order PLL with a perfect integrator is infinite. Indeed, the larger the VCO mistuning, $\Delta\omega$, the longer is the acquisition time. Viterbi derives the frequency acquisition time for a second-order PLL with a perfect integrator as

$$t_{\text{acq,freq}} \cong \tau_2\left(\frac{\Delta\omega}{K_\phi K_{\text{VCO}}} - \sin\theta_0\right) \text{sec}, \qquad (4\text{-}72)$$

where θ_0 is the initial phase difference in radians between the reference and VCO signals. When the loop is described by Eq. 4-59, the expression for the frequency acquisition time becomes

$$t_{\text{acq,freq}} \cong \tau_2\left[\frac{\Delta\omega}{K_\phi K_{\text{VCO}}(R_2/R_1)} - \sin\theta_0\right] \text{sec}. \qquad (4\text{-}73)$$

Acquisition

Analog phase-locked loops operating at a constant frequency or over a narrow frequency band can be designed so that no VCO tuning mechanism is required to acquire and to keep phase lock. Broadband PLLs are provided with some means of VCO tuning. In such cases a separate reactance control circuit is usually added to the VCO (see Fig. 4-18) so that tuning has little effect on loop parameters. Four basic approaches to VCO tuning are known to the writer.

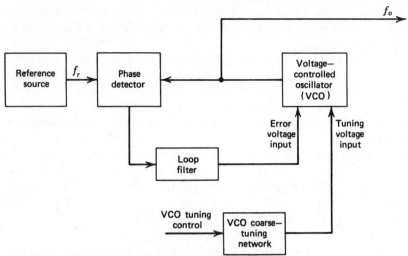

Figure 4-18. Externally tuned phase-locked loop.

In the approach considered first, tuning voltage is provided at all times so that the tuning error, $\Delta\omega$, is always smaller than the capture range. This approach is used in first-order loops, that is, when capture range is equal to hold-in range and removal of tuning voltage results instantaneously in asynchronous operation. A simple (and often the least expensive) way of implementing this approach utilizes resistive dividers, which are switched in and out by mechanical or electronic means (Fig. 4-19). This method is

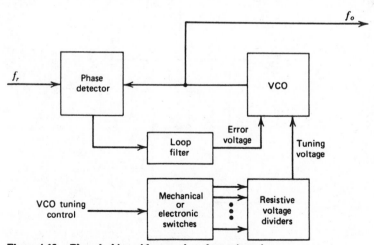

Figure 4-19. Phase locking with manual or electronic tuning.

particularly useful when only a few tuning voltage values have to be generated, as is the case in the stability analysis example presented in the next section.

In applications requiring many discrete values of tuning voltage, a digital frequency comparator and a digital-to-analog (D/A) converter are used in conjunction with resistive voltage dividers (Fig. 4-20). The resistive dividers provide a few widely spaced voltages, and the voltage increments in between are produced by the comparator-D/A converter combination. The comparator counts the frequencies of both phase-detector input signals and presents a difference frequency in the form of a binary-coded number, which is converted into a voltage by the D/A converter. This incremental voltage is added to the voltage generated by the resistive dividers. The binary-coded number from the comparator changes only when a new value of the difference frequency is detected by the comparator; hence the comparator-D/A converter combination always provides the VCO with a proper tuning voltage.

It was shown in this chapter that the capture range of a second-order loop with a passive phase-lag filter is smaller than the hold-in range. Whenever system requirements permit a high-loop gain (i.e., a large hold-in range), this characteristic is utilized in VCO tuning as follows. A course-tuning mechanism provides a tuning voltage in as few steps as necessary to keep the VCO tuning error smaller than the hold-in range, but larger than the capture range, and a capturing device operates in a manner that results in VCO acquisition. The output of this device drops to zero

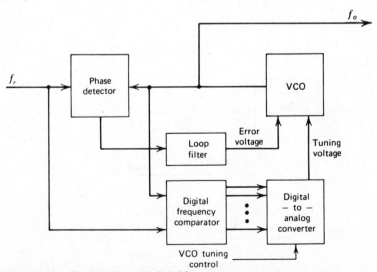

Figure 4-20.   Phase locking with digital frequency comparator.

soon after phase locking. In Fig. 4-21 such a capturing device is a frequency discriminator. Connected in parallel with the phase detector, it generates a correction voltage whenever the frequency difference between the phase-detector input signals is large. When this difference is reduced to a value smaller than the capture range, the phase detector takes over and the VCO locking takes place. The bandwidth of the discriminator is made large enough to result in negligible control voltage at the output of the device whenever frequency lock is achieved, independently of the discriminator center frequency drifts. Both of these functions can be combined in one circuit, as shown in Fig. 4-22. Such a circuit is the balanced discriminator, described in Chapter 6, which exhibits the desired two modes of operation: frequency discrimination and phase detection. Another capturing device is shown in Fig. 4-23. It consists of an ac detector, which senses asynchronous loop operation by detecting the beat note and produces a dc voltage; an amplifier, which amplifies this dc signal to the level adequate to trigger a sweep generator following the amplifier; and the sweep generator, which when triggered generates a sawtooth voltage at the level and sweep rate required by the loop parameters and system requirements. The sawtooth voltage is superimposed on the coarse-tuning voltage and sweeps the VCO frequency over a narrow band. When the VCO frequency comes so close to the reference frequency as to make $\Delta\omega$ less than the capture range, locking takes place. Right after the loop is synchronized, the signal triggering the sweep generator drops to zero, terminating the sweep.

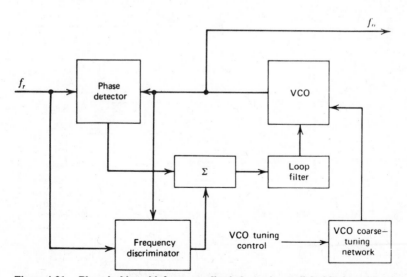

Figure 4-21. Phase locking with frequency discriminator in parallel with phase detector.

In phase-locked loops designed for a small capture range, large hold-in range, and external means of locking, VCO mistuning may be much larger than the loop bandwidth. Under these conditions the level of the beat note, when the loop is not locked, is very low and the beat note is a poor indication of asynchronism. In such cases the ac detector is replaced by an auxiliary phase detector, as shown in Fig. 4-24. When the loop is locked, the auxiliary phase-detector output is set to a maximum by adjusting

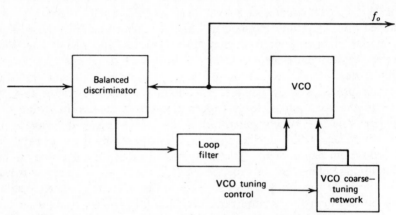

Figure 4-22.   Phase locking with frequency discriminator-phase detector.

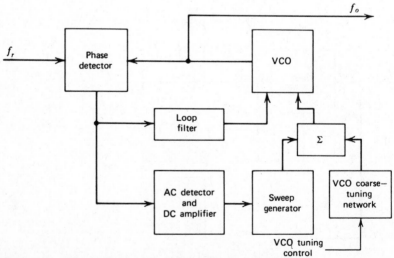

Figure 4-23.   Phase locking with ac detector and sweep generator.

appropriately the phase shift introduced by the variable phase shifter. The output of the inverter following the auxiliary phase detector is low. Hence no signal is present to trigger the sweep generator. As soon as the lock is broken, the auxiliary phase-detector output drops to a low-level difference beat note, the inverter output rises, and the sweep generator is triggered into a sweep mode. This approach of asynchronism detection can be used only with type 2 or type 3 PLLs.

The principle of operation of the VCO tuning circuits shown in Figs. 4-21 to 4-24 is based on effectively widening the loop bandwidth to increase the capture range to its maximum limit during the acquisition mode. The last approach to VCO tuning considered here consists of literally enlarging the loop bandwidth. This can be done by changing either the loop gain or the parameters of the loop filter. The loop gain can be increased by increasing the gain of a dc amplifier inserted in the loop for that purpose (see Fig. 4-25). A switching control signal can be obtained by any of the ways of asynchronism detection previously described. The loop filter is effectively removed from the loop if the resistor R_1 is shorted. This is accomplished by connecting two diodes in a parallel configuration across R_1 as in Fig. 4-26. When the loop falls out of synchronism, the ac

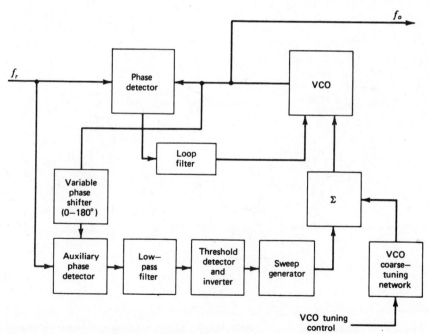

Figure 4-24.   Phase locking with auxiliary phase detector and sweep generator.

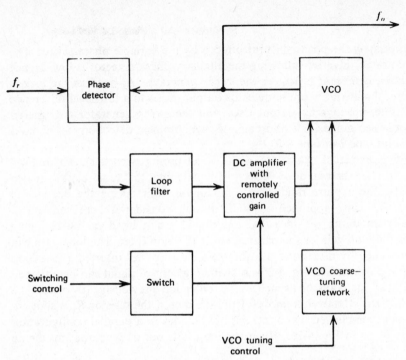

Figure 4-25. Phase locking with automatic widening of loop bandwidth by increasing loop gain.

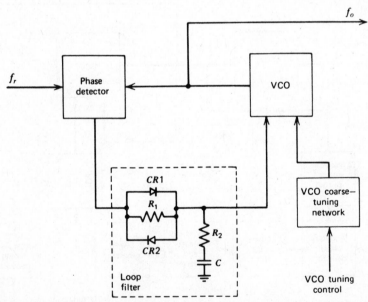

Figure 4-26. Phase locking with automatic widening of loop bandwidth by changing loop filter parameters.

274

beat note generated by the phase detector drives $CR1$ and $CR2$ into conduction; this provides a short across R_1, resulting in a wideband mode of operation. The diode approach is to be used only if it is known that the level of the beat note is high enough to drive the diodes into conduction.

Examples of circuit implementation of some of these tuning mechanisms are given in Chapter 6.

Stability of Analog Phase-Locked Loops

The condition for phase locking does not assure loop stability. A PLL can oscillate if at some frequency its open-loop gain is greater than unity and phase shift exceeds 180 degrees (see Section 4-1). The example of stability analysis presented below illustrates this point.

Consider the first-order PLL shown in Fig. 4-27. It is designed to combine a narrowband signal, varying in 10,000 frequency increments at 3 MHz, with a signal at 7 MHz varying in 10 steps. A PLL approach is chosen instead of a mixer to obtain an improved phase-noise spectrum of the 7 MHz signal far out from signal. A 50 dB improvement at 1 MHz offset frequency is expected. The measured worst-case VCO temperature drift is $\pm0.1\%$ over a specified temperature range. The VCO is coarse tuned in five bands, each 200 kHz wide; that is, five coarse-tuning voltages set the VCO frequency to the center of each 200 kHz band. The isolation amplifier is used to attenuate the 3 MHz signal leaking through the mixer, so that it is -80 dB below the desired signal at the final PLL output. The functions of the band-pass filter, low-pass filter, and dc amplifier will become evident as the material of this section is presented.

The worst case of VCO frequency drift with temperature is ±11 kHz,

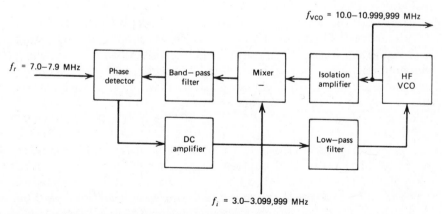

Figure 4-27. Phase-locked loop used in stability analysis.

and the range over which the VCO is to lock is ± 100 kHz; hence the maximum VCO mistuning, $\Delta\omega$, is

$$\Delta\omega = 2\pi(11 \times 10^3 + 100 \times 10^3) = 6.96 \times 10^5 \text{ rad/sec}.$$

Let

$$\Delta\omega_{\text{hold-in}} = 7 \times 10^5 \text{ rad/sec}.$$

A value of $K_{\text{VCO}} = 1.88 \times 10^5$ (rad/sec)/V is chosen on the basis of the VCO phase-noise requirement. Under these conditions, for phase locking to take place the phase-detector constant should be

$$K_\phi = \frac{\Delta\omega_{\text{hold-in}}}{K_{\text{VCO}}} = 3.72 \text{ V/rad}.$$

Assume that a sinusoidal phase detector is used whose gain constant, K'_ϕ, is equal to 0.64 V/rad. Hence a dc amplifier must be used. The amplifier gain should be

$$A = \frac{K_\phi}{K'_\phi} = 5.81 \quad \text{or} \quad 15.5 \text{ dB}.$$

The basic first-order PLL is stable unless the filtering provided in the forward and feedback paths introduces a large amount of phase shift. Let us consider first the band-pass filter following the mixer. To determine the phase characteristic of this circuit, we establish frequencies and levels of the mixer spurious outputs by performing intermodulation products analysis.

Assume that the mixer is a Watkins-Johnson model WJ-M1 operating at $+7$ dBm of the VCO power at f_{VCO} and 0 dBm injection signal power at f_i, as shown in Fig. 4-28a. The mixer insertion loss is 7 dB with the matching condition established in the circuit. The VCO signal leaking to the output and the low-order intermodulation products that fall in or close to the output frequency band are as follows: (a) f_{vco} at 10 to 11 MHz (-36 dB below the desired signal); (b) $2f_i$ at 6.0 to 6.2 MHz (-69 dB); (c) $3f_i$ at 9.0 to 9.3 MHz (-51 dB); (d) $2f_{\text{vco}} - 4f_i$ at 8.0 to 9.6 MHz (-82 dB); (e) $6f_i - f_{\text{VCO}}$ at 7.6 to 8.0 MHz (-86 dB), and $2f_{\text{vco}} - 5f_i$ at 5.0 to 6.5 MHz (-71 dB). The order of these intermodulation products is determined using the techniques described in Chapter 2. The level of the products and the signal at f_{vco} leaking to the output of the mixer are read from Fig. 2-6b for the model WJ-M1 mixer and for appropriate input power levels.

At first glance, one may disregard these spurious signals appearing at the phase-detector input because they are outside the 3 dB loop bandwidth and are attenuated by the loop at a 6 dB/octave rate before they appear at the final PLL output. Unfortunately this approach can lead to unpleasant surprises. A phase detector operates as a mixer with respect to spurious signals applied to either of its inputs, and there are likely to be intermodulation products of relatively low order that fall within the loop bandwidth and are not attenuated by the loop, even though the spurious signals producing these products are outside the loop bandwidth. The proper approach is to perform spurious analysis with respect to each of the spurious signals at the phase-detector inputs that can combine with the

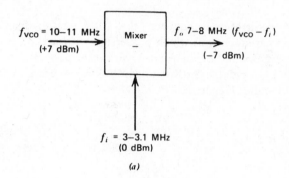

(a)

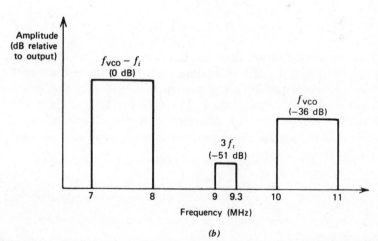

(b)

Figure 4-28. Analysis of the mixer spurious outputs. (a) Mixer block diagram; (b) close-in spurious outputs.

reference signal at f_r. In this example, though, it is assumed that to meet a -80 dB requirement only spurious signals higher than -65 dB should be considered. These are shown in Fig. 4-28b.

A preliminary filter analysis indicates that a two-pole filter attenuates the third harmonic of f_i by the required 14 dB, but a three-pole filter is needed to reduce the VCO signal leaking to the mixer output to below -65 dB. If one assumes a three-pole filter with a Butterworth performance, the in-band phase shift is expressed as (Ref. 13)

$$\Delta\theta = 114.6\left(\frac{\Delta f}{\Delta f_{3\,\text{dB}}}\right) \text{ degrees,} \tag{4-74}$$

where $\Delta\theta =$ filter passband phase shift Δf Hz away from the center frequency,

$\Delta f =$ difference between the filter center frequency and the frequency at which phase shift information is required (Hz),

$\Delta f_{3\,\text{dB}} =$ one half of the filter 3 dB bandwidth (Hz).

In this example, the 3 dB bandwidth is made equal to the mixer output frequency band. Hence

$$\Delta f_{3\,\text{dB}} = 500 \text{ kHz}$$

and

$$\Delta\theta = \left(\frac{114.6}{5\times 10^5}\right)\Delta f$$

$$= (2.292\times 10^{-4})\Delta f \text{ degrees.} \tag{4-75}$$

It was mentioned above that, for the purpose of stability analysis, a band-pass filter with a 3 dB bandwidth inserted into the feedback path of a PLL can be represented by a low-pass filter with half of this bandwidth located in the forward path of the PLL. Hence the $\Delta f_{3\,\text{dB}}$ in Eq. 4-74 is the 3 dB cutoff frequency of an equivalent low-pass filter, and Δf is the frequency at which phase shift information is required. We can write Eq. 4-75 for the low-pass equivalent as

$$\Theta(f) = (2.292\times 10^{-4})f \text{ degrees.} \tag{4-76}$$

For simplicity of presentation, we assume that the in-band insertion loss of this filter is negligible.

The 3 dB loop bandwidth of a first-order PLL is equal to $K_\phi K_{\text{VCO}}$, and the attenuation of the reference signal provided by the loop is 6 dB/

octave. Hence a PLL with a 3 dB bandwidth of 111.5 kHz reduces the phase-noise level of $0-7.0$ to 7.9 MHz signal at 1 MHz offset frequency by approximately 19 dB. The low-pass equivalent of the three-pole band-pass filter attenuates the noise by an additional 18 dB. To obtain a total of 50 dB attenuation, a low-pass filter has to be added in the forward path. (It is assumed that the VCO phase noise is below the requirement at the offset frequency of interest.) A low-pass filter with a cutoff frequency of 200 kHz is selected. The design of the filter is based on the information presented in Fig. 4-29. It provides the necessary 13 dB of noise attenuation.

Having defined the loop circuitry, we can perform a stability analysis. Figure 4-30 is an equivalent block diagram of the PLL considered in this example with important loop constants included in the corresponding functional blocks. To perform the stability analysis the magnitude and phase of

$$\frac{K_\phi K_{\text{VCO}}}{j\omega} F_1(j\omega) F_2(j\omega)$$

versus frequency are plotted. Since it is assumed that the in-band insertion loss of both filters is negligible,

$$\left| \frac{K_\phi K_{\text{VCO}}}{j\omega} F_1(j\omega) F_2(j\omega) \right| = \left(\frac{K_\phi K_{\text{VCO}}}{\omega} \right) \left[\frac{1}{\sqrt{1 + (\omega/\omega_{C1})^2}} \right] |F_2(j\omega)|$$

or

$$G(\omega) = 20 \log_{10} \left| \frac{K_\phi K_{\text{VCO}}}{j\omega} F_1(j\omega) F_2(j\omega) \right|$$

$$= 20 \log_{10} \left\{ \frac{1.115 \times 10^5}{f \sqrt{1 + [f/(2 \times 10^5)]^2}} |F_2(j\omega)| \right\} \text{ dB,} \qquad (4\text{-}77)$$

where

$|F_2(j\omega)| = 1$ for $0 < f < 5 \times 10^5$ Hz and decreases at a rate of 18 dB/octave thereafter.

The phase angle in terms of frequency is

$$\Theta_T(\omega) = \bigg/ \frac{K_\phi K_{\text{VCO}}}{j\omega} F_1(j\omega) F_2(j\omega)$$

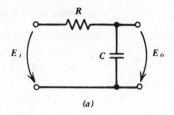

(a)

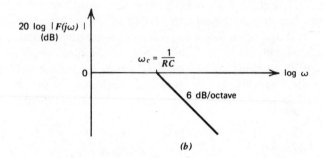

(b)

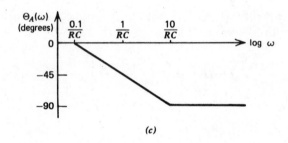

(c)

$$F(s) = \frac{1}{1 + RCs} \, .$$

$$F(j\omega) = \frac{1}{1 + jRC\omega} = |F(j\omega)| \, \underline{/\theta_A(\omega)}$$

$$|F(j\omega)| = \frac{1}{\sqrt{1 + (RC\omega)^2}} \, .$$

$$\Theta_A(\omega) = -\tan^{-1}(RC\omega) \text{ degrees.}$$

$$\omega_C = \frac{1}{RC} \, .$$

(d)

Figure 4-29. Low-pass filter. (*a*) **Schematic diagram;** (*b*) **amplitude response;** (*c*) **phase response;** (*d*) **related equations.**

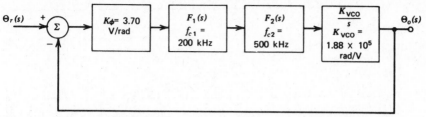

Figure 4-30. Equivalent block diagram of the phase-locked loop used in stability analysis.

or

$$\Theta_T(\omega) = \angle \frac{K_\phi K_{\text{VCO}}}{j\omega} + \angle F_1(j\omega) + \angle F_2(j\omega)$$

or

$$\Theta_T(\omega) = -\left\{ 90 + \tan^{-1}\left[(2\times 10^{-5})f\right] + (2.292\times 10^{-4})f \right\} \text{ degrees.} \quad (4\text{-}78)$$

The plot of loop gain versus frequency can be constructed in a simple manner. The value of $K_\phi K_{\text{VCO}}/\omega$ is computed for $f = 10^2$ Hz and plotted. A line is drawn from this point with a -6 dB/octave slope. At $f = 2\times 10^5$ Hz the slope is increased to -12 dB/octave because of the effect of the low-pass filter, and at $f = 5\times 10^5$ Hz the slope is changed to $-(12 + 18) = -30$ dB/octave to account for the attenuation provided by the band-pass filter.

Figure 4-31, solid curves, is a plot of $G(\omega)$ and $\Theta_T(\omega)$ versus frequency. At a -180 degree phase angle the gain is negative, but the system performance does not show the recommended gain margin of 10 dB. Under certain conditions, such as ambient temperature variations, the loop parameters may shift in direction so as to result in an unstable performance, and loop oscillations will take place. To perform reliably, the PLL has to be ac compensated. The way of performing ac compensation chosen by the writer is graphical. To achieve a gain margin of 10 dB a line is drawn parallel to the amplitude characteristic in the neighborhood of the frequency at which the phase shift is -180 degrees, so as to achieve the desired margin (dotted line). The line is extended to the second break frequency of the phase-lag filter used to ac-compensate the loop, and below that frequency rises at an additional 6 dB/octave until it crosses the solid line representing the plot of the uncompensated gain. The choice of the first and second break frequencies, $1/2\pi\tau_1$ and $1/2\pi\tau_2$, respectively, is

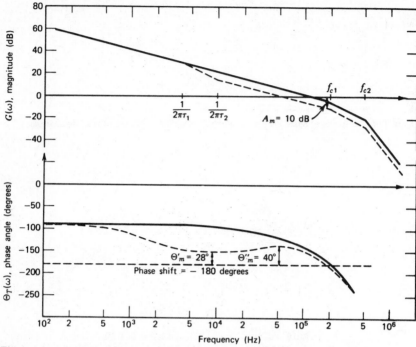

Figure 4-31.   Stability analysis, uncompensated and ac compensated phase-locked loop.

based on the following target specifications:

1. The amplitude response curve should drop between $1/2\pi\tau_1$ and $1/2\pi\tau_2$ by an additional 7 dB to achieve a gain margin of 10 dB.
2. The phase shift introduced by the ac compensation filter should not increase the total phase shift above -150 degrees at frequencies where the loop gain is positive (a phase margin of 30 degrees).
3. The capture range should be reduced as little as possible to achieve the fastest acquisition time.

Figure 4-12 is used to determine the phase-lag filter parameters.

The resultant plot of the gain and phase of the ac-compensated PLL considered in this example is shown in Fig. 4-31, dotted curves. It is obtained by trial and error. The gain and phase margins are 10 dB and 28 degrees, respectively. It is recommended that the interested student determine the impact of the addition of the low-pass and phase-lag filters on loop parameters, such as capture range and acquisition time, and try to improve upon it. What has to be changed or added to the original system design of this PLL to make it lock without providing any additional tuning voltage values?

4-3 Phase Noise

One of the requirements governing a PLL design involves phase noise, sometimes called phase jitter. This requirement is expressed in terms of either a total signal-to-noise ratio measured in a given bandwidth centered at some offset-from-signal frequency or a single sideband-phase-noise-to-signal ratio per Hz bandwidth specified over a wide range of offset frequencies. In either case a PLL noise analysis should start with a basic question: Over what region of offset frequencies is the phase noise specified?

It was demonstrated in Chapter 1 (also see Eqs. 4-17 and 4-20) that a phase-locked loop is effectively a low-pass filter with respect to the noise associated with the reference signal and a high-pass filter with respect to the VCO noise. It is very important, therefore, to know the differences in phase-noise spectra of reference sources, which are crystal-controlled oscillators, and VCOs, which are oscillators with one or two variable-reactance tuned circuits.

A typical example of these differences is shown in Fig. 4-32. Curve 1 is the phase-noise spectrum of the series CO-211 crystal oscillator, manufactured by Vectron Laboratories, at 5 MHz. Curve 2 represents this oscillator, designed for low noise operation, followed by a crystal filter. Notice that the noise of the oscillator with filter is almost 20 dB lower than the noise of the standard model at offset frequencies above 500 Hz. If the noise spectra of several crystal oscillators offered by various manufacturers are compared, significant differences, attributed to circuit design techniques, in units otherwise equal in performance are also found. Nevertheless, noise spectra of all high-quality crystal-controlled oscillators differ from a VCO noise spectrum in a similar manner.

For comparison both curves, 1 and 2 in Fig. 4-32, are normalized to 300 MHz by enhancing the noise by 35.5 dB, curves 3 and 4, respectively. Curve 5 is the noise spectrum of the 300 MHz LC voltage-controlled oscillator estimated in Chapter 2 and originally plotted in Fig. 2-36. Comparison of curves 3 and 4 with curve 5 reveals that the crystal oscillator with and without filter displays superior phase-noise performance close to the signal, which is explained by the filtering effect of the very-high-quality-factor resonators employed in this circuit. At offset frequencies far from the signal the VCO noise is significantly lower than the noise of the crystal oscillator even with the filter because of the noise enhancement associated with frequency multiplication. Even if the crystal oscillator displayed the best practically attainable noise spectrum at 5 MHz, curve 6, the spectrum would be degraded by 35.5 dB, curve 7, as the oscillator frequency was multiplied to 300 MHz, raising the noise 14.5 and 27.5 dB above the VCO noise at 200 kHz and 1 MHz offset frequency, respectively.

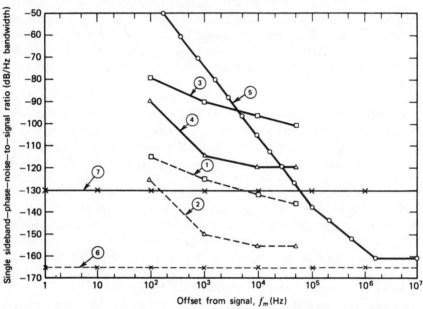

Figure 4-32. Differences in phase-noise spectra of a crystal-controlled and LC oscillators at uhf.

Phase-locked loop parameters are selected so as to arrive at the optimum phase-noise performance at the minimum cost. For example, if the phase-noise requirement called for low noise at offset frequencies close to signal, one would select a reference oscillator with the noise spectrum that satisfies the requirement and design a wideband loop to attenuate VCO noise in that region. If, on the other hand, the requirement called for a low-level far-out noise, the designer would develop a low-noise VCO, select a low-cost reference oscillator exhibiting a poor phase-noise characteristic, and use a narrowband PLL, assuming that the switching speed specification was not a limiting factor, to attenuate the noise of the reference signal to a level below the requirement. If the phase-noise requirement covered a wide band of offset frequencies, the best available reference oscillator would be selected, the best possible VCO would be developed, and the loop bandwidth would be chosen so as to result in optimum phase-noise performance of the system.

A few practical design hints of a general nature are given below to serve as a guide.

A PLL is effectively an AM-to-PM converter. Any noise appearing as a voltage at the error voltage input of the VCO is converted by the latter into phase noise. Hence the error voltage input is the most sensitive point in a PLL system, if no dc amplifier is used preceding the VCO. Any noise picked up at the error voltage input or leaking to it is acted upon by the PLL as defined by Eqs. 4-17 to 4-20; however, the higher K_{VCO}, the higher is the resultant level of phase noise at the final PLL output. It is desirable to select K_{VCO} as small as possible; this means that, for a given loop gain, K_ϕ should be as large as possible. Good design practice is to keep the loop gain, $K_\phi K_{VCO}$, small and to develop a VCO that exhibits the best frequency stability with temperature and time and the best tuning accuracy economically feasible.

Inserting a dc amplifier in the forward path moves the point of highest sensitivity from the VCO error voltage input to the dc amplifier input, because for a PLL to function properly the dc amplifier bandwidth is made larger than the loop bandwidth, and at the same time increases the sensitivity by a factor that is equal to the amplifier in-band gain. In low-noise PLLs, therefore, it is not advisable to use high-gain dc amplifiers in the forward path unless analysis shows that the phase-noise requirement can be satisfied with an ample margin when practical power levels of pickup and leakage are assumed. Power levels below -165 dBm/Hz should not be considered practical.

Phase-locked loops that are electronically tuned over a broad frequency band suffer from noise present at the input of the coarse-tuning reactance control circuit. The gain constant of this circuit may be several megahertz per volt at the low end of the band at uhf and several tens of megahertz at shf, which is more than enough to convert the thermal noise present at the output of the dc power supply into a high-level phase noise at the PLL output. (The problem associated with VCO coarse tuning is considered in the section on phase noise in oscillators and a numerical example is given in Chapter 2.) This noise is reduced in magnitude, but not eliminated, by dividing the VCO frequency band into several subbands and by switching a fixed capacitance in the VCO tuned circuit corresponding to each subband. Under these conditions the gain constant of the coarse-tuning network is reduced because of a reduction in the range over which the variable-reactance element, such as a voltage-controlled capacitance, is tuned. A detailed description of a VCO band-switching circuit is given in Chapter 6.

The noise contributions of rf amplifiers and mixers located in the feedback path of a PLL, such as shown in Fig. 4-27, can be made negligible by designing for levels of rf signal that give the required signal-to-noise ratio at the input of each device. (See Chapter 2 on noise in amplifiers and mixers.) The filtering effect of a PLL on this noise can be estimated from examining the loop transfer functions, Eqs. 4-17 through 4-20.

A band-pass, band-reject, high-pass, or low-pass filter inserted in the appropriate place in the loop, as in the example cited in the section on stability, (p. 275), may be used to increase the roll-off slope of the PLL characteristic from 6 to 12 dB/octave. The designer should be aware of the fact that under certain conditions a filter may enhance, instead of reducing, the phase noise of a signal. This noise degradation can be prevented by achieving a low-loss filter design and by proper matching of the filter to the driving source and load impedances.

An example of PLL phase-noise analysis was given in Chapter 2. Although a digital PLL was used in that example, all of the important points considered there, such as the derivation of the reference source noise appearing at the final PLL output and of the VCO noise, apply to an analog PLL as well and hence are not repeated here.

4-4 Spurious Outputs

The techniques of identification, elimination, and measurement of spurious signals generated by circuits such as mixers, multipliers, and dividers are described in detail in Chapters 2 and 6. Because of the variety of circuits comprising a phase-locked loop, PLL spurious outputs manifest themselves as AM, PM, or signals unrelated to the final PLL output. Whether injected into the loop as part of the reference signal spectrum or generated by the PLL circuitry, these signals are acted upon by the PLL as defined by the loop transfer functions and the Bode plot of loop gain versus frequency. To avoid repetition, only a few cases of special interest are considered here.

Amplitude Modulation

The most significant source of spurious amplitude modulation in a PLL is the dc power supply. Amplitude modulation results when a signal at the line frequency or its harmonics leaks to, or is picked up by, circuits, such as rf amplifiers, that effectively are amplitude modulators. Although the leakage path is usually through the dc power supply, a direct pickup of the power supply transformer field is possible and should be considered if the level of AM sideband components exceeds the requirement.

It is costly to perform filtering of a signal at 60 Hz or at harmonics of 60 Hz on an individual circuit basis. Depending on the place in the loop at which AM is taking place, the loop may provide some attenuation of the sideband components. However, usually the steps taken to prevent spurious AM consist of (*a*) selecting a low-ripple power supply, and (*b*) enclosing the power supply transformer in a Mu-metal can. The 250 μV peak-to-peak ripple is a common power supply requirement and 20-μV-ripple power supplies have been used in hf synthesizers exhibiting better than -100 dB AM sideband performance. Care should be exercised in routing harnesses of cables carrying high currents at the line frequency so as not to generate high-intensity fields in the vicinity of circuits susceptible to AM. If necessary, these harnesses should be enclosed in jackets made of low-permeability material, such as the type MS zip-on jacket manufactured by Zippertubing Company.

Phase Modulation

Phase modulation of the output signal is a more serious problem in phase-locked loops because a PLL, referred to the VCO error voltage or coarse-tuning voltage inputs, is a highly sensitive phase modulator. Any ac signal appearing at these inputs produces a high-level phase modulation of the PLL output. The spurious-output problem is more severe in PLLs that are electronically tuned over a broad frequency band, as was the case also with phase noise.

The steps taken to reduce levels of PM sidebands are (*a*) designing for a low value of K_{VCO}; (*b*) selecting a low-ripple power supply; (*c*) using low-noise voltage regulators, such as type μA723 (Part No. U6A7723393 manufactured by Fairchild Semiconductor Company), to provide additional filtering of the dc power line directly at the critical circuits (the VCO and coarse-tuning network); (*d*) providing adequate rf shielding for the circuits susceptible to PM; (*e*) placing the phase detector and coarse-tuning network either in the VCO module or in close proximity to it, so that the error and tuning voltage leads (preferably rf coaxial cables) are short and introduce negligible pickup levels of spurious signals; and (*f*) enclosing the power supply transformer in a Mu-metal can.

Spurious Signals Unrelated to the Final PLL Output

The mechanism of generation of spurious signals unrelated to the final output in analog PLLs is best demonstrated with an example. Consider the PLL previously used in the stability analysis with some functional blocks removed, as shown in Fig. 4-33, because they do not contribute to this discussion. Assume that the loop is a part of a microwave synthesizer and that, therefore, the requirement for spurious signals at the PLL output is

−100 dB. To begin with, assume also that no isolation amplifier is used. The output signal power of the PLL is +7 dBm at f_{VCO}. This is also the power of the VCO signal applied to the local oscillator (LO) port of the hf mixer, Watkins-Johnson model WJ-M1. The power at f_i applied to the signal port of the mixer is 0 dBm. The mixer provides a signal-to-LO-port isolation of 65 dB at f_i, which is 3 MHz. Hence the leakage level of the signal at f_i at the LO port is −65 dBm. To reduce this signal, which is only −72 dB below the output signal, to −100 dB, an isolation amplifier with a forward gain of unity and an output-to-input loss of 28 dB at 3 MHz must be inserted between the mixer and hf VCO. A one-stage tuned amplifier will provide the required isolation.

The leakage problem becomes severe if all frequencies of the PLL signals are multiplied by a factor of 10^2, as is shown in Fig. 4-34. A uhf

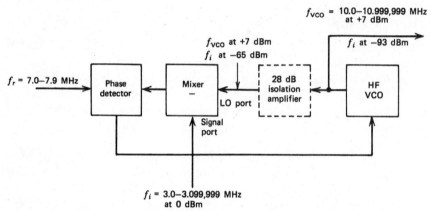

Figure 4-33.   HF phase-locked loop, analysis of spurious outputs.

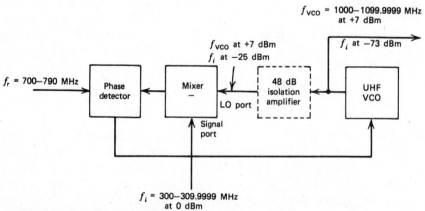

Figure 4-34.   UHF phase-locked loop, analysis of spurious outputs.

mixer, Lorch Electronics model FC301, provides only 25 dB of isolation at 300 MHz between the signal and LO ports. On the assumption that at uhf the spurious-output requirement is -80 dB, the isolation amplifier output-to-input loss should be 48 dB. This is costly to implement even if accounted for in the original design. The addition of the circuit would be significantly more costly to implement, however, if the designer discovered the problem during his final testing of the PLL.

4-5 Example of an Analog PLL Design

An hf phase-locked loop was partially designed in the section of this chapter on PLL stability. Two functions performed by the PLL, frequency addition of two rf signals and phase-noise spectrum improvement of one of the signals, governed the PLL design during the first design phase. An example of spurious-output analysis was given in Section 4-4. In this example the loop analysis is continued to complete the description of the PLL performance.

The following loop parameters were previously computed for the first-order loop:

$$K_\phi = 3.72 \text{ V/rad},$$

$$K_{\text{VCO}} = 1.88 \times 10^5 \text{ (rad/sec)/V},$$

$$K_\phi K_{\text{VCO}} = 7 \times 10^5 \text{ rad/sec},$$

$$\Delta\omega_{\text{max}} = 6.96 \times 10^5 \text{ rad/sec},$$

$$\Delta\omega_{\text{capture}} = \Delta\omega_{\text{hold-in}} = K_\phi K_{\text{VCO}} = 7 \times 10^5 \text{ rad/sec},$$

$$\Delta f_{\text{capture}} = \Delta f_{\text{hold-in}} = \frac{K_\phi K_{\text{VCO}}}{2\pi} = 111.4 \text{ kHz},$$

$$B_{3\,\text{dB}} = \frac{K_\phi K_{\text{VCO}}}{2\pi} = 111.4 \text{ kHz}.$$

Notice that an approximate value of $B_{3\,\text{dB}}$ could be obtained also from Fig. 4-31 since $B_{3\,\text{dB}}$ is equal to the loop crossover frequency, that is, the frequency at which the loop gain is unity.

Originally the loop was designed so that the tuning voltage set the VCO frequency within the capture range. When this is the case, the acquisition time for the PLL without ac compensation is

$$t_{\text{acq, phase}} = \frac{2}{K_\phi K_{\text{VCO}} \cos\varepsilon_{ss}} \log_e\left(\frac{2}{\gamma_{\text{lock}}}\right),$$

where

$$\varepsilon_{ss} = \sin^{-1}\left(\frac{\Delta\omega_{max}}{K_\phi K_{VCO}}\right)$$

$$= 83.9 \text{ degrees} = 1.464 \text{ rad},$$

so that

$$\cos\varepsilon_{ss} = 0.1063.$$

Letting

$$\gamma_{lock} = 0.1 \text{ rad},$$

we obtain

$$t_{acq, phase} = 8.1 \ \mu sec.$$

This is a broadband, high-loop-gain PLL, and the phase acquisition time of the loop is very short.

By adding a phase-lag network for the purpose of ac compensation we introduce the following performance changes. The capture range no longer equals the hold-in range. The new value of $\Delta\omega_{capture}$ is

$$\Delta\omega_{capture} \cong \left(\frac{\tau_2}{\tau_1}\right)\Delta\omega_{hold\text{-}in}.$$

From Fig. 4-31,

$$\frac{1}{2\pi\tau_1} = 4\times10^3, \qquad \tau_1 = 3.98\times10^{-5} \text{ sec},$$

$$\frac{1}{2\pi\tau_2} = 10^4, \qquad \tau_2 = 1.59\times10^{-5} \text{ sec},$$

$$\Delta\omega_{capture} \cong (0.4)\Delta\omega_{hold\text{-}in} = 2.8\times10^5 \text{ rad/V},$$

$$\Delta f_{capture} \cong (0.4)\Delta f_{hold\text{-}in} = 44.56 \text{ kHz}.$$

The 3 dB bandwidth computed from Eq. 4-45 is equal to 50.9 kHz. (When determined from Fig. 4-31, it is approximately 53 kHz.) Notice that with the addition of a band-pass and low-pass filters it is the Bode plot of loop gain versus frequency, and no longer the equation for the 3 dB bandwidth, that contains all the pertinent information describing the phase-noise and spurious-signal filtering characteristics of the PLL.

Equation 4-54 for the frequency acquisition time of the second-order PLL can be used only if

$$\Delta\omega_{max} \leqslant \Delta\omega_{pull\text{-}in},$$

where $\Delta\omega_{pull\text{-}in}$ is defined by Eq. 4-53. In this example $F(0)=1$; hence

$$\omega_n = \left(\frac{K_\phi K_{VCO}}{\tau_1}\right)^{1/2} = 1.33 \times 10^5 \text{ rad/sec},$$

$$\zeta = \frac{1}{2}\left(\frac{1}{\tau_1 K_\phi K_{VCO}}\right)^{1/2}(1+\tau_2 K_\phi K_{VCO}) = 1.145,$$

$$\Delta\omega_{pull\text{-}in} \cong \sqrt{2}\left[2\zeta\omega_n K_\phi K_{VCO} F(0) - \omega_n^2\right]^{1/2} = 6.25 \times 10^5 \text{ rad/sec},$$

and

$$\Delta f_{pull\text{-}in} = \frac{\Delta\omega_{pull\text{-}in}}{2\pi} \cong 100 \text{ kHz}.$$

By adding one more value of tuning voltage, we divide the 1 MHz output frequency band into six subbands, each approximately 166.7 kHz wide. Under these conditions, the greatest VCO mistuning, $\Delta f'_{max}$, is $(166.7/2+11)=94.35$ kHz, and VCO capture is feasible with some cycles skipped before locking takes place. The frequency acquisition time, Eq. 4-54, is

$$t_{acq,freq} \cong 4\frac{(\Delta f'_{max})^2}{B_n^3} \cong 47.8 \ \mu sec$$

since

$$B_n = \frac{\omega_n}{2}\left(\zeta + \frac{1}{4\zeta}\right) \cong 90.6 \text{ kHz}.$$

The total acquisition time is

$$t_{acq,total} = t_{acq,freq} + t_{acq,phase} \cong 60.0 \ \mu sec.$$

If, instead of adding another value of tuning voltage, a capturing circuit is used, the acquisition time is affected in the following manner. The maximum VCO sweep repetition rate is

$$\left(\frac{dE}{dt}\right)_{\text{VCO}} = \left(\frac{1}{K_{\text{VCO}}}\right)\left(\frac{\omega_n^2}{2}\right) \cong 4.69 \times 10^4 \text{ V/sec}.$$

For a linear sweep of magnitude E V and period T sec,

$$\left(\frac{dE}{dt}\right)_{\text{VCO}} = \frac{E}{T}.$$

Let $E = 10$ V; then $T = 213$ μsec. Hence if it is assumed that VCO locks during the first period of the sweep, the longest acquisition time is

$$t_{\text{acq, total}} = T = 213 \text{ } \mu\text{sec}.$$

The actual acquisition time may be significantly longer than 213 μsec because of the initial jump in the VCO tuning voltage at the start of the sweep, which can throw the VCO off frequency and make the initial VCO mistuning larger than the estimated value, so that the VCO acquires lock only after a few cycles of the sweep repetition rate. If the fastest possible acquisition time were required, the designer would have to add enough tuning voltage values to bring the VCO frequency within the capture range each time the PLL frequency was changed. For $\Delta f_{\text{capture}} = 44.56$ kHz the maximum subband will be equal to $2(44.56 - 11) = 67.12$ kHz. Hence 15 values of tuning voltage would have to be provided.

References

1. Savant, C. J., Jr. *Basic Feedback Control System Design* (New York: McGraw-Hill Book Company, 1958).

2. Gruen, W. J. "Theory of AFC Synchronization," *Proceedings of the IRE*, August 1953, pp. 1043–1048.

3. Richman, D. "Color-Carrier Reference Phase Synchronization Accuracy in NTSC Color Television," *Proceedings of the IRE*, January 1954, pp. 106–134.

4. Jaffe, R. and E. Rechtin. "Design and Performance of Phase-Lock Circuits Capable of Near-Optimum Performance over a Wide Range of Input Signal and Noise Levels," *IRE Transactions—Information Theory*, March 1955, pp. 66–76.

5. Gilchriest, C. E. "Application of the Phase-Locked Loop to Telemetry as a Discriminator or Tracking Filter," *IRE Transactions on Telemetry and Remote Control*, June 1958, pp. 20–35.

6. Rey, T. J. "Automatic Phase Control: Theory and Design," *Proceedings of the IRE*, October 1960, pp. 1760–1771.

7. Viterbi, A. J. "Phase-Locked Loop Dynamics in the Presence of Noise by Fokker-Planck Techniques," *Proceedings of the IEEE*, December 1963, pp. 1737–1753.

8. Gardner, F. M. *Phaselock Techniques* (New York: John Wiley and Sons, 1966).

9. Viterbi, A. J. *Principles of Coherent Communication* (New York: McGraw-Hill Book Company, 1966).

10. Tausworth, R. C. "Theory and Practical Design of Phase-Locked Receivers," *Technical Report* 32-819 (Pasadena, Calif.: Jet Propulsion Laboratory, California Institute of Technology, February 15, 1966).

11. Viterbi, A. J. "Acquisition and Tracking Behavior of Phase-Locked Loops," *Jet Propulsion Laboratory External Publication* 673, July 14, 1959.

12. Richman, D. "APC Color Sync for NTSC Color Television," *IRE Convention Record*, Part 4, 1953.

13. Howard W. Sams & Co., Inc. *Reference Data for Radio Engineers*, 6th ed., first printing, 1975.

5 Digital Phase-Locked Loops

The discussion of digital phase-locked loops in this chapter is based on the assumption that the reader has studied the material presented in Chapters 1 (pp. 31 through 36), 2, and 4 and is thoroughly familiar with the principles governing a PLL operation in general and a digital PLL in particular.

Because of the close similarity between analog and digital PLLs, which is demonstrated below, Chapter 5 presents mainly equations necessary for analysis and design of a digital PLL. To avoid repetition, only comments that refer specifically to a digital PLL operation are included. The reader will find an extensive discussion of various subjects common to both PLLs, such as acquisition, in Chapter 4.

5-1 Performance of a Digital Phase-Locked Loop

Figure 5-1 is the general block diagram of a digital PLL. The VCO signal at f_{out} is divided by N_2 and downconverted before it is applied to a variable-ratio divider. Phase comparison is accomplished at f_ϕ which is derived from a reference signal at f_{ref} divided by N_1. The functions that each circuit in Fig. 5-1 performs were described in Chapters 1 and 2, except for the band-reject filter, and are briefly summarized below.

The fixed-ratio divider, N_1, is used because f_ϕ is often in the neighborhood of 10 kHz and it is not economical to build stable low-noise crystal-controlled oscillators at these frequencies. (The reference oscillator frequency is usually between 1 and 10 MHz.) The term "phase comparator" replaces "phase detector" to point out that a digital sample-and-hold device is preferred to a sinusoidal phase detector because it provides more than 80 dB suppression of the reference signal at f_ϕ, which is an important factor when f_ϕ is close to the loop 3 dB bandwidth and the loop is not capable of attenuating the signal at f_ϕ by the required amount. The fact that this phase comparator operates over a phase range equal to approximately $\pm\pi$ rads is another important advantage because it results in a higher effective phase-comparator gain constant as compared to the gain

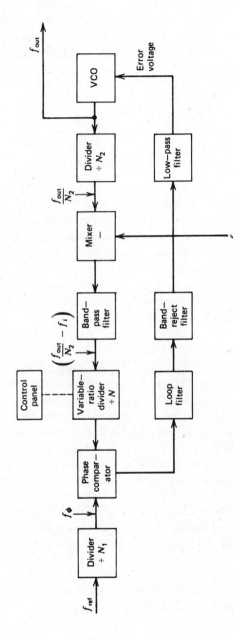

Figure 5-1. Digital phase-locked loop.

295

constant associated with a sinusoidal phase detector. The loop filter ac stabilizes the PLL while the band-reject and low-pass filters attenuate the reference signal at f_ϕ and its harmonic leaking to the output of the phase comparator. The other fixed-ratio divider, N_2, reduces the frequency of the VCO signal so that after the downconversion the frequency is within the operating range of the variable-ratio divider, N. Finally, the band-pass filter following the mixer attenuates the sum and higher-order intermodulation products of the two mixer input signals.

The frequency of the output signal, f_out, is a function of f_ref and f_i, Eq. 1-14:

$$f_\text{out} = N_2(Nf_\phi + f_i),$$

$$f_\phi = \frac{f_\text{ref}}{N_1}.$$

Hence f_out is varied in N_2Nf_ϕ frequency increments as N is changed in an appropriate manner. The output frequency will also vary in $N_2\Delta f_i$ frequency increments if f_i is varied in Δf_i increments. It seems that by making f_ϕ arbitrarily small, one can design a digital PLL capable of generating any desired number of frequency increments. In many practical cases, however, this is not so. Some considerations limiting the minimum value of f_ϕ are given in Chapter 1, p. 32. More is said on this subject in the section on the design of digital PLLs in the present chapter.

Transfer Functions

Figure 5-2 is a linearized equivalent block diagram of a digital PLL. It differs from the equivalent block diagram of an analog PLL, Fig. 4-7, in two respects: (a) the feedback function is no longer unity but is equal to $1/NN_2$, and (b) the phase comparator is represented by πK_ϕ because the phase range over which a sample-and-hold device operates is $\pm\pi$ rads. (In cases where the phase-comparator range is less than $\pm\pi$ rads, an appropriate correction should be introduced into the following equations. See Chapter 6 for a detailed description of a digital phase comparator.) The divide-by-N_1 circuit is omitted from the block diagram because it does not affect the operation of the PLL and can be dealt with separately.

According to the feedback control systems analogy, the forward gain is $\pi K_\phi K_\text{VCO} F(s)/s$ and the open-loop gain is $[(\pi K_\phi)(K_\text{VCO}/NN_2)F(s)]/s$. The system transfer function is

$$\frac{\Theta_o(s)}{\Theta_r(s)} = \frac{\pi K_\phi K_\text{VCO} F(s)}{s + (\pi K_\phi)(K_\text{VCO}/NN_2)F(s)}. \tag{5-1}$$

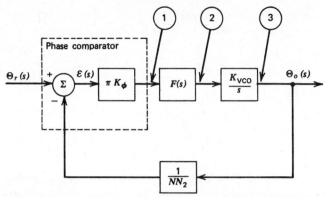

Figure 5-2. Equivalent block diagram of the digital phase-locked loop shown in Fig. 5-1.

Similarly, the transfer functions from points 1, 2, and 3 to the output are as follows:

$$\text{transfer function from point 1 to output} = \frac{K_{\text{VCO}}F(s)}{s+(\pi K_\phi)(K_{\text{VCO}}/NN_2)F(s)}, \tag{5-2}$$

$$\text{transfer function from point 2 to output} = \frac{K_{\text{VCO}}}{s+(\pi K_\phi)(K_{\text{VCO}}/NN_2)F(s)}, \tag{5-3}$$

$$\text{transfer function from point 3 to output} = \frac{1}{1+\left[(\pi K_\phi)(K_{\text{VCO}}/NN_2)F(s)/s\right]}. \tag{5-4}$$

Let $\pi K_\phi = \alpha_\phi$ and $K_{\text{VCO}}/NN_2 = \alpha_{\text{VCO}}$; then

$$\text{forward gain} = NN_2\left[\frac{\alpha_\phi\alpha_{\text{VCO}}F(s)}{s}\right], \tag{5-5}$$

$$\text{open-loop gain} = \frac{\alpha_\phi\alpha_{\text{VCO}}F(s)}{s}. \tag{5-6}$$

Similarly, the transfer functions are

$$\frac{\Theta_o(s)}{\Theta_r(s)} = (NN_2)\left[\frac{\alpha_\phi\alpha_{\text{VCO}}F(s)}{s+\alpha_\phi\alpha_{\text{VCO}}F(s)}\right], \tag{5-7}$$

$$\text{transfer function from} = (NN_2)\left[\frac{\alpha_{VCO}F(s)}{s + \alpha_\phi \alpha_{VCO}F(s)}\right], \qquad (5\text{-}8)$$
point 1 to output

$$\text{transfer function from} = (NN_2)\left[\frac{\alpha_{VCO}}{s + \alpha_\phi \alpha_{VCO}F(s)}\right], \qquad (5\text{-}9)$$
point 2 to output

$$\text{transfer function from} = \frac{1}{1 + \left[\alpha_\phi \alpha_{VCO}F(s)/s\right]}. \qquad (5\text{-}10)$$
point 3 to output

Comparing Eqs. 5-7 to 5-10 with Eqs. 4-17 to 4-20, respectively, we see a one-to-one correspondence between the expressions characterizing an analog and a digital PLL (except for the factor NN_2, which is a constant with respect to the Laplace variable, s). The similarity of the expressions is one of the considerations on which an equivalent analog model of a digital PLL, used in obtaining various loop parameters, is developed. It is also assumed in developing the model that the sampling, or the phase-comparator, frequency, f_ϕ, is much higher than the loop 3 dB bandwidth.

First-Order Phase-Locked Loop

In analyzing a digital PLL, it is convenient to consider the system referred to the phase-comparator frequency, f_ϕ. Under these conditions the block diagram in Fig. 5-2 can be modified as shown in Fig. 5-3, where $\alpha_\phi = \pi K_\phi$ and $\alpha_{VCO} = K_{VCO}/NN_2$. When $F(s) = 1$, the following expressions are obtained for a first-order digital PLL on the basis of the material presented in Chapter 4. The forward gain is

$$\begin{array}{l}\text{forward gain of} \\ \text{first-order digital PLL} = \dfrac{\alpha_\phi \alpha_{VCO}}{s}, \\ \text{referred to } f_\phi \end{array} \qquad (5\text{-}11)$$

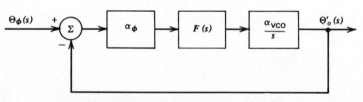

Figure 5-3. Equivalent block diagram of the PLL referred to the phase-comparator frequency; time delays of the phase comparator and variable-ratio divider are omitted.

and so is the open-loop gain. The system transfer function referred to f_ϕ is

$$\left[\frac{\Theta'_o(s)}{\Theta_\phi(s)}\right]_{\text{first order}} = \frac{\alpha_\phi \alpha_{\text{VCO}}}{s + \alpha_\phi \alpha_{\text{VCO}}}. \tag{5-12}$$

Notice that all we have to do to express these functions referred to the VCO frequency, f_{out}, is to multiply Eqs. 5-11 and 5-12 by NN_2. The expression for the open-loop gain is the same in both cases, as one would anticipate.

The phase error function is

$$\mathcal{E}(s) = \frac{s\Theta_\phi(s)}{s + \alpha_\phi \alpha_{\text{VCO}}}, \tag{5-13}$$

so that, according to the analog PLL analogy, the steady-state phase error resulting from a step change of input phase of magnitude $\Delta\theta_\phi$ or from a VCO phase change of magnitude $\Delta\theta_{\text{out}}$ is

$$\varepsilon_{ss} = \lim_{s \to 0} \frac{s\,\Delta\theta_\phi}{s + \alpha_\phi \alpha_{\text{VCO}}} = \lim_{s \to 0} \frac{s(\Delta\theta_{\text{out}}/NN_2)}{s + \alpha_\phi \alpha_{\text{VCO}}} = 0, \tag{5-14}$$

and the steady-state error resulting from a step change in input frequency of magnitude $\Delta\omega_\phi$ or from a VCO frequency change of magnitude $\Delta\omega_{\text{out}}$ is

$$\varepsilon_{ss} = \lim_{s \to 0} \frac{\Delta\omega_\phi}{s + \alpha_\phi \alpha_{\text{VCO}}} = \lim_{s \to 0} \frac{(\Delta\omega_{\text{out}}/NN_2)}{s + \alpha_\phi \alpha_{\text{VCO}}},$$

$$= \frac{\Delta\omega_\phi}{\alpha_\phi \alpha_{\text{VCO}}} = \frac{\Delta\omega_{\text{out}}}{NN_2 \alpha_\phi \alpha_{\text{VCO}}} \text{ rad.} \tag{5-15}$$

The first-order digital PLL response to a linear change in input frequency at a rate of $d\Delta\omega_\phi/dt$ is

$$\varepsilon_{ss} = \lim_{s \to 0} \frac{d\Delta\omega_\phi/dt}{s^2 + s\alpha_\phi \alpha_{\text{VCO}}} = \infty. \tag{5-16}$$

The 3 dB bandwidth of the first-order digital PLL is

$$B_{3\text{dB}} = \frac{\alpha_\phi \alpha_{\text{VCO}}}{2\pi}. \tag{5-17}$$

Notice that B_{3dB} varies with N because $\alpha_{VCO} = K_{VCO}/NN_2$. As is shown later, B_{3dB} of a second-order digital PLL is approximately proportional to $(1/N)^{1/2}$. Hence, if it is important to keep the 3 dB bandwidth relatively constant, a second-order PLL is preferred to a first-order PLL. The expression for noise bandwidth is obtained by replacing K_ϕ with α_ϕ and K_{VCO} with α_{VCO} in Eq. 4-29.

The hold-in range referred to the phase-comparator frequency is

$$\Delta\omega_{\phi,\text{hold-in}} = \alpha_\phi \alpha_{VCO},$$

and, referred to the VCO frequency (since $\Delta\omega_\phi = \Delta\omega_{\text{out}}/NN_2$), it is

$$\Delta\omega_{\text{out,hold-in}} = \alpha_\phi K_{VCO} \text{ rad/sec.} \tag{5-18}$$

The corresponding expressions for the capture range are

$$\Delta\omega_{\phi,\text{capture}} = \Delta\omega_{\phi,\text{hold-in}} \text{ rad/sec,}$$

$$\Delta\omega_{\text{out,capture}} = \Delta\omega_{\text{out,hold-in}} \text{ rad/sec.} \tag{5-19}$$

The time to acquire phase lock, when $\Delta\omega$ is made smaller than $\Delta\omega_{\text{capture}}$, is

$$t_{\text{acq,phase}} \cong \frac{2}{\alpha_\phi \alpha_{VCO} \cos\varepsilon_{ss}} \log_e\left(\frac{2}{\gamma_{\text{lock}}}\right) \text{ sec,} \tag{5-20}$$

where

$$\gamma_{\text{lock}} = \text{specified deviation from the steady-state error}$$
$$\text{at the time equal to } t_{\text{acq,phase}} \text{ (rads).}$$

It was demonstrated in Chapter 4 that, although a basic first-order PLL is unconditionally stable, practical forms of PLLs often require ac compensation. In digital PLLs the delays introduced by the phase comparator and digital frequency dividers must be included in the equations of the open-loop phase. Figure 5-4 is the equivalent block diagram of a digital PLL used in stability analysis. It includes delays introduced by the phase comparator and dividers. The phase comparator, a sample-and-hold circuit, can be described by the equation (Ref. 1, Section 7.4)

$$G_\phi(s) = \frac{\alpha_\phi(1 - e^{-T_\phi s})}{s}, \tag{5-21}$$

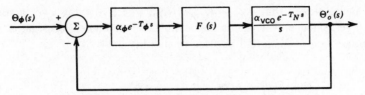

Figure 5-4. Equivalent block diagram of the PLL referred to the phase-comparator frequency; time delays of the phase comparator and variable-ratio divider are included.

where

$$T_\phi = \text{sampling period of the}$$
$$\text{sample-and-hold circuit,}$$
$$\text{that is, } T_\phi = 1/f_\phi.$$

When the phase comparator is followed by a low-pass filter that provides a large attenuation of the signal at f_ϕ and its harmonics, $G_\phi(s)$ can be simplified to

$$G_\phi(s) \cong \alpha_\phi e^{-T_\phi s}. \qquad (5\text{-}22)$$

The transfer function of the variable-ratio digital divider is

$$G_N(s) = \frac{e^{-T_N s}}{N}, \qquad (5\text{-}23)$$

where T_N = the time delay of the divider.
Similarly, for the fixed-ratio divider.

$$G_{N_2}(s) = \frac{e^{-T_{N_2} s}}{N_2}. \qquad (5\text{-}24)$$

In a synchronous digital divider the time delay through the circuit is less than the period of the input frequency. Hence, when N and N_2 are large, the effect of the dividers on loop phase shift can be neglected. However, these delays must be included in stability analysis when the frequency of the input signals to either of the dividers is only one or two orders of magnitude higher than the phase-comparator frequency. When these delays are included, the expressions for the magnitude and phase angle of the open-loop gain are as follows:

$$G_T(\omega) = 20\log_{10}\left|\frac{\alpha_\phi \alpha_{VCO}}{j\omega}\right| = 20\log_{10}\left(\frac{\alpha_\phi \alpha_{VCO}}{\omega}\right) \text{ dB}, \qquad (5\text{-}25)$$

$$\Theta_T(\omega) = \left(-\frac{\pi}{2}\right) + (-T_\phi\omega) + (-T_N\omega) + (-T_{N_2}\omega) \text{ rad}, \qquad (5\text{-}26)$$

respectively. Notice that $\alpha_{VCO} = K_{VCO}/NN_2$, where N varies over a specified range. Since a PLL tends to be unstable at higher loop gains, the minimum value of N should be used in stability analysis.

Second-Order Phase-Locked Loop

When the passive phase-lag filter described by Eq. 4-41 is used, the loop parameters referred to the phase-comparator frequency are as follows.

The open-loop gain and the forward gain of a second-order digital PLL are

$$\frac{\alpha_\phi \alpha_{VCO}(\tau_2/\tau_1)[s+(1/\tau_2)]}{s[s+(1/\tau_1)]}, \qquad (5\text{-}27)$$

where τ_1 and τ_2 are defined in Fig. 4-12. The transfer function is

$$\left[\frac{\Theta'_o(s)}{\Theta_\phi(s)}\right]_{\text{second order}} = \frac{\alpha_\phi \alpha_{VCO}(1/\tau_1)(1+\tau_2 s)}{s^2+(1/\tau_1)(1+\alpha_\phi \alpha_{VCO}\tau_2)s+(\alpha_\phi \alpha_{VCO}/\tau_1)}, \qquad (5\text{-}28)$$

and, when expressed in terms of the loop natural frequency, ω_n, and damping factor, ζ, it is

$$\left[\frac{\Theta'_o(s)}{\Theta_\phi(s)}\right]_{\text{second order}} = \frac{s\omega_n[2\zeta-(\omega_n/\alpha_\phi \alpha_{VCO})]+\omega_n^2}{s^2+2\zeta\omega_n s+\omega_n^2}, \qquad (5\text{-}29)$$

where

$$\omega_n = \left(\frac{\alpha_\phi \alpha_{VCO}}{\tau_1}\right)^{1/2} \text{rad/sec} \qquad (5\text{-}30)$$

and

$$\zeta = \tfrac{1}{2}\left(\frac{1}{\tau_1 \alpha_\phi \alpha_{VCO}}\right)^{1/2}(1+\tau_2 \alpha_\phi \alpha_{VCO}). \qquad (5\text{-}31)$$

The 3 dB bandwidth is

$$B_{3dB} = \frac{\omega_n}{2\pi}\left(d+\sqrt{d^2+1}\,\right)^{1/2} \text{Hz}, \qquad (5\text{-}32)$$

where

$$d = \left[2\zeta^2 + 1 - \frac{\omega_n}{\alpha_\phi \alpha_{\mathrm{VCO}}} \left(4\zeta - \frac{\omega_n}{\alpha_\phi \alpha_{\mathrm{VCO}}} \right) \right].$$

Notice that because ω_n is proportional to $(1/N)^{1/2}$ the 3 dB bandwidth also varies approximately as $(1/N)^{1/2}$.

The noise bandwidth is

$$B_n = \frac{\omega_n}{2} \left(\zeta + \frac{1}{4\zeta} \right) \text{ Hz.} \tag{5-33}$$

The phase error function is

$$\mathcal{E}(s) = \frac{s(1 + \tau_1 s)\Theta_\phi(s)}{\tau_1 s^2 + (1 + \alpha_\phi \alpha_{\mathrm{VCO}} \tau_2)s + \alpha_\phi \alpha_{\mathrm{VCO}}}. \tag{5-34}$$

For a step phase and a ramp frequency input, the steady-state error is zero and infinite, respectively. For a step change in input frequency of magnitude $\Delta\omega_\phi$ or from a VCO frequency change of magnitude $\Delta\omega_{\mathrm{out}}$, the steady-state phase error is

$$\varepsilon_{ss} = \frac{\Delta\omega_\phi}{\alpha_\phi \alpha_{\mathrm{VCO}}} = \frac{\Delta\omega_{\mathrm{out}}}{NN_2 \alpha_\phi \alpha_{\mathrm{VCO}}} = \frac{\Delta\omega_{\mathrm{out}}}{\alpha_\phi K_{\mathrm{VCO}}} \text{ rad.} \tag{5-35}$$

The critical value of rate of change of reference frequency above which the loop will no longer stay locked is

$$\left(\frac{d\Delta\omega_\phi}{dt} \right)_{f_\phi} = \omega_n^2. \tag{5-36}$$

Referred to the phase-comparator frequency, the maximum rate at which VCO frequency can be swept in an attempt to achieve locking, for locking to be guaranteed, is

$$\left(\frac{d\Delta\omega_\phi}{dt} \right)_{\mathrm{VCO}} < \frac{\omega_n^2}{2}. \tag{5-37}$$

Referred to the VCO frequency, the maximum rate at which VCO may be swept is

$$\left(\frac{d\Delta\omega_{\mathrm{out}}}{dt} \right)_{\mathrm{VCO}} < \frac{NN_2 \omega_n^2}{2} \text{ rad/sec}^2. \tag{5-38}$$

The maximum rate at which VCO control voltage may be swept is

$$\left| \frac{dE}{dt} \right|_{\text{max}} = \frac{NN_2\omega_n^2}{2K_{\text{VCO}}} \text{ V/sec.} \tag{5-39}$$

The hold-in range in terms of the phase-comparator frequency is

$$\Delta\omega_{\phi,\text{ hold-in}} = \alpha_\phi\alpha_{\text{VCO}} \text{ rad/sec}$$

and in terms of the VCO frequency is

$$\Delta\omega_{\text{out, hold-in}} = \alpha_\phi K_{\text{VCO}} \text{ rad/sec.} \tag{5-40}$$

The capture range of a second-order digital PLL is always smaller than the hold-in range. Referred to the phase comparator and VCO frequencies, the capture range is

$$\Delta\omega_{\phi,\text{ capture}} \cong \alpha_\phi\alpha_{\text{VCO}}\left(\frac{\tau_2}{\tau_1}\right)\text{rad/sec,}$$

$$\Delta\omega_{\text{out, capture}} \cong \alpha_\phi K_{\text{VCO}}\left(\frac{\tau_2}{\tau_1}\right)\text{rad/sec,} \tag{5-41}$$

respectively.

When the VCO mistuning, $\Delta\omega$, is smaller than the capture range, the time to acquire phase lock is approximately the time given by Eq. 5-20. When $\Delta\omega$ is larger than $\Delta\omega_{\text{capture}}$, locking is still possible with some cycles of VCO frequency skipped as long as $\Delta\omega$ is smaller than $\Delta\omega_{\text{pull-in}}$, where the pull-in range referred to the phase comparator and VCO frequencies is

$$\Delta\omega_{\phi,\text{ pull-in}} \cong \sqrt{2}\left[2\zeta\omega_n\alpha_\phi\alpha_{\text{VCO}}F(0) - \omega_n^2\right]^{1/2}\text{ rad/sec,}$$

$$\Delta\omega_{\text{out, pull-in}} \cong \sqrt{2}\,(NN_2)\left[2\zeta\omega_n\alpha_\phi\alpha_{\text{VCO}}F(0) - \omega_n^2\right]^{1/2}\text{ rad/sec,} \tag{5-42}$$

respectively.

Equation 5-42 is a good approximation of the pull-in range for $\omega_n F(0)/\alpha_\phi\alpha_{\text{VCO}} < 0.4$ (moderate or high loop gains), but is very poor for $\omega_n F(0)/\alpha_\phi\alpha_{\text{VCO}} > 0.5$.

An approximate expression for the time required for the loop to frequency-lock is

$$t_{\text{acq, freq}} \cong \frac{4(\Delta f_{\text{out}}/NN_2)^2}{B_n^3} \text{ sec,} \tag{5-43}$$

where $\Delta f_{out}/NN_2 =$ VCO frequency mistuning referred to the phase-comparator frequency (Hz),

$B_n =$ noise bandwidth (Hz).

The total time to acquire both frequency and phase lock, assuming that

$$\Delta\omega_{capture} < \Delta\omega < \Delta\omega_{pull-in},$$

is

$$t_{acq,\,total} \cong t_{acq,\,freq} + t_{acq,\,phase}, \tag{5-44}$$

where $t_{acq,\,phase}$ is given by Eq. 5-20.

The stability analysis of a second-order digital PLL is performed in the same manner as the analysis of an analog PLL. The recommended approach is to make Bode plots of the magnitude and phase angle of the open-loop gain versus frequency. As was demonstrated in Chapter 4, a Bode plot of the magnitude of the open-loop gain contains design information valuable in loop ac compensation and in evaluation of PLL performance with respect to phase noise and spurious outputs.

The expressions for the magnitude and phase angle, including the delays of the phase comparator and digital frequency dividers, are as follows:

$$G_T(\omega) = 20\log_{10}\left(\frac{\alpha_\phi \alpha_{VCO}}{\omega}\right) + 20\log_{10}|F(j\omega)| \text{ dB} \tag{5-45}$$

and

$$\Theta_T(\omega) = \left(-\frac{\pi}{2}\right) + \left(-T_\phi\omega\right) + \left(-T_N\omega\right) + \left(-T_{N_2}\omega\right) + \underline{/F(j\omega)} \text{ rad.} \tag{5-46}$$

Second-Order Phase-Locked Loop with a Perfect Integrator

It was shown in Chapter 4 that a true phase-locked loop is a PLL whose loop filter is a perfect integrator with a transfer function given by Eq. 4-56:

$$F(s) = \frac{s+a}{s}.$$

A practical filter with a transfer function approximating Eq. 4-56 was described in Fig. 4-17. The transfer function of this circuit, Eq. 4-59, is

$$F(s) = \left(\frac{R_2}{R_1}\right)\left[\frac{s+(1/\tau_2)}{s}\right].$$

Utilizing this filter, one obtains the following expressions for loop parameters referred to the phase-comparator frequency. The forward gain, as well as the open-loop gain, is

$$\frac{\alpha_\phi \alpha_{VCO}(R_2/R_1)[s+(1/\tau_2)]}{s^2}.$$

The transfer function of the system is

$$\left[\frac{\Theta'_o(s)}{\Theta_\phi(s)}\right]_{\text{perfect integrator}} = \frac{\alpha_\phi \alpha_{VCO}(R_2/R_1)[s+(1/\tau_2)]}{s^2 + \alpha_\phi \alpha_{VCO}(R_2/R_1)s + (\alpha_\phi \alpha_{VCO}/\tau_2)(R_2/R_1)}.$$

$$(5\text{-}47)$$

Expressed in terms of the loop natural frequency and damping factor, it is

$$\left[\frac{\Theta'_o(s)}{\Theta_\phi(s)}\right]_{\text{perfect integrator}} = \frac{2\zeta\omega_n s + \omega_n^2}{s^2 + 2\zeta\omega_n s + \omega_n^2},\qquad (5\text{-}48)$$

where

$$\omega_n = \left[\frac{\alpha_\phi \alpha_{VCO}}{\tau_2}\left(\frac{R_2}{R_1}\right)\right]^{1/2} \text{ rad/sec} \qquad (5\text{-}49)$$

and

$$\zeta = \tfrac{1}{2}\left[\alpha_\phi \alpha_{VCO}\tau_2\left(\frac{R_2}{R_1}\right)\right]^{1/2} \qquad (5\text{-}50)$$

The 3 dB bandwidth is

$$B_{3dB} = \frac{\omega_n}{2\pi}\left[2\zeta^2 + 1 + \sqrt{(2\zeta^2+1)^2 + 1}\,\right]^{1/2} \text{ Hz}, \qquad (5\text{-}51)$$

and the noise bandwidth is

$$B_n = \frac{\alpha_\phi \alpha_{VCO}(R_2/R_1) + (1/\tau_2)}{4} \text{ Hz}. \qquad (5\text{-}52)$$

The phase error function is

$$\mathcal{E}(s) = \frac{s^2\Theta_\phi(s)}{s^2 + \alpha_\phi\alpha_{\text{VCO}}(R_2/R_1)s + (\alpha_\phi\alpha_{\text{VCO}}/\tau_2)(R_2/R_1)}. \qquad (5\text{-}53)$$

The steady-state error resulting from a step change in input phase or frequency of magnitude $\Delta\theta_\phi$ and $\Delta\omega_\phi$, respectively, is zero. For a ramp change in input frequency, that is, for $\Theta_\phi(s) = (d\Delta\omega_\phi/dt)/s^3$, the steady-state phase error is

$$\varepsilon_{ss} = \left(\frac{R_1}{R_2}\right)\left[\frac{\tau_2(d\Delta\omega_\phi/dt)}{\alpha_\phi\alpha_{\text{VCO}}}\right] \text{ rad}. \qquad (5\text{-}54)$$

When an excitation is applied at Doppler rates,

$$\frac{d\Delta\omega_\phi}{dt} > \left(\frac{R_2}{R_1}\right)\left(\frac{\alpha_\phi\alpha_{\text{VCO}}}{\tau_2}\right),$$

the loop will fall out of lock. The maximum rate at which VCO frequency can be swept in order for acquisition to be certain is

$$\left(\frac{d\Delta\omega_\phi}{dt}\right)_{\text{VCO,max}} = \frac{1}{2\tau_2}\left(4B_n - \frac{1}{\tau_2}\right) \text{ rad/sec}^2,$$

where B_n is given by Eq. 5-52. Referred to the VCO frequency, this quantity becomes

$$\left(\frac{d\Delta\omega_{\text{out}}}{dt}\right)_{\text{VCO,max}} = \frac{NN_2}{2\tau_2}\left(4B_n - \frac{1}{\tau_2}\right) \text{ rad/sec}, \qquad (5\text{-}55)$$

and since

$$\left(\frac{d\Delta\omega_{\text{out}}}{dt}\right)_{\text{VCO}} = K_{\text{VCO}}\frac{dE}{dt},$$

we have

$$\left|\frac{dE}{dt}\right|_{\text{max}} = \frac{NN_2}{2K_{\text{VCO}}\tau_2}\left(4B_n - \frac{1}{\tau_2}\right) \text{ V/sec}. \qquad (5\text{-}56)$$

The pull-in range of a second-order digital PLL with a perfect integrator is infinite. The larger the VCO mistuning, however, the longer the acquisition time is. An approximate expression for the frequency acquisition time is

$$t_{\text{acq,freq}} \cong \tau_2 \left(\frac{\Delta\omega_\phi}{\alpha_\phi \alpha_{\text{VCO}}} - \sin\theta_0 \right) \text{ sec,} \qquad (5\text{-}57)$$

where θ_0 is the initial phase difference between the reference and VCO signals referred to the phase-comparator frequency, f_ϕ. When the loop filter is described by Eq. 4-59, the approximate expression for the frequency acquisition time becomes

$$t_{\text{acq,freq}} \cong \tau_2 \left[\frac{\Delta\omega_\phi}{\alpha_\phi \alpha_{\text{VCO}}(R_2/R_1)} - \sin\theta_0 \right] \text{ sec.} \qquad (5\text{-}58)$$

The total acquisition time is

$$t_{\text{acq,total}} \cong t_{\text{acq,freq}} + t_{\text{acq,phase}}, \qquad (5\text{-}59)$$

where $t_{\text{acq,phase}}$ is given by Eq. 5-20.

5-2 Design Considerations

It was pointed out in Chapter 2 that, as various synthesizer requirements are introduced, the complexity of a frequency synthesis approach satisfying these requirements increases. It was demonstrated later in Chapter 2 that a single-loop synthesizer, Fig. 2-1, cannot satisfy a typical specification. In particular, it was shown in Figs. 2-63 and 2-64 that in a single-loop digital synthesizer generating a signal at 220 to 300 MHz in 1 kHz frequency increments the 1 MHz reference source and VCO noise can be reduced to a level below a typical phase-noise requirement, Fig. 2-62, only if the loop bandwidth is made 20 Hz and 5 kHz, respectively, values that cannot be achieved simultaneously. It was also shown in Chapter 2, Fig. 2-65, that in a two-loop approach the factor by which the reference signal noise is multiplied is reduced by three orders of magnitude; this permits the loop bandwidth to be 5 kHz, making the VCO noise negligible at the offset from signal frequencies of interest (400 to 600 Hz). The two-loop synthesizer, however, falls short of meeting the phase-noise requirement by approximately 9 dB when it is designed to operate with a 1 MHz reference source. Adding another digital PLL does not improve the phase-noise performance of the synthesizer because the minimum multiplication factor for the 1 MHz reference phase noise is already achieved in the two-loop approach, and it is this noise, multiplied to 220 to 300 MHz, that is dominant at the synthesizer output.

To simplify the analysis, it is assumed that the level of the residual synthesizer noise is low and can be neglected.

A number of steps can be taken to resolve the conflict between the phase-noise and other requirements. First of all, the phase-noise requirement itself should be reviewed. Usually, this requirement is prepared by members of the receiver-transmitter group (or some other in-house systems design group, depending on the synthesizer application), who might have added to the requirement an excessively large safety margin without being aware of the cost increment due to that margin. Second, the system of which the synthesizer is a part should be studied again to determine whether or not the system as a whole can be designed so as to result in a relaxed synthesizer phase-noise requirement. Only after these first two investigative steps prove fruitless should the designer look for a better reference oscillator. If such an oscillator is found, it will significantly increase the cost of the synthesizer. Finally, the two-loop synthesizer could be redesigned as shown in Fig. 5-5, making it operate with a 5 MHz reference source such as the Vectron series CO-211 followed by a narrow-band crystal filter. (For the phase-noise performance of this oscillator, see curve 2, Fig. 4-32.) It can be shown, however, that by making such a change, one increases the cost, size, and power consumption of the

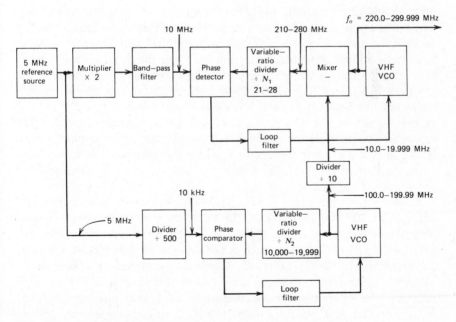

Figure 5-5.   Two-loop frequency synthesizer, 1 kHz frequency increments.

synthesizer, shown in Fig. 2-65, by more than 10 percent each. This is a typical synthesizer performance-cost trade-off problem.

The purpose of this discussion is to demonstrate design problems and their solutions, and not to discredit the usefulness of single-loop synthesizers. Such synthesizers are currently utilized in many applications and display satisfactory performance with respect to the requirements of these applications. In general, a single-loop synthesizer is used with great success whenever the phase-noise requirement is not stringent or whenever it specifies far-out noise only, and there is no conflict between switching time and generation of frequency increments. In this presentation we describe occasions when a single-loop synthesizer cannot be used and try to justify the selection of a two- or multiloop approach.

Figure 5-5 is a block diagram of a synthesizer generating 1 kHz increments at vhf and displaying excellent close-to-signal phase-noise performance, together with a 5 msec switching time. Notice that the two-loop approach permits the loop bandwidth of the 220 to 300 MHz PLL to be adjusted to any desired value necessary to attenuate VCO noise without degrading the close-in noise of the synthesizer as long as the stability conditions are satisfied. Hence, by developing a VCO with good far-from-signal phase-noise characteristics, a stringent far-from-signal phase-noise specification can also be satisfied by this system.

Before further changes in requirements are considered, one more comment should be made with respect to the synthesizer described in Fig. 5-5. The signal at the sum of two input frequencies falls in the output frequency band of the mixer. Hence a multipole voltage-tuned band-pass filter would have to be added to the block diagram following the mixer. It may be less expensive to increase the frequency of the injection signal generated by the auxiliary PLL from 10.0–19.999 to 40.0–49.999 MHz and to use a fixed-frequency band-pass filter (which is cheaper and presents fewer problems) at the output of the mixer. This, of course, is possible only if other low-order intermodulation products fall outside the mixer output band. It is left to the interested student to evaluate the proposed changes of the block diagram in Fig. 5-5 and to show that they do result in elimination of the voltage-tuned filter.

Assume now that the 1 kHz frequency-increment requirement is changed to 100 Hz. In this case the tendency is to decrease the phase-comparator frequency of the 100.0 to 199.99 MHz PLL from 10 kHz to 1 kHz (see Fig. 5-6). This approach is not acceptable, however, if a 5 msec switching time is to be achieved. A possible approach that satisfies all of the requirements considered above is shown in Fig. 5-7. Instead of reducing the phase-comparator frequency to 1 kHz, another PLL-frequency divider pair, which provides 100 Hz frequency increments without

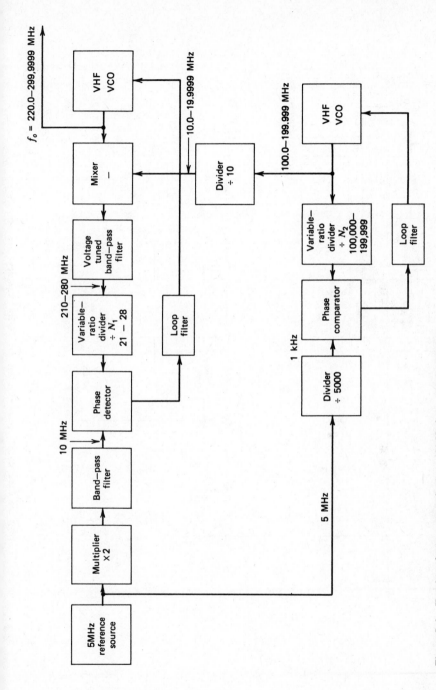

Figure 5-6. Two-loop frequency synthesizer, 100 Hz frequency increments.

311

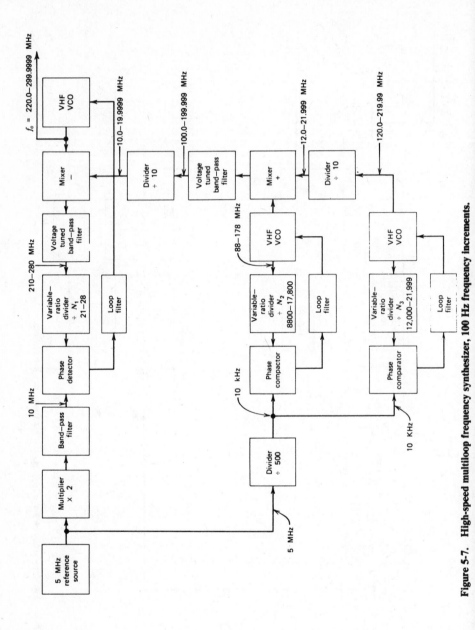

Figure 5-7. High-speed multiloop frequency synthesizer, 100 Hz frequency increments.

degrading phase noise and switching time, is added to the synthesizer. Although two voltage-tuned band-pass filters are used to attenuate spurious outputs of both mixers, the writer feels that an attempt directed toward replacing both voltage-tuned filters with fixed-frequency filters by selecting appropriate mixer input frequencies will prove fruitful.

The next topic to be considered is leakage of the phase-comparator input signals at f_ϕ to the VCO error voltage input. Keeping this leakage under control is a serious design limiation in digital PLLs. It will be demonstrated that the problem of leakage at f_ϕ is quite severe in a single-loop digital synthesizer but is conveniently dealt with in a multiloop synthesizer.

Consider the single-loop synthesizer shown in Fig. 2-1. The output and the phase-comparator frequencies of this system are 200.0 to 299.99 MHz and 1 kHz, respectively. Assume that the requirement for spurious outputs is -80 dB and that the VCO gain constant, K_{VCO}, is 3.14×10^6 (rad/sec)/V. Using Eq. 2-17, we compute the allowable frequency deviation, Δf_{peak}, as

$$-80 = 10\log_{10}\left(\frac{\Delta f_{peak}}{2f_m}\right)^2,$$

where

$$f_m = f_\phi = 10^3 \text{ Hz.}$$

Hence $\Delta f_{peak} = 0.2$ Hz or $\Delta\omega_{peak} = 2\pi \Delta f_{peak} = 1.256$ rad/sec. This is equivalent to the maximum allowable peak amplitude of a spurious signal at 1 kHz of $\Delta\omega_{peak}/K_{VCO} = 4 \times 10^{-7}$ V peak appearing at the VCO error voltage input when the loop is locked.

The next step is to estimate the value of ripple voltage at the phase-comparator output. Figure 5-8a is a simplified schematic diagram of a sample-and-hold phase comparator. The reference voltage, $e_r(t)$, is a sawtooth wave with a period, T_r, as shown in Fig. 5-8b. The sampling signal, $e_s(t)$, is a pulse train generated in the process of frequency division of the VCO signal. The repetition rate and the width of the pulse are T_s and T_w, and the driving source and load resistances are R_g and R_l, respectively. The hold capacitance is C_{hold}, and the isolation resistance of C_{hold} is R_C. When the loop is in synchronism, $T_r = T_s = 1/f_\phi$.

Assume that the initial charge-discharge period took place before t_0 and that C_{hold} is discharged to E_1 as shown in Fig. 5-8b. When the switch closes again at t_0, C_{hold} charges to E_2. During the time that the switch is open, t_1, to t_2, C_{hold} discharges through R_C and R_l back to E_1, and the charge-discharge cycle repeats.

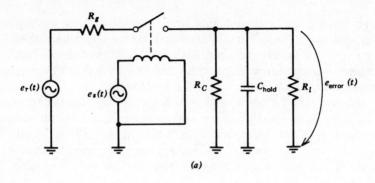

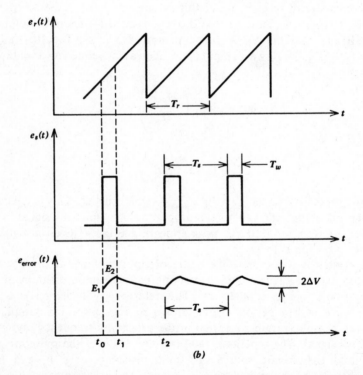

Figure 5-8. Timing diagram of a sample-and-hold phase comparator. (a) Simplified schematic diagram; (b) voltage waveforms.

314

The equivalent circuits of the phase comparator during charging and discharging periods are shown in Fig. 5-9a and b. To minimize the peak ripple voltage at the sampling frequency, which to the first-order approximation is $\Delta V = (E_2 - E_1)/2$, the following ideal conditions should be approached as closely as possible:

$$R_g C_{\text{hold}} = 0 \quad \text{or} \quad R_g = 0, \tag{5-60}$$

$$R_l C_{\text{hold}} = \infty \quad \text{or} \quad R_l = \infty, \tag{5-61}$$

$$T_w = 0. \tag{5-62}$$

In practical systems R_g and R_l may be 100 Ω and 10 MΩ, respectively. The width of the sampling pulse, T_w, depends on the nature of the variable-ratio divider. The divider whose operation is based on the pulse-swallowing techniques described in Chapter 6 generates an output pulse that has a width equal to approximately five times the period of the input-to-the-divider signal, so that in the example cited $T_{w,\text{max}} \cong 4.5 \times 10^{-9}$ sec. This is negligible compared to T_s (1 msec).

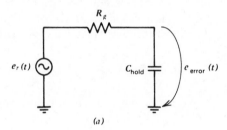

(a)

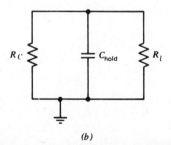

(b)

Figure 5-9. Phase comparator during sampling and hold periods. (a) Simplified equivalent circuit during sampling period; (b) simplified equivalent circuit during hold period.

The ripple voltage at the output of the phase comparator can be determined in a number of ways. In this discussion it is assumed that R_C is much larger than the load resistance and can be neglected. With only R_l present in the circuit, the change in the voltage across C_{hold} during the discharge period is approximated by

$$\Delta V \cong \frac{i_l T_s}{2 C_{hold}} \text{ V peak,} \tag{5-63}$$

where

$$i_l = \text{load current.}$$

Equation 5-63 indicates that, for a given T_s, the value of the hold capacitor should be as large as possible. Unfortunately, there is a practical limit on the maximum value of the hold capacitor. The equivalent circuit of a sample-and-hold device during the charging period, Fig. 5-9a, is a low-pass filter. For the phase comparator not to affect such loop parameters as capture range, the cutoff frequency of this circuit, $f_{C, hold}$, should be higher than the loop bandwidth.

Instead of performing an extensive PLL design to obtain the various loop parameters used in this example, let us assume that the loop bandwidth is one half of the sampling frequency, or 500 Hz. Let us choose the minimum value of the phase-comparator cutoff frequency as one and a half times the loop bandwidth, that is, let $f_{C, hold} = 750$ Hz; then

$$f_{C, hold} = \frac{1}{2\pi R_g C_{hold}} = 750 \text{ Hz,}$$

and for $R_g = 100 \ \Omega$, the maximum value of the hold capacitor is

$$C_{hold} \leqslant 2 \mu F.$$

For $R_l = 10$ MΩ and a maximum error voltage of 25 V (which is approximately the maximum phase-comparator output if a 28 V dc power supply is used), the load current is 2.5 μA, so that the peak ripple at the sampling frequency is

$$\Delta V \cong \frac{2.5 \times 10^{-6} \times 10^{-3}}{2 \times 2 \times 10^{-6}} \cong 625 \ \mu V \text{ peak.}$$

The allowable peak amplitude of the spurious signal at f_ϕ is 0.4 μV peak. Hence approximately 63 dB of attenuation at f_ϕ should be provided by the PLL and external filters.

A first-order PLL with a 3 dB bandwidth of 500 Hz attenuates the spurious signal at 1 kHz by only 6 to 9 dB, so that the external filters have to provide an additional 54 to 57 dB. This is difficult to achieve without affecting the capture range and stability of the PLL. A 90 dB spurious-output specification would increase the attenuation requirement for external filters to 64 to 67 dB.

Consider now Fig. 5-5. Although the value of the maximum allowable peak amplitude of the spurious signal at f_ϕ has not changed for the 220 to 300 MHz loop, the frequency at which phase comparison is done has been increased to 10 MHz, that is, many octaves above the loop bandwidth. At this frequency one would use a sinusoidal phase detector, which attenuates input signals significantly less than its sample-and-hold counterpart. Nevertheless, a PLL with a 500 Hz bandwidth provides more than 80 dB of attenuation at 10 MHz, and an external rf low-pass filter can be designed to attenuate the signal at 10 MHz by an additional 50 to 60 dB, if necessary, without affecting either the capture range or the stability of the PLL.

The leakage problem is made simple in the auxiliary PLL as well because (a) the phase-comparator frequency is 10 times as high as in the previous case, so that a PLL with a 500 Hz bandwidth attenuates a signal at 10 kHz by approximately 27 dB, and (b) the frequency divider (divide-by-10) provides an additional 20 dB of attenuation. Thus, to satisfy a -80 dB requirement, the external filters in the auxiliary PLL have to attenuate the signal at 10 kHz only by 16 dB; this can be easily accomplished by utilizing the Twin-T band-reject filter described in Figs. 5-10, 5-11, and 5-12. The advantage of using a Twin-T filter is that, for a given attenuation at the sampling frequency, a Twin-T introduces a significantly smaller phase shift within the PLL bandwidth, thus affecting the loop stability to a lesser degree than a low-pass filter.

Since not only the signal at the sampling frequency but also its harmonics are present at the phase-comparator output, the Twin-T is usually followed by a low-pass filter.

The problem of leakage at f_ϕ in a multiloop synthesizer is further reduced.

It would add little to the subject to discuss here such topics as acquisition, loop stability, and optimum design of digital PLLs. These are dealt with in Chapters 1, 2, and 4. Indeed, such considerations as optimum loop bandwidth, power supply isolation, and problems associated with broadband VCO tuning and with a dc amplifier inserted in the forward path apply equally to both analog and digital PLLs. One problem characteristic of a digital PLL, not considered before, results from a leakage (by way of the dc power supply) of high-rise-time transients generated in digital circuits, such as frequency dividers, to analog circuits, such as VCOs, capable of converting these transients into amplitude or phase modulation of the synthesizer output signal. It is highly recommended, therefore, to use separate power supplies for analog and digital sections of the synthesizer. Only the analog circuitry power supply needs to have good regulation and low ripple characteristics.

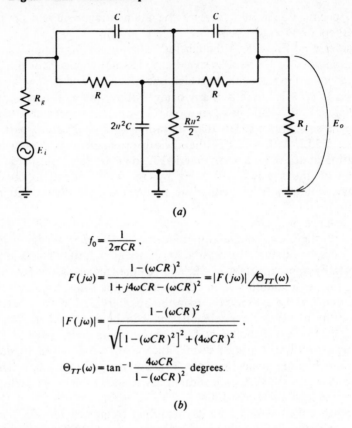

$$f_0 = \frac{1}{2\pi CR} \,,$$

$$F(j\omega) = \frac{1-(\omega CR)^2}{1+j4\omega CR-(\omega CR)^2} = |F(j\omega)| \underline{/\Theta_{TT}(\omega)}$$

$$|F(j\omega)| = \frac{1-(\omega CR)^2}{\sqrt{\left[1-(\omega CR)^2\right]^2+(4\omega CR)^2}} \,,$$

$$\Theta_{TT}(\omega) = \tan^{-1}\frac{4\omega CR}{1-(\omega CR)^2} \text{ degrees.}$$

(b)

Figure 5-10. Twin-T filter. (a) Schematic diagram; (b) related equations for $n=1$.

It is clear from the forgoing discussion that there are many reasons for making the phase-comparator frequency as high as possible. Phase noise, switching time, and spurious FM at f_ϕ are among the most important considerations. Ground loop currents, which present a problem below approximately 5 kHz, are another reason for making f_ϕ high. In low-spurious-output digital synthesizers it is advisable to keep the phase-comparator frequency above 10 kHz.

It is left to the interested student to show that the loop parameters of a PLL utilizing a frequency multiplier, instead of a divider, can be easily derived using the techniques demonstrated in this chapter.

Reference

1. Monroe, A. J. *Digital Processes for Sampled Data Systems* (New York: John Wiley and Sons, 1962).

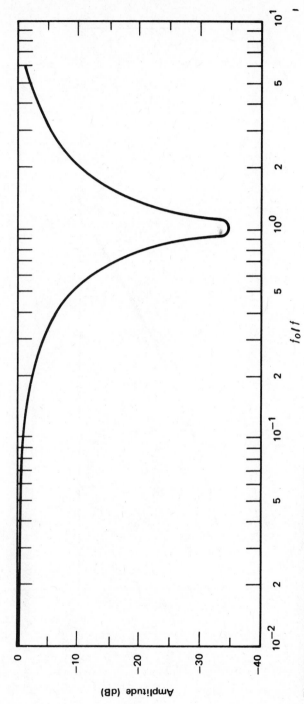

Figure 5-11. Amplitude response of the Twin-T filter.

319

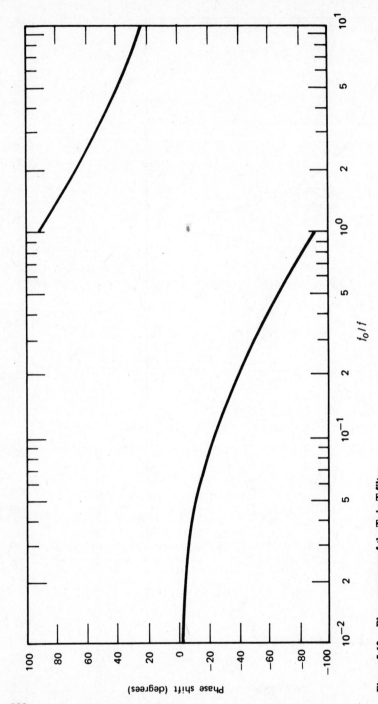

Figure 5-12. Phase response of the Twin-T filter.

320

6 Basic Circuits

In Chapter 1 we presented various configurations of basic building blocks, such as frequency multipliers, frequency dividers, and mixers, which constitute frequency synthesizers. Here we show how the functions of these basic blocks can be implemented.

6.1 Tuned Amplifiers

An rf tuned amplifier is probably the most commonly used circuit in frequency synthesis. It provides power gain and, frequently, filtering of unwanted signals.

Figure 6-1 is the schematic diagram of a single-stage rf tuned amplifier.

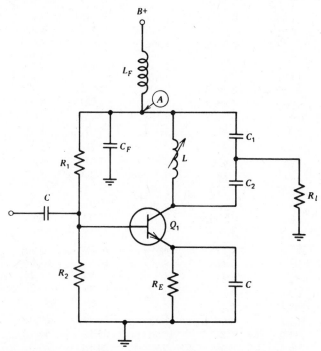

Figure 6-1.   Radio-frequency tuned amplifier; schematic diagram.

The resistors, R_1, R_2, and R_E, are used for dc biasing of the transistor, Q_1; the inductor L and capacitors C_1 and C_2 constitute the tuned circuit of the amplifier; the inductor L_F and capacitor C_F are used as a low-pass filter for power supply isolation (C_F also shorts point A to ground); and the reactance of capacitors C's is negligibly small at the signal frequency.

The design of a tuned amplifier is best demonstrated with a practical example. Assume that the prime function of the amplifier is to provide a 10 dB rf gain at 100 MHz at the specified power levels, which are as follows:

Input power, $P_{in} = 1$ mW (or 0 dBm; see Table 6-1);

Table 6-1. Conversion from dBm to Power and Voltage into a Given Load Resistance

0 dBm = 1 mW		50 Ω	500 Ω	1 kΩ	5 kΩ	10 kΩ
	Power	Volts	Volts	Volts	Volts	Volts
DBM	(mW)	(rms)	(rms)	(rms)	(rms)	(rms)
0	1.000	0.2236	0.7071	1.000	2.236	3.162
+1	1.259	0.2509	0.7934	1.122	2.509	3.548
+2	1.585	0.2815	0.8902	1.259	2.815	3.981
+3	1.995	0.3158	0.9987	1.412	3.158	4.467
+4	2.512	0.3544	1.121	1.585	3.544	5.012
+5	3.162	0.3976	1.257	1.778	3.976	5.623
+6	3.981	0.4461	1.411	1.995	4.461	6.310
+7	5.012	0.5006	1.583	2.239	5.006	7.079
+8	6.310	0.5617	1.776	2.512	5.617	7.943
+9	7.943	0.6301	1.993	2.818	6.301	8.913
+10	10.00	0.7071	2.236	3.162	7.071	10.00
+11	12.59	0.7934	2.509	3.548	7.934	11.22
+12	15.85	0.8902	2.815	3.981	8.902	12.59
+13	19.95	0.9987	3.158	4.467	9.987	14.12
+14	25.12	1.121	3.544	5.012	11.21	15.85
+15	31.62	1.257	3.976	5.623	12.57	17.78
+16	39.81	1.411	4.461	6.310	14.11	19.95
+17	50.12	1.583	5.006	7.079	15.83	22.39
+18	63.10	1.776	5.617	7.943	17.76	25.12
+19	79.43	1.993	6.301	8.913	19.93	28.18
+20	100.0	2.236	7.071	10.00	22.36	31.62
+21	125.9	2.509	7.934	11.22	25.09	35.48
+22	158.5	2.815	8.902	12.59	28.15	39.81
+23	199.5	3.158	9.987	14.12	31.58	44.67
+24	251.2	3.544	11.21	15.85	35.44	50.12
+25	316.2	3.976	12.57	17.78	39.76	56.23
+26	398.1	4.461	14.11	19.95	44.61	63.10
+27	501.2	5.006	15.83	22.39	50.06	70.79
+28	631.0	5.617	17.76	25.12	56.17	79.43
+29	794.3	6.301	19.93	28.18	63.01	89.13
+30	1 W	7.071	22.36	31.62	70.71	100.0

Output power, $P_o = 10$ mW ($+10$ dBm) into a 50 Ω load;

Power supply voltage, $B+ = +15$ V dc.

Assume also that the power dissipated in the tuned circuit of the amplifier is less than 10 % of the power delivered to the load, so that the total power delivered by the amplifier, P_T, is 11 mW.

A familiar equation for the power output of a class A amplifier (Ref. 1) is

$$P_T = \frac{V_m^2}{2(R_l)_T} \text{ W,} \tag{6-1}$$

where V_m = peak value of the sinusoidal voltage developed at the collector;

$(R_l)_T$ = total rf load seen by the transistor.

This expression, however, is based on somewhat ideal conditions of 50 % efficiency, which is a theoretical limit, and on design parameters implemented as computed mathematically. Neither is achievable in most practical cases. To develop the desired output power the amplifier circuit is designed to operate at higher power levels, that is, lower efficiency. The recommended equation relating the output power to the collector load is

$$P_T \cong \frac{V_{CE}^2}{8(R_l)_T} \text{ W,} \tag{6-2}$$

where V_{CE} = collector-to-emitter voltage (see Fig. 6-2).

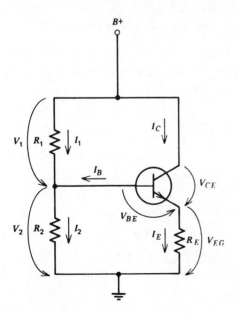

Figure 6-2. Direct current bias of a class A transistor amplifier.

Let $V_{CE} = 12$ V dc; then

$$(R_l)_T \cong \frac{(12)^2}{8 \times 11 \times 10^{-3}} = 1.64 \text{ k}\Omega.$$

Assume that the transistor may be driven from cutoff to saturation. Then the collector dc bias current is

$$I_C = \frac{B+}{(R_l)_T}. \qquad (6\text{-}3)$$

Hence

$$I_C = \frac{15}{1.64 \times 10^3} = 9.2 \text{ mA}.$$

At this stage of the design it is advisable to compute the collector dc power dissipation and compare it with the maximum allowable dissipation of the transistor at the highest specified temperature.

Selection of a transistor is based on the rf power gain requirement. The maximum available power gain (MAPG) of a transistor is

$$\text{MAPG} = 10 \log_{10} \left[\frac{|Y_{21}|^2}{4 g_{11} g_{22}} \right] \text{dB}, \qquad (6\text{-}4)$$

where Y_{ij}, used also in Eq. 6-16 later in this section, are transistor small-signal admittance parameters at the frequency of operation, and g_{ij} is the real part of Y_{ij}.

For good dc bias temperature stability the emitter resistor, R_E, should be at least one order of magnitude larger than the emitter-base junction resistance, r_e, which at room temperature is

$$r_e = \frac{25}{I_E \text{ (mA)}} \Omega. \qquad (6\text{-}5)$$

Hence

$$r_e = \frac{25}{9.2} = 2.7 \ \Omega,$$

whereas the emitter resistance is

$$R_E = \frac{V_{EG}}{I_E} \cong \frac{(B+) - V_{CE}}{I_C} \Omega \qquad (6\text{-}6)$$

or

$$R_E = \frac{15-12}{9.2 \times 10^{-3}} = 326 \ \Omega.$$

For the same reason the bias current flowing in R_1 and R_2, I_2, should be significantly larger than the base current, I_B. Assume that the minimum dc current gain of the transistor, h_{FE}, is 30; then the maximum base current is

$$(I_B)_{max} = \frac{I_C}{(h_{FE})_{min}} \ \text{mA} \qquad (6\text{-}7)$$

or

$$(I_B)_{max} = \frac{9.2 \times 10^{-3}}{30} = 0.31 \ \text{mA}.$$

Let $I_2 = 6(I_B)_{max} = 2 \ \text{mA}$; then

$$R_2 = \frac{V_{BG}}{I_2} = \frac{V_{BE} + V_{EG}}{I_2} \ \Omega \qquad (6\text{-}8)$$

or

$$R_2 = \frac{3.7}{2 \times 10^{-3}} = 1.85 \ \text{k}\Omega,$$

and

$$R_1 = \frac{V_1}{I_1} = \frac{(B+) - V_{BG}}{I_2 - I_B} \qquad (6\text{-}9)$$

or

$$R_1 = \frac{15 - 3.7}{(2 + 0.31) \times 10^{-3}} = 5.32 \ \text{k}\Omega.$$

Since the computed values for R_E, R_1, and R_2 are nonstandard, the closest standard values are chosen as 330 Ω, 5.1 kΩ, and 1.8 kΩ, respectively.

Having decided upon the transistor biasing, we design the collector tuned circuit. An equivalent of this circuit is shown in Fig. 6-3a, where the output transistor resistance is R_o, R_c is the equivalent parallel resistance of the coil, L, and R_l' is the load reflected across the tuned circuit. At present R_o is not included in the analysis because all that we are designing for is the specified rf output power of $+10$ dBm. If the maximum rf power or the amplifier bandwidth were the governing consideration, R_o would be of prime importance.

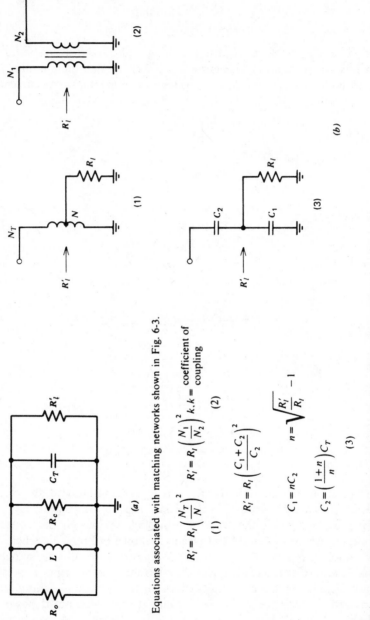

(2)

(1)

(3)

(b)

(a)

Equations associated with matching networks shown in Fig. 6-3.

$$R_i' = R_l \left(\frac{N_T}{N}\right)^2 \qquad R_i' = R_l \left(\frac{N_1}{N_2}\right)^2 \, k. \; k = \begin{array}{l}\text{coefficient of}\\ \text{coupling}\end{array}$$

(1)   (2)

$$R_i' = R_l \left(\frac{C_1 + C_2}{C_2}\right)^2$$

$$C_1 = nC_2 \qquad n = \sqrt{\frac{R_i'}{R_l}} \; - 1$$

$$C_2 = \left(\frac{1+n}{n}\right) C_T$$

(3)

Figure 6-3. Single-pole tuned circuit. (a) Equivalent circuit; (b) matching networks.

It was originally assumed that the rf power dissipated in the coil itself was 10% of the total power delivered by the stage; hence

$$R_c \gtrless 10 R_l',$$

$$(R_l)_T = 1.64 \text{ k}\Omega,$$

and, for

$$R_l' = 1.1 \, (R_l)_T = 1.1 \times 1.64 = 1.8 \text{ k}\Omega,$$

$$R_c \gtrless 11 \, (R_l)_T = 11 \times 1.64 = 18 \text{ k}\Omega.$$

The total capacitance, C_T, which tunes with the inductance L to f_o is chosen to be 12 pF, that is,

$$C_T = \frac{C_1 C_2}{C_1 + C_2} = 12 \text{ pF.} \tag{6-10}$$

Then

$$L = \frac{1}{(2\pi f_o)^2 C_T} \text{ H} = \frac{1}{(6.28 \times 10^8)^2 \times 12 \times 10^{-12}} = 0.21 \, \mu\text{H,} \tag{6-11}$$

and the required quality factor of the inductance is

$$Q_c = \frac{R_c}{2\pi f_o L} = \frac{18 \times 10^3}{6.28 \times 10^8 \times 0.21 \times 10^{-6}} \simeq 130, \tag{6-12}$$

which can be easily obtained at this frequency.

A capacitance-matching network is selected as shown in Fig. 6-1. The equations used to compute the values of C_1 and C_2 are given in Fig. 6-3b. Hence

$$C_1 = C_2 \left(\sqrt{\frac{1.8 \times 10^3}{50}} - 1 \right) = 5 C_2,$$

$$C_2 = \left(\tfrac{6}{5}\right) \times 12 \times 10^{-12} = 14.4 \text{ pF,}$$

$$C_1 = 5 \times 14.4 \times 10^{-12} = 72 \text{ pF.}$$

The standard values of C_1 and C_2 are 75 and 15 pF, respectively.

One has to exercise care in using the equations associated with Fig. 6-3 at frequencies above 10 MHz. First of all, these equations, which are only approximations to more complicated expressions, though accurate enough for most practical purposes, may give unsatisfactory results when the ratio R_l'/R_l is high. Secondly, a design may result in requirements that cannot be met in practice, such as tapping a coil which consists of three wire turns each $\frac{1}{8}$ in. in diameter at one sixth of the turn from ground, or that are costly to implement as, for example, nonstandard capacitance values. Finally, one has to verify that the matching circuit can be implemented with the components available on the market. It would be difficult to make the circuit in Fig. 6-3*b*-3 work at 100 MHz if C_1 were a 200 pF capacitor in a wire-lead configuration. The self-resonance of such a component would be below the operating frequency, and it would display an inductive rather than a capacitive reactance.

If a bandwidth requirement, in addition to rf output power, governed the design, the design steps would be modified. For example, if the tuned circuit were to attenuate a spurious signal 16 MHz away from 100 MHz by 18 dB, the 3 dB bandwidth of the circuit, B_{3dB}, would have to be no wider than 4 MHz. (The frequency response of a single-pole tuned circuit has a 6 dB/octave slope.) In such a case, one would first find the collector load associated with the 4 MHz bandwidth as

$$Q_l = \frac{f_0}{B_{3dB}} \tag{6-13}$$

or

$$Q_l = \frac{10^8}{4 \times 10^6} = 25$$

and

$$(R_l)_T = 2\pi f_o L Q_l = 6.28 \times 10^8 \times 0.21 \times 10^{-6} \times 25 = 3.3 \text{ k}\Omega \tag{6-14}$$

and then determine the bias parameters of the transistor. In case of a high transistor output resistance, R_o, the problem is easy to solve, though sometimes at the price of an increase in the power supply cost. When $R_o \gg (R_l)_T$ we compute, Eq. 6-2,

$$V_{CE} = \sqrt{8 \times 11 \times 10^{-3} \times 3.3 \times 10^3} = 17 \text{ V},$$

which indicates that the power supply voltage has to be increased from 15 to 20 V dc. The dc collector current then is, Eq. 6-3,

$$I_C = \frac{20}{3.3 \times 10^3} = 6.1 \text{ mA.}$$

The value of R_o between 1 and 2 kΩ is not uncommon at high frequencies and cannot be neglected in many practical cases. One way of attacking the problem is to tap the coil, thus effectively increasing R_o reflected across the tuned circuit so as to make it negligible compared to the required $(R_l)_T$, and to select the transistor dc bias for the specified output rf power and the load seen by the transistor. The exercise carried out below demonstrates this design approach.

Assume that $R_o = 2$ kΩ and that, when reflected across the tuned circuit it is equal to $10 \times (R_l)_T$ or to 33 kΩ. To achieve this transformation the coil should be tapped at

$$N = N_T \sqrt{\frac{2 \times 10^3}{33 \times 10^3}} = (0.25)N_T,$$

which can be easily implemented. Under this condition the effective collector load seen by the transistor is

$$R_{\text{col}} = (R_l)_T \left(\frac{N}{N_T}\right)^2 \tag{6-15}$$

$$= 3.3 \times 10^3 (0.25)^2 = 200 \ \Omega$$

and

$$V_{CE} = \sqrt{8 R_{\text{col}}, P_T}$$

$$= \sqrt{8 \times 200 \times 11 \times 10^{-3}} = 4.2 \text{ V dc.}$$

Similarly,

$$I_C = \frac{B+}{R_{\text{col}}},$$

$$= \frac{15}{200} = 75 \text{ mA.}$$

Hence imposing a composite bandwidth and rf output power requirement on a single-stage amplifier may result in a significant increase in the power supply current drain.

Before completing the design it is advisable to analyze the potential stability of the rf amplifier. The expression of Stern's K factor, Ref. 2, is used for this purpose:

$$K = \frac{2(g_{11} + G_s)(g_{22} + G_l)}{|Y_{12}Y_{21}| + \text{Re}\{Y_{12}Y_{21}\}}, \qquad (6\text{-}16)$$

where the driving source and load admittances at the frequency of operation are G_s and G_l, respectively. The amplifier is stable for all values of G_s and G_l that result in $K > 1$. The amplifier will oscillate if $K < 1$. Equation 6-16 is useful in determining the range of design values of G_s and G_l for a given transistor and frequency.

The task of selecting a transistor that satisfies the gain and power requirements stated above and of performing a stability analysis is left to the interested student.

6.2 Radio-Frequency Mixers

Intermodulation products analysis of an rf mixer and selection of input frequencies that minimize the number of close-to-band and in-band spurious outputs were considered in Chapter 2. Here we discuss the important differences between various mixer configurations to assist the reader in selecting the mixer element, dc bias, and circuit configuration that most closely satisfy his requirements.

Figure 6-4 is the functional block diagram of an rf mixer. Two signals at f_1 and f_2 are applied to the input ports of a nonlinear element via the input matching networks. The desired output signal at f_3, which is equal to $f_1 + f_2$ or $f_1 - f_2$, is delivered to a load via the load matching network. The performance characteristics of a mixer ideal for frequency synthesis are as follows: (a) a transfer function that contains no terms higher than the second order (i.e., the mixer nonlinear element exhibits a square-law characteristic); (b) a conversion gain, instead of a conversion loss; (c) a high degree of isolation between any two ports of the device; and (d) a wideband operation (often, several octaves). Notice that the dynamic range of the mixer, so important in receiver design, plays no role in frequency synthesis because, in all practical cases, the variation in the level of both input rf signals is several decibels only. Similarly, with the exception of cases in which the mixer is followed by a high-multiplication-ratio frequency multiplier, the noise figure of the device is of little importance because the levels of rf power dealt with are high relative to the thermal noise level.

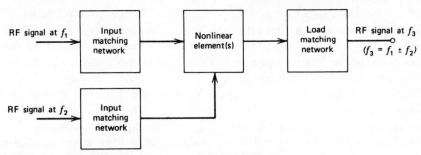

RF signal at f_1 → Input matching network → Nonlinear element(s) → Load matching network → RF signal at f_3

($f_3 = f_1 \pm f_2$)

RF signal at f_2 → Input matching network →

Figure 6-4. Radio-frequency mixer; functional block diagram.

Transistor Mixer

Figure 6-5 is a schematic diagram of an rf transistor mixer with both input signals injected into the base. The dc bias design and load matching are conducted in the same manner as was described for a class A rf amplifier in Section 6-1. An approximate rf gain of the circuit may be computed by assuming that an equivalent configuration consists of a diode mixer with a

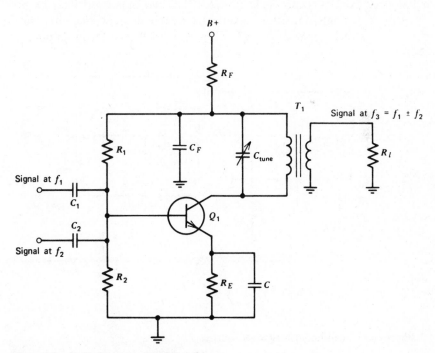

Figure 6-5. Radio-frequency transistor mixer.

10 dB loss followed by an rf amplifier, tuned to the output frequency, whose gain can be calculated if f_{max} of the transistor and the collector load are known. In other words, in some cases a transistor mixer can be designed to have conversion gain. The input impedance matching is accomplished with two capacitors, C_1 and C_2, which couple both input signals to the base of Q_1. Input port isolation is achieved by making the impedances of C_1 and C_2 adequately high at f_2 and f_1, respectively. Hence, if the isolation requirement is stringent, the advantage of a conversion gain provided by the transistor may be nullified by a light coupling of the driving sources to the mixer element.

To make this mixer operate over several octaves, the input- and load-matching techniques utilized in the circuit, as well as the circuit itself, would have to be modified to account for rf gain decrease with frequency and changes in the input and output impedances of the transistor. Probably such modifications also would result in conversion gain reduction.

Since a transistor does not exhibit a square-law characteristic, all intermodulation products, as well as the two input signals and their harmonics, are present at the output of a single-transistor mixer.

A better circuit is shown in Fig. 6-6, in which two transistors are used in a balanced configuration. This circuit offers two important features not present in a single-transistor mixer: the matching networks, the transformers T_1 and T_2, provide 20 to 30 dB isolation between the input ports of

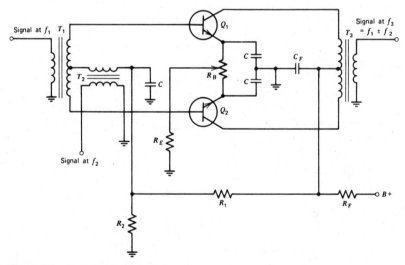

Figure 6-6. **Transistor balanced mixer.**

the mixer with relatively small loss, and (what is more important) the input signal at f_2 and all even harmonics of the signal are balanced out at the output of the mixer. The degree of balancing depends on the rf balancing of T_1 and T_2. (The rf gains of the two transistors at f_2 are equalized by adjusting R_B.) Whenever the phase-noise requirement permits, the balanced transistor mixer can be operated at low rf levels of both input signals, thus saving power and facilitating the rf shielding task.

Field-Effect Transistor Mixer

The drain current of a field-effect transistor (FET) is nearly a square-law function of gate-to-source voltage:

$$I_D \cong I_{DSS} \left(1 - \frac{V_{GS}}{V_P} \right)^n, \tag{6-17}$$

where $n \cong 2$,

$I_{DSS} = $ dc drain saturation current at $V_{GS} = 0$ and $V_{DS} > V_P$,

$V_P = $ pinchoff voltage; the value of V_{GS} at which $I_D = 0$.

It is expected, therefore, that, when properly biased, an FET mixer will display better performance with respect to spurious outputs than a transistor mixer.

The best bias for square-law operation is at half of the minimum pinchoff voltage, that is,

$$V_{GS,\text{bias}} = - \frac{V_{P(\text{min})}}{2} . \tag{6-18}$$

This results in a maximum allowable rf signal swing at the gate of the FET before the FET is driven into saturation or cutoff. Figure 6-7 shows an FET dc bias circuit and demonstrates a graphical approach to selecting the value of the source resistor, R_S. At no time is the FET to be driven into saturation or cutoff because, if it were driven into either of these modes of operation, it would no longer operate as a square-law device, that is, it would display the same spurious-output performance as any other switch. In other words, the following relationship is to be satisfied at all times:

$$V_{GS,\text{bias}} + E_s + E_{LO} < V_{P(\text{min})}, \tag{6-19}$$

where E_s and E_{LO} are peak voltages of the rf and local oscillator (LO)

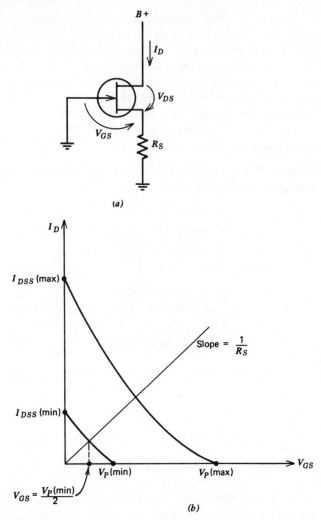

Figure 6-7. FET mixer, operating bias. (*a*) Schematic diagram; (*b*) transfer characteristic and selection of source resistor.

signals (see Fig. 6-8). To indicate a preferred condition, namely, that one input signal, the local oscillator, be higher in amplitude than the other, receiver terminology is used. Under this condition the conversion transconductance is (Ref. 3)

$$g_c = \frac{E_{LO} I_{DSS}}{V_P^2}, \tag{6-20}$$

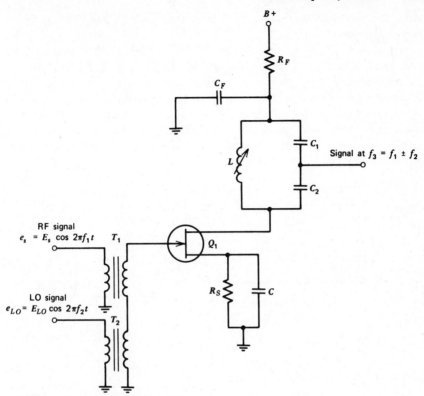

Figure 6-8. Radio-frequency FET mixer.

and the rf term of the drain current at the sum and difference frequency is

$$I_{d,f_3} = \frac{E_{LO}E_sI_{DSS}}{V_P^2}\cos 2\pi(f_1 \pm f_2)t. \qquad (6\text{-}21)$$

The conversion voltage gain at $f_1 \pm f_2$ is

$$k_V = g_c R_l', \qquad (6\text{-}22)$$

where R_l' = effective drain load at $f_1 + f_2$ or at $f_1 - f_2$.

For the conversion gain to be greater than unity, $R_l' > 1/g_c$. It is evident that, when one desires to operate at a high signal voltage, the local oscillator level must be reduced to satisfy Eq. 6-19, with a consequent reduction in conversion transconductance and, therefore, voltage gain.

The minimum drain-to-source supply voltage, V_{DS}, should be greater than the pinchoff voltage. It is experimentally adjusted for maximum conversion gain.

A typical FET mixer circuit is shown in Fig. 6-8. Both rf input signals are injected into the gate via the transformers, T_1 and T_2. The gate-to-source bias of $V_{P(min)}/2V$ is provided by the resistor R_S. The output circuit is tuned to f_3. The load transformation is accomplished by a capacitive divider, C_1 and C_2. Both rf input signals and their harmonics are present at the output. Analogously to a transistor mixer, some degree of balancing of one of the input signals is provided by the balanced FET mixer circuit shown in Fig. 6-9.

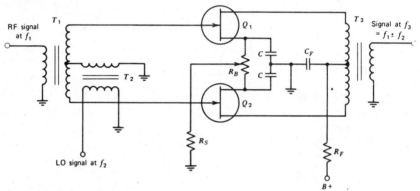

Figure 6-9.   FET balanced mixer.

An FET mixer, as well as its transistor counterpart, is most effective in applications where the frequency band of operation is narrow.

Diode Mixers

The most commonly used mixing element is a semiconductor diode. It requires no external dc bias, it is a broadband high-frequency device, and it displays very low $1/f$ noise when a Schottky barrier type of diode is employed. Because of very poor input-output port isolation and high levels of spurious outputs, however, a diode mixer is seldom used in a single-element configuration.

Single- and double-balanced mixers are periodic sampling switches whose open and close states are determined by the polarity of the local oscillator signal. In a single-balanced configuration, Fig. 6-10a, the two diodes, $CR1$ and $CR2$, are switched on on a one-half cycle of the LO signal, at which time rf power at the output signal frequency, f_3, is transmitted to the load. On the other half cycle the diodes are switched off, and the mixer presents a high impedance to the transmission path. The local oscillator switching action is such that the unfiltered output spectrum of the mixer theoretically does not contain the LO signal, its even-order harmonics and all intermodulation products arising from the even-order harmonics of the LO signal, such as $f_1 + 2f_2$, $f_1 + 4f_2$, and so forth. In a practical circuit the degree of balancing depends on diode match for equal forward resistance and shunt capacitance, dc bias level (if any is used), and rf balance at LO frequency provided by the transformers. In a double-balanced configuration, Fig. 6-10b, on one cycle of LO signal the diodes $CR1$ and $CR2$ are switched on, while $CR3$ and $CR4$ are off. On the other half-cycle of LO signal the diodes $CR3$ and $CR4$ are on and $CR1$ and $CR2$ are off. Hence almost all available signal power at f_1 is delivered to the load. A typical insertion loss of a double-balanced mixer varies from 6 to 8 dB, depending on the frequency of operation. The resulting output spectrum theoretically should not contain both input signals, all their even-order harmonics and all intermodulation products arising from the even-order harmonics of either input signal, such as $2f_1 + 5f_2$, $f_1 - 4f_2$, $2f_1 + 2f_2$, and so on. The extent to which spurious signals are balanced out is determined by the degree of balance provided by the transformers, T_1 and T_2, the degree of match in the diode quad with respect to both input signals, the proper phase relation, the input-output port isolation provided by the layout of the circuit at the frequencies of operation, and the level of input signal at f_1.

It is desirable to operate a diode mixer in a linear region where, theoretically, for each decibel of signal power reduction the nf_1 intermodulation product decreases by $n-1$ dB in relation to the desired output (Ref. 4). This permits the designer to select an input signal power that results in satisfactory levels of intermodulation products. In some applications the dependence of spurious outputs on the input signal power is a disturbing factor. In design cases where a high-rf-level operation is mandatory (as, e.g., in ultralow-phase-noise systems), an extension of the mixer linear range is achieved by using improved double-balanced mixers. D. Cheadle (Ref. 4) classifies various double-balanced configurations as follows. Class I mixers use one diode in each balanced arm as shown in Fig. 6-10b. This class of mixers requires +7 dBm nominal LO power and has a 1 dB conversion compression point typically between 0 and +5 dBm of signal power. Class II mixers use either two diodes in series or a precision resistor in series with a diode in each balanced arm. This type of mixer has a better-balanced ring and, therefore, offers higher port-to-port isolation and lower levels of intermodulation products. Class II mixers operate at +13 to +20 dBm LO power and exhibit 1 dB compression points at +8 to +12 dBm of signal power. Class III double-balanced mixers utilize a precision resistor paralleled by a capacitor in series with a diode in each balanced arm. They require +20 to +27 dBm nominal LO power and have a 1 dB compression point at +15 dBm or higher.

It should be pointed out that, by operating a diode mixer at a high LO power, one simultaneously increases the levels of LO fundamental and harmonic power appearing at the output of the mixer. In some applications this may not be desirable. It is advisable, therefore, to study the intermod-

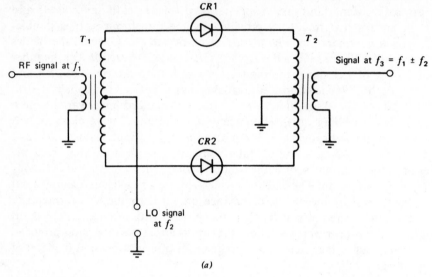

(a)

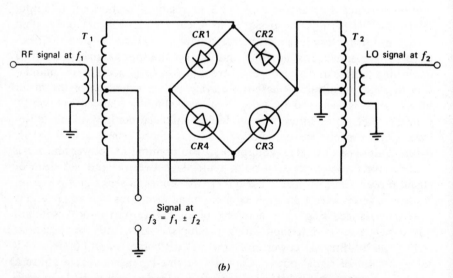

(b)

Figure 6-10.   Diode mixers. (a) Single-balanced mixer; (b) double-balanced mixer.

ulation products tables, such as Fig. 2-6b which are available from manufacturers of mixers, to determine whether or not the improvement achieved by selecting a high-LO-power mixer really helps in coping with the spurious-output problem. The designer should be also aware that the cost of a class III mixer is an order of magnitude higher than that of a class I mixer. In some cases, therefore, it may be less expensive to modify the frequency plan of a system (e.g., by providing an additional up- or down-conversion) and use inexpensive components than to rely on a high-cost mixer operation.

Parametric Mixers

A parametric mixer, often referred to in literature as a parametric or varactor frequency converter, is basically a back-biased varactor diode driven by an rf power source known as a pump. When, in addition to the pump, a signal is applied to the varactor, rf currents are generated at the sum and the difference of all the integral harmonics of the pump and signal frequencies, that is, at $mf_p \pm nf_s$, where f_p and f_s are the pump and signal frequencies, respectively. If the varactor is current driven, that is, if the pump current through the diode is sinusoidal, a true square-law characteristic is displayed by the varactor (Refs. 5 and 6). The parametric mixer is a low-noise device. It may have a conversion gain, and it can be made to operate over a broad band of frequencies (Ref. 7).

In spite of such a wide collection of desirable characteristics, a parametric mixer cannot compete in general types of applications with a low-level double-balanced diode mixer for two reasons: (1) the rf power required to pump the varactor is relatively high ($+20$ to $+25$ dBm), and (2) a parametric mixer is prone to instabilities, which may present severe design problems. It is recommended, therefore, that one consider using a parametric mixer only if extremely low levels of intermodulation products are to be obtained and a diode mixer falls short of meeting the requirement.

In discussing transistor, FET, and diode mixers no distinction was made between the performance of these circuits in either the upper- or the lower-sideband mode because they operate in the same manner in both modes. Unfortunately, the performance of parametric mixers depends not only on whether the mixer output signal is at $f_p + f_s$ or $f_p - f_s$ but also on whether the output frequency is greater or smaller than the signal frequency. For this reason it is convenient to separate parametric mixers into four groups (see Table 6-2) in order that the important differences between mixers belonging to various groups can be easily pointed out. The following discussion is based on the material presented by P. Penfield and R. P. Rafuse in *Varactor Applications* (Ref. 8).

Table 6-2. Four Groups of Parametric Frequency Converters

Upper-sideband upconverter	Upper-sideband downconverter
$f_o = f_p + f_s$	$f_o = f_s - f_p$
$f_o > f_s$	$f_o < f_s$
Lower-sideband upconverter	Lower-sideband downconverter
$f_o = f_p - f_s$	$f_o = f_p - f_s$
$f_o > f_s$	$f_o < f_s$

Seven assumptions are made by Penfield and Rafuse in deriving expressions for the input and output impedances and the gains of parametric mixers:

1. The varactor equivalent circuit consists of a voltage-variable capacitance, $C(v)$, in series with a resistor, R_s

2. The varactor is pumped at a frequency f_p such that its incremental elastance is

$$s(t) = \sum_{k=-\infty}^{\infty} S_k e^{jk\omega_p t}, \qquad (6\text{-}23)$$

where $\omega_p = 2\pi f_p$ and $S_2 = 0$.

3. A linear operation is ensured, that is, the voltages and currents are small at all frequencies except nf_p, where $n = 1, 2, 3, \ldots$.

4. Only one varactor diode is used.

5. Losses in the coupling networks are negligible.

6. The varactor is open-circuited at all frequencies except the pump, signal, and output frequencies.

7. The input and output networks are tuned to their corresponding frequencies, so as to tune out the varactor average elastance, S_0.

The model used here does not constitute a very accurate characterization of the operation of a parametric mixer, particularly at very high frequencies. For example, it would be difficult to implement assumption 6 in a practical current-driven varactor circuit. However, results obtained from an analysis based on these assumptions can serve as a guide to parametric mixer design and development.

Before proceeding with the discussion of parametric mixers, several quantities must be described in some detail. One of these, the varactor diode capacitance, is expressed as

$$C(v) = C_{\min}\left(\frac{\phi - V_B}{\phi - v}\right)^n, \qquad (6.24)$$

where C_{min} is the varactor capacitance at the breakdown voltage of V_B, ϕ is the diode contact potential (approximately 0.7 V), v is an externally applied bias voltage, and $n = \frac{1}{2}$ for an abrupt-junction diode. Notice that, since both V_B and v are reverse-bias voltages, they are negative quantities.

The diode elastance is

$$S(v) = \frac{1}{C(v)}. \tag{6-25}$$

The elastance can be expressed in terms of the diode bias voltage as

$$S(v) = S_{max}\left(\frac{\phi - v}{\phi - V_B}\right)^{1/2}, \tag{6-26}$$

where

$$S_{max} = \frac{1}{C_{min}}. \tag{6-27}$$

The pumped figure of merit for a varactor is $m_1\omega_c$, where m_1 and ω_c are the varactor modulation ratio and angular cutoff frequency, respectively. In terms of the diode parameters it is

$$m_1\omega_c = \frac{|S_1|}{R_s}, \tag{6-28}$$

where $S_1 =$ the fundamental component in the expression of the diode elastance, Eq. 6-23.

A pumped varactor with a higher value of $m_1\omega_c$ is capable of better noise performance and more gain. For simplicity of presentation, in all equations that follow we use $m_1 f_c$, where f_c is the diode cutoff frequency in hertz.

The constant term and the magnitude of the fundamental component of elastance, S_0 and $|S_1|$, often appear in design equations of parametric mixers. In the case of sinusoidal pumping, these quantities can be easily determined with the help of Eq. 6-26 if the varactor dc bias voltage and the amplitude of the rf pump signal are known. S_0 is the elastance at the dc bias voltage of V_0, and $|S_1|$ is equal to one-half the amplitude of the elastance variation due to pumping (see Fig. 6-11).

The available power from a source is the maximum power that can be delivered by the source into any impedance that terminates it. For a sinusoidal source the available power is

$$P_a = \frac{V_{oc}^2}{4\text{Re}\{Z_o\}}, \tag{6-29}$$

where V_{oc} is the open-circuit voltage, and $\text{Re}\{Z_o\}$ is the real part of the output impedance of the source. It is assumed that $\text{Re}\{Z_o\} > 0$. Hence the available power gain of a parametric mixer with a positive real part of the output impedance may be expressed as

$$G_a = \frac{\text{available power from mixer at } f_p \pm f_s}{\text{available power from signal source at } f_s}. \qquad (6\text{-}30)$$

Under some conditions parametric mixers exhibit input and output impedances with negative real parts. In such cases Eqs. 6-29 and 6-30 no longer apply. To overcome the difficulty of defining a general form of the mixer power gain, the concept of exchangeable power is introduced. The exchangeable power of a source is

$$P_e = \frac{V_{oc}^2}{4\,\text{Re}\{Z_o\}}, \qquad (6\text{-}31)$$

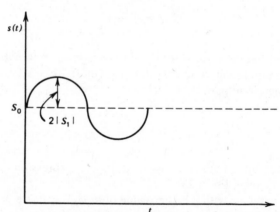

Figure 6-11. Varactor elastance waveform at sinusoidal pumping. Reprinted from *Varactor Applications*, Paul Penfield, Jr. and Robert P. Rafuse, by permission of the M.I.T. Press, Cambridge, Massachusetts.

where the real part of the source impedance, $\text{Re}\{Z_o\}$, can be positive or negative. The exchangeable power gain of a parametric mixer is

$$G_e = \frac{\text{exchangeable power from mixer at } f_p \pm f_s}{\text{exchangeable power from signal source at } f_s}. \qquad (6\text{-}32)$$

Notice that for $\text{Re}\{Z_o\} > 0$ the exchangeable gain is equal to the available gain of the mixer. This concept of rf power gain is used throughout the section on parametric mixers.

Figure 6-12 is a block diagram of a parametric mixer. The driving source impedance, Z_g, tunes out the varactor average elastance, S_0, at f_s. The load impedance, Z_l, tunes out the diode average elastance at f_o. The input and output impedances of the pumped varactor circuit at the frequencies of interest are Z_{in} and Z_{out}, respectively.

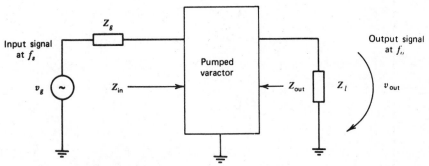

Figure 6-12. Parametric mixer.

Figure 6-13 is an equivalent schematic diagram of a current-pumped parametric mixer. In this particular circuit

$$Z_g = R_g + j\left(\omega_s L_1 - \frac{1}{\omega_s C_1}\right) \tag{6-33}$$

and

$$Z_l = R_l + j\left(\omega_o L_3 - \frac{1}{\omega_o C_3}\right), \tag{6-34}$$

where $\omega_s = 2\pi f_s$ and $\omega_o = 2\pi f_o$. Tuning of the mixer input and output is accomplished by satisfying the following relationships:

$$f_s = \frac{1}{2\pi\sqrt{L_1[C_1 C_0/(C_1 + C_0)]}} \tag{6-35}$$

and

$$f_o = \frac{1}{2\pi\sqrt{L_3[C_3 C_0/(C_3 + C_0)]}}, \tag{6-36}$$

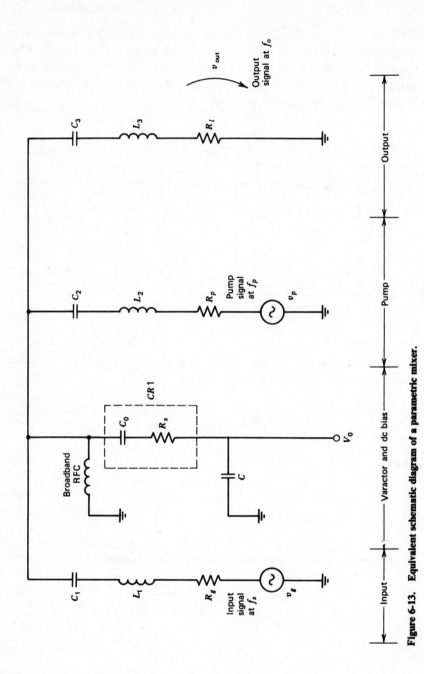

Figure 6-13. Equivalent schematic diagram of a parametric mixer.

344

where C_0 is the varactor capacitance at a dc bias voltage of V_0. For a sinusoidal current pumping

$$f_p = \frac{1}{2\pi\sqrt{L_2[\,C_2C_0/(C_2+C_0)\,]}} \,. \tag{6-37}$$

For the upper-sideband up- and downconverters the input and output impedances of the pumped varactor circuit and the mixer exchangeable gain are as follows:

$$Z_{in} = \left[R_s + \frac{(m_1 f_c R_s)^2}{f_s f_o (R_s + R_l)} \right] + \frac{1}{j2\pi f_s C_0} \,, \tag{6-38}$$

$$Z_{out} = \left[R_s + \frac{(m_1 f_c R_s)^2}{f_s f_o (R_s + R_g)} \right] + \frac{1}{j2\pi f_o C_0} \,, \tag{6-39}$$

and

$$\frac{1}{G_e} = \left(\frac{f_s}{f_o} \right)\left(\frac{R_s + R_g}{R_g} \right) + \left(\frac{f_s}{m_1 f_c} \right)^2 \frac{(R_s + R_g)^2}{R_s R_g} \,. \tag{6-40}$$

It is assumed in deriving Eqs. 6-38, 6-39, and 6-40 that both input and output circuits are tuned, and the varactor is pumped sinusoidally.

The gain of an upper-sideband downconverter is always less than unity; however, this should not be considered an important disadvantage unless it is lower than -8 dB. Both configurations exhibit good stability and are not prone to parametric oscillations.

The exchangeable gain can be maximized in the upper-sideband up- and downconverters by selecting R_g so that

$$R_g = R_s\sqrt{1 + \frac{(m_1 f_c)^2}{f_s f_o}} \tag{6-41}$$

Under this condition, the gain is

$$G_{e,\,max} = \frac{(m_1 f_c)^2}{f_s^2\left\{ 1 + \sqrt{1 + [\,(m_1 f_c)^2/f_s f_o\,]}\, \right\}^2} \,, \tag{6-42}$$

and the real part of the mixer output impedance is

$$R_{\text{out}} = R_s \sqrt{1 + \frac{(m_1 f_c)^2}{f_s f_o}} \qquad (6\text{-}43)$$

If we make $R_l = R_{\text{out}}$, the real part of the mixer input impedance, R_{in}, is equal to R_g and the mixer is matched at the input and output.

For the lower-sideband up- and downconverters the input and output impedances and exchangeable gain are as follows:

$$Z_{\text{in}} = \left[R_s - \frac{(m_1 f_c R_s)^2}{f_s f_o (R_s - R_l)} \right] + \frac{1}{j2\pi f_s C_0}, \qquad (6\text{-}44)$$

$$Z_{\text{out}} = \left[R_s - \frac{(m_1 f_c R_s)^2}{f_s f_o (R_s + R_g)} \right] + \frac{1}{j2\pi f_o C_0}, \qquad (6\text{-}45)$$

and

$$\frac{1}{G_e} = \left(\frac{f_s}{m_1 f_c} \right)^2 \frac{(R_s + R_g)^2}{R_s R_g} - \left(\frac{f_s}{f_o} \right) \left(\frac{R_s + R_g}{R_g} \right). \qquad (6\text{-}46)$$

The exchangeable gain of the lower-sideband up- and downconverters may be made large but only at the expense of possible instabilities. Moreover, the gain of the lower-sideband downconverter is greater than unity only if

$$f_s f_o < (m_1 f_c)^2. \qquad (6\text{-}47)$$

However, this is also the condition at which a potential instability can occur.

The exchangeable gain can be maximized in both configurations if

$$R_g = R_s \sqrt{1 - \frac{(m_1 f_c)^2}{f_s f_o}} \qquad (6\text{-}48)$$

Then

$$G_{e,\text{max}} = \frac{(m_1 f_c)^2}{f_s^2 \left\{ 1 + \sqrt{1 - \left[(m_1 f_c)^2 / f_s f_o \right]} \right\}^2}, \qquad (6\text{-}49)$$

and

$$R_{\text{out}} = R_s \sqrt{1 - \left[(m_1 f_c)^2 / f_s f_o \right]} \ . \tag{6-50}$$

By making $R_l = R_{\text{out}}$, we provide input and output matching.

For an abrupt-junction varactor diode that is fully pumped with a sinusoidal current, $m_1 = 0.25$. The dc bias voltage is

$$V_0 = 0.375(\phi - V_B) - \phi, \tag{6-51}$$

and the required rf pump power is

$$P_p = 0.5 P_n \left(\frac{f_p}{f_c} \right)^2, \tag{6-52}$$

where the nominal power, P_n, is

$$P_n = \frac{(-V_B - v_{\min})^2}{R_s} ; \tag{6-53}$$

and $v_{\min}$ is the minimum instantaneous diode voltage. This voltage is equal to the dc bias voltage minus the peak rf voltage across the varactor.

6-3 Frequency Multipliers

Depending on the frequency of operation, multipliers are classified as low and high rf power devices. At frequencies at which rf power gain is cheap the governing design consideration is the phase-noise performance of the device; that is, given a multiplication number and the power efficiency associated with it, the rf drive power at the frequency of the input wave is determined by the required signal-to-noise ratio at the output of the multiplier. It usually is sufficient to drive such a device at a low rf input power (less than a hundred milliwatts). At microwave frequencies signal amplification is expensive at present, so that all rf power amplification is accomplished at the input frequency. Under this condition the multiplier operates at a high rf power (several watts).

Both types of multipliers can be implemented with transistor or step recovery diode circuits when small multiplication numbers are employed, but step recovery diodes are preferred when the multiplication number is large and/or when high efficiency is of prime importance. Transistor multipliers are inexpensive, are simple to implement, and do not suffer from instabilities. Step recovery diode multiplier circuits are more complicated and may present instability problems because the capability of operating in a subharmonic parametric oscillation mode is inherent to the step recovery diode (Refs. 9 and 10). Both the transistor and the step recovery diode low-rf-power multipliers are described below. The interested reader is referred to Refs. 11 and 12 for numerous examples of high-power multiplier design.

Transistor Frequency Multiplier

Transistors biased for class C operation are successfully used as low-harmonic-number frequency multipliers. The constituent parts of the circuit are a transistor biased in class C, a tuned circuit that is resonant at the output frequency, and a band-pass filter (see Fig. 6-14). When an input signal at f_{in} is applied to the base of the transistor, the transistor conducts during a portion of the cycle of the input wave. The collector current approximates a half-wave rectifier waveform. The collector tuned circuit provides a high load at the frequency of the desired harmonic component, and the band-pass filter attenuates the unwanted harmonics present at the output of the tuned circuit.

The efficiency of such a multiplier depends on the conduction angle of the collector current, so that the circuit is optimized by adjusting the dc transistor bias in such a way as to result in the maximum rf power output

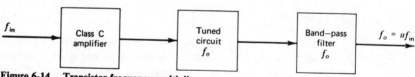

Figure 6-14.   Transistor frequency multiplier.

at the selected harmonic component of the input frequency. Unfortunately, this is not an efficient way of high-order harmonic generation because the harmonics of a half-wave rectifier waveform are proportional to the reciprocal of the multiplication number squared and drop off very rapidly with the multiplication number. Significantly higher efficiency at high harmonic numbers is achieved with a circuit that converts a sine wave into a periodic fast-rise-time pulse.

There are devices that convert a sine wave into a pulse train. At frequencies below approximately 1 MHz the saturable inductor multiplier would be a recommended choice. At frequencies above 1 MHz the step recovery diode exhibits superior performance.

Saturable Inductor Multiplier

The magnetic flux density in an inductor changes with a varying current conducted by the inductor. The change in magnetic flux induces a voltage across the inductor that is proportional to the rate of change of flux (Ref. 13), that is,

$$e(t) = Ri(t) + \frac{d\lambda}{dt}, \qquad (6\text{-}54)$$

where R = resistance of the winding,
   $i(t)$ = current in the inductor,
   λ = magnetic flux density,
$d\lambda/dt$ = rate of change of magnetic flux density.
In the case considered,

$$Ri(t) \ll \frac{d\lambda}{dt}$$

and

$$e(t) \cong \frac{d\lambda}{dt}. \qquad (6\text{-}55)$$

If the inductor is a coil that is wound on a ferromagnetic material core, such as 4-79 molybdenum permalloy prepared in thin sheets and wound on a core bobbin, no change in flux is observed above a certain value of current, I_s—the core is in saturation, and the induced voltage is zero. Hence, if the current in such an inductor is a sine wave, the inductor is periodically driven into and out of saturation and the induced voltage is a periodic pulse, as shown in Fig. 6-15.

It should be obvious that for high multiplication numbers the best core material is one which is characterized by a square hysteresis loop with as short a transition time as possible, that is, the highest rate of change of flux during the transition phase (see Fig. 6-16).

This method of sine wave-to-pulse conversion is successfully utilized in the design of saturable inductor multipliers. As shown in Fig. 6-17, a saturable inductor, driven into saturation by an amplifier-driver, generates a train of positive and negative pulses. A diode clipping circuit removes negative pulses. A pulse amplifier may or may not be used, depending on the multiplication number and the desired signal-to-noise ratio at the output of the multiplier. Finally, the band-pass filter at the output of the multiplier attenuates all unwanted harmonic components of the input wave.

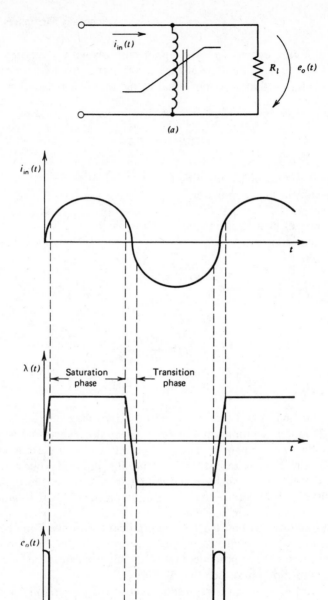

Figure 6-15. Saturable inductor. (a) Schematic diagram; (b) waveforms.

350

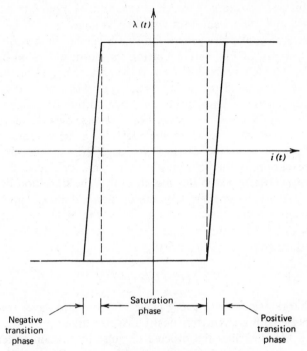

Figure 6-16. Hysteresis loop of a ferromagnetic material.

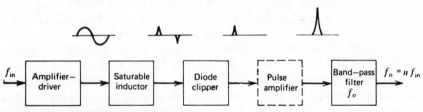

Figure 6-17. Saturable inductor multiplier.

With new developments in the manufacture of square-hysteresis-loop materials for high-speed computers constantly taking place, it is hard to set a high-frequency limit for the saturable inductor multiplier. The writer chose 1 MHz because from this frequency on, well into the microwave region, the step recovery diode multiplier exhibits excellent efficiency for both low and high multiplication numbers. This circuit is treated next.

Step Recovery Diode Multiplier

The step recovery diode (SRD) is effectively a two-impedance-state electronic switch. When forward biased, the diode exhibits a low impedance (large diffusion capacitance), and in a reverse bias mode the diode impedance is high and almost constant with bias voltage.

The mechanism of pulse generation in a SRD can be described utilizing the circuit shown in Fig. 6-18a. During the positive portion of the input wave cycle, the SRD is conducting and the voltage across the diode is low. At this time a charge of minority carriers is injected into the active region of the diode junction. When on the negative portion of the cycle a reverse bias voltage is applied to the diode, heavy reverse current supplied by the carriers in the junction flows in the SRD for a time comparable to the carriers' lifetime until the carriers are depleted. As the charge at the junction reaches zero, the field built into the junction causes the current of the minority carriers to stop flowing abruptly. The transition time is many orders of magnitude smaller than the carriers' lifetime. Since the diode current passes through the inductor L, when the SRD is switched off the energy stored in the inductor, $\frac{1}{2}LI^2$, is transferred to the load, R_l', in the form of a transient current (Ref. 14):

$$i_l(t) = Ie^{-(R_l/L)t}. \qquad (6\text{-}56)$$

This discharge occurs once every cycle of the input wave, resulting in a periodic pulse with a characteristically fast rise time.

Figure 6-18b describes the process of pulse generation in terms of the waveforms associated with the impulse generator. For simplicity of presentation, the effects of stray capacitances, lead inductances, and imperfections in the diode are neglected in the sketch of the waveforms.

A SRD frequency multiplier consists of an amplifier-driver, which delivers the required rf power to the diode; an input matching network, which downconverts a high output impedance of the amplifier to match the impedance of the diode (usually 20 to 30 Ω); an impulse generator, which converts the input sine wave into a periodic pulse; an output matching network, which upconvers the low diode impedance; and a band-pass filter, which rejects unwanted harmonic components of the input wave. Figure 6-19 shows the order in which these circuits are interconnected.

A recommended procedure consists of designing the multiplier circuits in the following order:

1. The output band-pass filter.
2. The impulse generator.
3. The amplifier-driver.
4. The input and output matching networks.

Only steps 2 and 4 are considered here. Filter design is outside the scope of this work, and rf power amplifier design was treated at the beginning of this chapter.

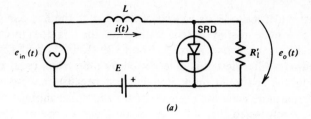

(a)

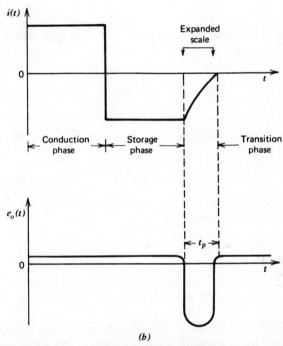

(b)

Figure 6-18. Impulse generator. (*a*) Schematic diagram; (*b*) diode current and load voltage, one cycle of the input wave. Courtesy Hewlett-Packard Company.

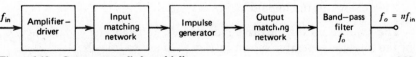

Figure 6-19. Step recovery diode multiplier.

353

The selection of a diode, as well as the design of the impulse generator and matching networks, is based mainly on the material presented in Hewlett-Packard *Application Notes* 913 and 920 (Refs. 11 and 12).

The band-pass filter design is determined by the system requirement for spurious outputs. It should be completed first because, having computed the filter loss, a more accurate estimate of the multiplier input rf power can be made. It is suggested that the filter be designed to operate from a 50Ω driving source and into a 50Ω load impedance for ease of alignment and testing of the filter and for ease of matching the SRD to the filter.

Six diode parameters are important in low rf power design:

1. The reverse bias capacitance, C_R, which determines the energy in the impulse and impedance of the diode. The impedance of C_R at the output frequency is

$$X_o = \frac{1}{2\pi f_o C_R}. \tag{6-57}$$

2. The minority carrier's lifetime, τ, which determines the loss occurring during forward charge storage. Select the diode so that

$$\tau \gg \frac{1}{2\pi f_{in}}.$$

3. The transition time, t_t, which sets the upper output frequency limit. Choose the diode with

$$t_t \leqslant \frac{1}{f_o}.$$

4. The series forward conduction resistance, R_S. Select a diode with R_S as small as possible to minimize the loss occurring in the diode input circuit.

5. The package inductance, L_p, which should be as small as possible to minimize the portion of energy stored in the inductor L that is not transferred to the load. At least, make sure that

$$L_p < \frac{X_o}{2\pi f_o}.$$

6. Of lesser importance in low rf power multipliers: the diode breakdown voltage, V_{BR}, which sets the maximum limit on the amplitude of and the energy in the impulse.

The expressions for the drive inductor, L, and shorting capacitance, C_T, of the impulse generator shown in Fig. 6-20 are as follows:

$$L \cong \frac{2.31 \times 10^{-2}}{C_R f_o^2} \tag{6-58}$$

and

$$C_T \cong \frac{1.265 \times 10^{-2}}{L f_{in}^2}, \tag{6-59}$$

where L is in henries, C_T and C_R are in farads, f_{in} and f_o are in hertz, and C_R is the diode capacitance at a reverse bias of 10V. The shorting capacitance is used to provide an ac short at all harmonics of f_{in} up to and above the output frequency by tuning out the drive inductance at f_{in}. The rf capacitor should be a good one with a self-resonant frequency well above f_o. The shorting capacitance also makes the input impedance of the impulse generator, R_{in}, purely resistive. An approximate expression for R_{in}

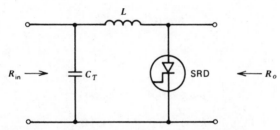

Figure 6-20. Impulse generator. Courtesy of Hewlett-Packard Company.

is

$$R_{in} = \frac{f_{in}}{2\pi f_o^2 C_R} \Omega. \tag{6-60}$$

The drive inductance and shorting capacitance can be expressed in terms of the pulse width, t_p, as

$$L \cong \left(\frac{t_p}{\pi}\right)^2 \frac{1}{C_R} \text{H} \tag{6-61}$$

and

$$C_T \cong \frac{C_R}{(2 f_{in} t_p)^2} \text{F}. \tag{6-62}$$

For most of the energy stored in L to be delivered at f_o, the parameters of the impulse generator should be chosen so that the value of the pulse width is $1/2f_o < t_p < 1/f_o$.

The pulse width is affected by the impulse generator load as $t_p \cong 2\pi\zeta R_l' C_R$, where ζ is the damping factor and R_l' is the effective diode load. For high efficiency ζ should be equal to 0.3. Hence, having determined t_p from the output frequency requirement and having selected a diode (i.e., having determined C_R), one may compute the effective diode load as

$$R_l' \cong \frac{t_p}{0.6\pi C_R}\,\Omega. \tag{6-63}$$

An empirically derived expression for the diode efficiency is

$$\eta \cong 100\left(\frac{kf_{\text{in}}}{f_o}\right)\% \tag{6-64}$$

where

$$k = 2 \quad \text{for} \quad f_o < 5 \text{ GHz},$$

$$k = 1 \quad \text{for} \quad f_o > 5 \text{ GHz}.$$

Hence, for a given required output power, P_o, and multiplication number, f_o/f_{in}, the power to be delivered by the amplifier-driver is

$$P_{\text{in}} \cong \left(\frac{f_o}{kf_{\text{in}}}\right)P_o$$

+ losses in the matching networks and band-pass filter.   (6-65)

Practical circuits are designed for larger P_{in} than the value given in Eq. 6-65 to account for losses due to imperfect matching between the diode and associated circuitry and for the approximate nature of the expressions used in the analysis.

The SRD should be dc biased for optimum efficiency. Biasing can be achieved by utilizing an external power supply voltage or a self-bias resistor. The advantages of self-biasing are low cost, temperature compensation of the diode (discussed below), and self-adjustment for different input power levels. The value of the bias resistor is computed from the expression

$$R_B \cong \frac{2\tau}{\pi (f_o/f_{\text{in}})^2 C_R} \ \Omega. \tag{6-66}$$

The resistor can be placed across the shorting capacitor as shown in Fig. 6-21. If optimum performance is required, a circuit adjustment of the bias resistor should be made for each diode as shown in the same figure, where $R_1 \cong 0.5R_B$ and $R_2 \cong 0.75R_B$. If no capacitive coupling is provided by the input matching network, a dc decoupling capacitor, C, should be added to the circuit. This capacitor should provide an ac short at the input frequency.

The carriers' lifetime of the SRD varies with temperature. It was determined empirically that τ increases with temperature at a rate that can vary between $\frac{1}{2}$ and $1\%/°C$. A bias resistor with a negative temperature coefficient that is equal to the temperature coefficient of the SRD lifetime, such as a silicon resistor (sensistor), is used in the circuit to temperature-

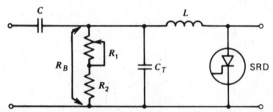

Figure 6-21. Step recovery diode, self-biasing. Courtesy of Hewlett-Packard Company.

compensate the diode. The resistor body should be in good thermal contact with the diode for proper temperature compensation.

Having established parameters of the impulse generator and estimated the required input rf power, one proceeds with the amplifier-driver design. In most practical circuits the power gain requirement dictates utilization of a collector load that is greater than the input impedance of the impulse generator. Similarly, the band-pass filter will require a driving source impedance greater than the output impedance of the impulse generator. To overcome this complication input and output matching networks are employed. The two types of matching networks described below may not present the best possible choice with respect to particular amplifier-driver and band-pass filter designs; however, practically any networks can be used as long as they match a low diode impedance to high driving source and load impedances.

Input matching is accomplished with an aid of the parallel capacitance-series inductance circuit shown in Fig. 6-22. The values of components for $R_g/R_{\text{in}} > 10$ are

$$L_m = \frac{\sqrt{R_g R_{in}}}{2\pi f_{in}} \text{ H} \tag{6-67}$$

and

$$C_m = \frac{1}{2\pi f_{in}\sqrt{R_g R_{in}}} \text{ F,} \tag{6-68}$$

where R_{in} is given by Eq. 6-60.

The output matching network consists of three components (see Fig. 6-22). The inductance L_n is

$$L_n = L, \tag{6-69}$$

and the two capacitances, C_n and C_c, are determined from the following two equations:

$$C_n + C_c = \frac{1}{(2\pi f_o)^2 L_n}, \tag{6-70}$$

since

$$f_o = \frac{1}{2\pi\sqrt{L_n(C_n + C_c)}},$$

and

$$C_n = C_c \left[\sqrt{\frac{R_l}{R_l'}} - 1 \right], \tag{6-71}$$

where R_l' is described by Eq. 6-63.

The multiplier is tuned by varying C_m, C_n, and C_c for maximum power output at the required harmonic component of the input signal.

Power losses in the multiplier come from numerous sources. The input matching network, C_m and L_m, the drive inductance, L, and the series forward resistance of SRD, R_S, introduce losses in the circuit at the input frequency. The reverse bias diode capacitance, C_R, the diode package inductance, L_p, and the output matching network, L_n, C_n, and C_c, introduce losses at the output frequency. All capacitors and inductors external to the diode should be high-quality-factor components with self-resonant frequencies well above the frequency of operation. Particular care should be exercised in designing the impulse generator because even small losses in this circuit have a significant effect on the efficiency of the multiplier.

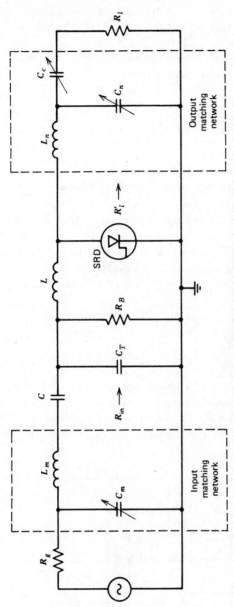

Figure 6-22. Impulse generator, input and output matching networks. Courtesy of Hewlett-Packard Company.

Losses due to the minority carriers' recombination occur during the forward charge storage phase. These losses can be minimized by selecting a diode with a long carrier lifetime, τ, such that $2\pi f_{in}\tau \gg 1$.

Transition losses occur during the time when the diode is opening. The multiplier efficiency due to transition losses alone is

$$\eta_t \cong \frac{100}{1+\left(V_{BR}/2f_{in}t_p^2\right)}\%.\tag{6-72}$$

Two basic problems inherent in a high-multiplication-number circuit, such as the step recovery diode multiplier, manifest themselves at the initial stage of the design: (a) suppression of unwanted harmonic components of the input signal, and (b) low signal-to-noise ratio at the output of the device due to poor efficiency. What is usually done to overcome these problems is to multiply the frequency of a signal in a number of steps, filtering out the unwanted harmonics and amplifying at each step. This approach cannot be used, however, when one multiplies by a prime number.

None of these problems is associated with the phase-locked loop frequency multiplier, which is the subject of the next discussion.

Phase-Locked Loop Frequency Multiplier

The functional block diagram of a PLL multiplier is shown in Fig. 6-23. The PLL utilizes a frequency divider in the feedback path. Multiplication numbers of 1000 and higher with 90 dB (or better) suppression of undesired harmonic components of the input signal can be achieved in most

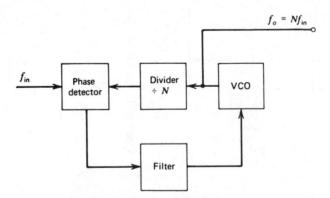

Figure 6-23.   Phase-locked loop frequency multiplier.

practical cases without any design problems or excessively high cost. Unless one deals with signals exhibiting extremely low noise spectra, the close-in noise of the input signal is multiplied up to the final frequency without being further degraded by the noise of the PLL circuits. Far-out noise may be reduced depending on the PLL loop bandwidth and VCO noise. Hence the PLL multiplier also acts as a narrow band-pass filter with respect to the phase noise and unwanted harmonics of the input signal. Another important advantage of this circuit as compared to a conventional multiplier is that the multiplication number can easily be changed, either manually or electronically, over a wide range without any elaborate tuning arrangements. The design of this type of PLL is considered in Chapter 5 on digital phase-locked loops.

6-4 Frequency Dividers

The types of circuits that perform frequency division are numerous. At present none satisfies all of the basic requirements, which are (a) a high frequency of operation (presently, up to 15 GHz); (b) a broadband operation (frequently, over an octave band); (c) a variable division ratio (manually or electronically controlled); (d) a high division ratio (of the order of 10^5); (e) zero rf output in the absence of an rf input signal; and (f) low cost. The four approaches to frequency division presented in this section are, therefore, evaluated in terms of the basic requirements, and the limitations of each approach are pointed out.

Regenerative Frequency Divider

The regenerative divider is one of the circuits used in early synthesizers. The functional block diagram of the divider is shown in Fig. 6-24. The key element of this circuit is a mixer that accepts an rf signal, whose frequency is divided, at the local oscillator port and its own output, multiplied by $N-1$, at the signal port. The frequency of the divider output signal, which is also the mixer output, is therefore

$$f_o = f_{in} - f = f_{in} - f_{in}\left(\frac{N-1}{N}\right)$$

or

$$f_o = \frac{f_{in}}{N}.$$ (6-73)

For the oscillations to be sustained, the loop gain should be greater than unity. It can be shown (Ref. 15) that the only stable output frequency at which the divider can operate is f_{in}/N.

The advantage of using this circuit is that theoretically it can be made to divide at a very high frequency because both mixers and frequency multipliers presently operate at and above 15 GHz. Another advantage is that in the absence of an input signal no rf output is generated by the divider. This feature is very useful in detecting circuit failures.

Detailed analysis of this circuit, however, discloses that many more functional blocks must be added to the block diagram of Fig. 6-24 to

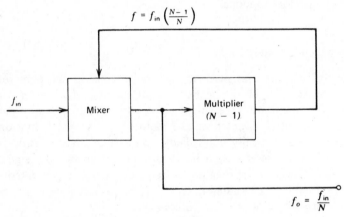

Figure 6-24.   Regenerative frequency divider.

implement the regenerative divider, making this approach relatively expensive. (See the detailed block diagram of the divider, Fig. 6-25.) Another disadvantage is that it may be necessary to introduce a large transient voltage into the circuit to start the divider (Ref. 15). The reason for the lack of a self-starting characteristic is obvious. To operate properly, the multiplier times $N-1$ has to be hard-driven. However, the mixer cannot provide an output signal without the multiplier operating properly.

The narrow frequency range over which regeneration takes place is another important disadvantage of this approach. Even a 10% range is difficult to achieve without tuning some of the divider circuits at the same time. The high cost, narrow band of operation, low division ratio, and difficulty of changing the division ratio make the regenerative divider inferior by far to other means of frequency division at frequencies below 600 MHz.

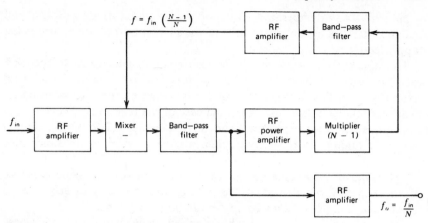

Figure 6-25. Regenerative divider; detailed block diagram.

Locked-Oscillator Frequency Divider

Oscillator injection locking (divide-by-1 frequency division) has been known for some time; see Refs. 16, 17, and 18. By injecting into an oscillator an rf signal whose frequency is approximately a Nth harmonic of the free-running oscillator frequency, one can force the oscillator to lock to a subharmonic of the injected signal.

It is convenient to discuss the divider behavior in terms of the oscillator mistuning, Δf (i.e., the difference between the oscillator free-running and locked frequencies), and the capture range, which is defined as the maximum oscillator mistuning at which locking takes place for a given injected rf power, division ratio, and bandwidth of the oscillator tuned circuit. In general, for a given circuit configuration and oscillator mistuning, the higher the division ratio, N, or the figure of merit of the tuned circuit, Q, the higher is the rf power that has to be injected into the oscillator to maintain constant capture range. Similarly, for a given N and Q, a large capture range requires a high injection rf power.

It is important to know that, since the circuit is forced to oscillate at a frequency different from the resonant frequency of the oscillator tuned circuit and since the locked frequency is often varied over a specified band, the phase of the output signal varies approximately as (Ref. 17)

$$\theta = \text{arc } \tan\left(-\frac{2Q\Delta f}{f_T}\right), \tag{6-74}$$

where $f_T =$ the resonant frequency of the tuned circuit.

Similarly, any temperature drift of the resonant frequency, f_T, will manifest itself as a slow phase drift of the divider output wave even if the input frequency is kept constant.

This approach has no lower or upper frequency limit because, theoretically, an oscillator operating at any frequency can be locked to a signal source by injection locking. Another advantage of the locked-oscillator divider is relatively low cost. As in the case of the regenerative divider, the circuit operates over a narrow band of frequencies (12 to 15 %), though electronic tuning can be easily implemented, and with low division ratios only.

The major disadvantage of the locked-oscillator divider is the presence of an output rf signal even if the driving circuitry fails to generate an injection signal. This makes equipment failure detection difficult.

A typical oscillator circuit is shown in Fig. 6-26. It is a grounded-base Colpitts-type oscillator, described in Section 6-5. If the reactance of the capacitor C_b is selected so that the capacitor no longer provides a good ac short from the base of the transistor to ground, and if an external signal whose frequency is harmonically related to the oscillator frequency is injected into the transistor base, the oscillator will lock in frequency to the injected signal.

The design of the divider is carried out as if it were a Colpitts-type oscillator. When the circuit is made to operate reliably at the selected frequency, the value of C_B is reduced to permit locking at the desired rf injection power. To increase the locking range, a load resistor, R_l, is inserted across the tuned circuit of the oscillator, L_1 and C_1, effectively increasing the bandwidth of the tuned circuit, that is, the capture range.

Other types of oscillators, such as a unijunction FET (Refs. 15, 19, and 20), a tunnel diode (Ref. 21), and an astable multivibrator (Refs. 15 and 22), have been successfully employed as frequency dividers at various frequencies.

Digital Frequency Dividers

Current advancements in the technology of digital frequency division have brought to the market fixed-ratio broadband dividers that divide by 2 to 10 and operate at frequencies above 1 GHz. They are available in dual-in-line or flat-pack integrated circuit packages and at low cost. Fairchild, Motorola, and Texas Instruments are a few of the companies that manufacture these devices.

Having such a variety of divider circuits available, the designer can combine a few of these devices in series to increase the division ratio to any desired number. Today it would be hard to find a good reason for designing one's own fixed-ratio divider at frequencies below 1 GHz.

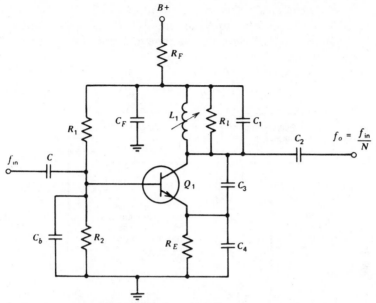

Figure 6-26.   Locked oscillator frequency divider, low division ratio.

At the same time a giant step has been taken in developing digital techniques of variable-ratio frequency division (presently up to 400 MHz) and in manufacturing devices to implement these techniques. One approach of variable-ratio frequency division, namely pulse swallowing, is considered here (also see Refs. 23 and 24). Others are described in application notes published by and available from semiconductor manufacturers.

The advantage of the pulse-swallowing technique is that only one high-speed integrated circuit is utilized in the divider, while the remaining divider circuitry operates at a frequency that is one tenth of the frequency of the input wave. This saves dc power, reduces cost, and makes the layout of the divider relatively noncritical.

The divider is defined by the three basic blocks shown in Fig. 6-27. The high-frequency two-modulo prescaler divides the frequency of the input wave by either the upper or the lower modulo count, N_U or N_L, respectively. The division ratio of the prescaler is controlled by a signal generated by the swallow counter. The swallow and program counters are preset by externally applied control logic to count from an "initial" state to a "zero" state. Initially the modulo control signal from the swallow counter is low, and the prescaler divides by N_U. When the swallow counter reaches its "zero" state, it generates a high modulo control signal, which changes the division ratio of the prescaler to N_L and inhibits the swallow counter from further counting. Program counters continue to count to their "zero" state,

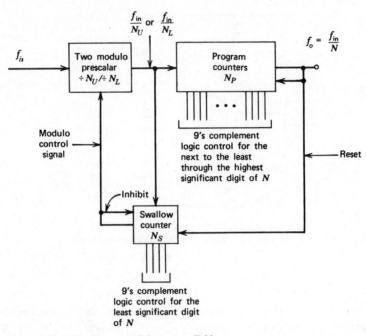

Figure 6-27. "Swallow-count" frequency divider.

at which time an output pulse is generated. The output pulse then resets both swallow and program counters to the "initial" state, and the cycle repeats.

The equation describing the operation of a pulse-swallowing divider is

$$N = N_U N_S + N_L N_P = x_1 x_2 \cdots x_n, \qquad (6\text{-}75)$$

where x_1 = the most significant digit of N,

x_n = the least significant digit of N.

The swallow counter is used for the least significant digit, and the program counters are used for the more significant digits, of N.

Figure 6-28 is a detailed block diagram of a pulse-swallowing frequency divider. It is drawn in such a manner as to point out the high- and low-frequency sections of the divider and the interface circuitry between them.

Figure 6-29 is a schematic diagram of a three-significant-digit variable-ratio divider utilizing Fairchild EC and TT logic. The vhf prescaler, type 95H90 (U1), is programmed to divide by 10 or 11. The swallow and program counters, U3 and U4–U5 respectively, are standard TTL/MSI BCD decade counters, type 9310, and U2 is a quad of two-input NAND gates, type 9002. The resistors R_1 and R_2 shift the threshold level of U1 in order to reduce the required rf power of the input wave to about +5 dBm. The transistor-diode-resistors combination, Q_1, CR1, R_3, and R_4, provides ECL-to-TTL interface, and resistors R_5 and R_6 provide TTL-to-ECL interface. The division ratio, N, can be varied from 90 to 999 at the frequency of an input signal as high as 200 MHz.

The operation of the divider is best demonstrated with a numerical example. Let the division ratio be equal to 208. Hence $N_S = 8$, and, for $N_U = 11$ and $N_L = 10$,

$$N_P = \frac{N - 11 N_S}{10} = 12.$$

Figure 6-30 shows the important waveforms and count ratios. The waveform in Fig. 6-30a is the input sine wave, described as a row of lines to show the number of cycles passed in a given time interval. The waveform in Fig. 6-30b is a train of pulses at the output of the prescaler when measured at the collector of Q_1. The pulses are numbered to indicate that for the first 8 cycles of f' (i.e., during the time the swallow counter is counting) the prescaler divides by 11, and during the consecutive 12 cycles of f', when the swallow counter is inhibited and only program counters are counting, the prescaler divides by 10. The waveform in Fig. 6-30c, the modulo control voltage, shows when switching from N_U to N_L occurs. Finally, the waveform in Fig. 6-30e is the output pulse, which appears once every $88 + 120 = 208$ cycles of the input wave. The 9's complement control logic required to achieve a division ratio of 208 is shown in Fig. 6-29.

Figure 6-31 is a schematic diagram of a four-significant-digit variable-ratio circuit dividing by 90 through 9999. The only difference between this circuit and the ones previously discussed is an additional 9310 counter to account for an extra digit of N. By adding 9310's and providing appropriate connections, the circuit can be easily designed to divide by five and six significant digits.

The advantages of this approach are numerious. The divider can be made to operate over more than an octave band if proper coupling and bypass capacitors are employed. Variable-division ratios as high as 10^6 can be easily achieved in this circuit. The divider generates no output in the absence of an input rf signal. The cost of the constituent parts of the divider is very low, considering the many desirable features that this approach offers to the designer.

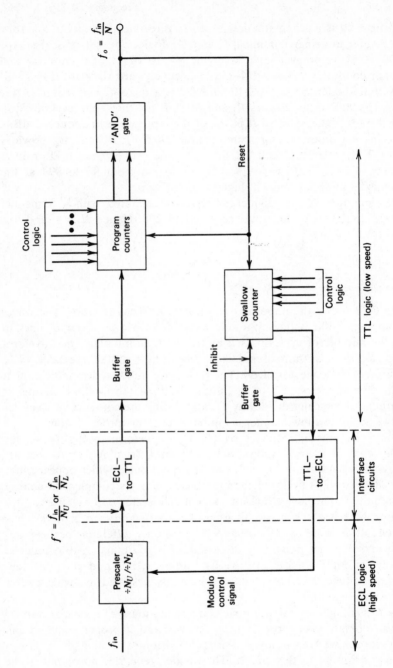

Figure 6-28. Swallow-count divider; detailed block diagram.

368

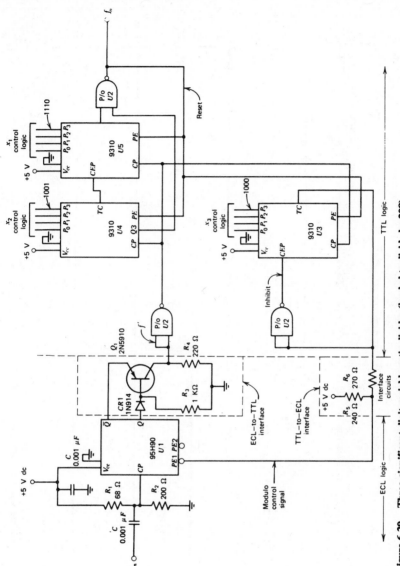

Figure 6-29. Three-significant-digit variable-ratio divider (loaded to divide by 208).

369

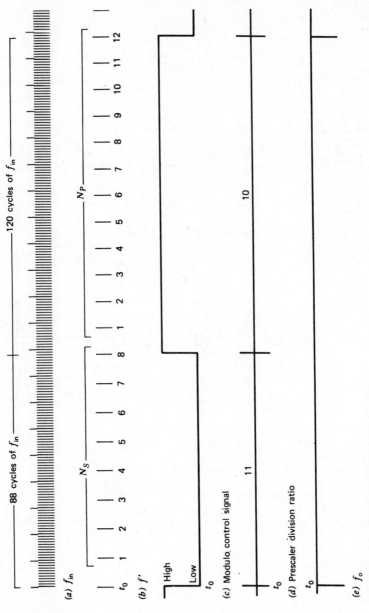

Figure 6-30. Divider waveforms and count ratios.

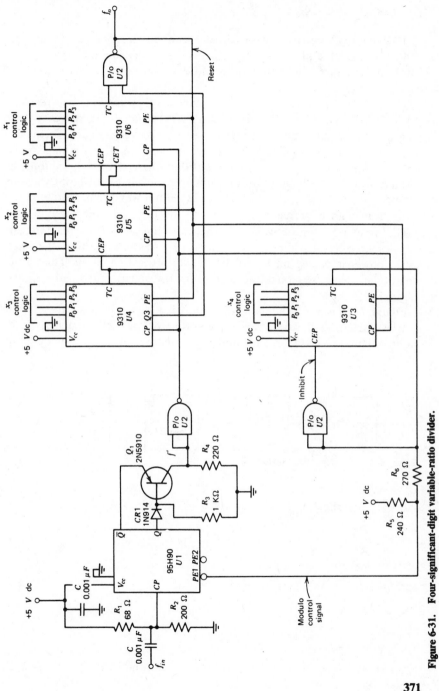

Figure 6-31. Four-significant-digit variable-ratio divider.

Divide-by-2 Carrier-Storage Frequency Divider

The divide-by-2 carrier-storage divider can be described as a charge-controlled switch and input/output networks that isolate the signal source from the load (see Fig. 6-32). An amplifier-driver provides the required rf power.

The principal element of the charge-controlled switch is a step recovery diode, which closes when the voltage across the diode drives it into forward conduction and opens when reverse recovery occurs. Frequency division is due to a relaxation process in which a given voltage pulse formed across the diode during the transition phase effectively inhibits the formation of a similar pulse on the subsequent cycle of the input wave by changing the recovery delay (Ref. 10). The output waveform, therefore, is a pulse train with a period equal to twice the period of the input wave.

The input isolation network prevents the wave at f_o from reaching the signal source by providing an ac short at f_o. At the same time the input signal power is passed on to the diode with little attenuation. The output isolation network separates the load from the signal source in a similar manner.

Figure 6-33 is the schematic diagram of a divide-by-2 carrier-storage divider excluding the amplifier-driver (Ref. 6-25). The elements of the input isolation network, L_1, C_1, and C_2, are chosen so as to provide an ac

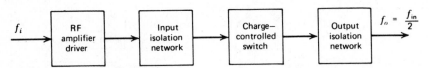

Figure 6-32.   Carrier-storage divider.

short at f_o by selecting L_1 and C_1 to resonate in series at f_o:

$$L_1 = \frac{1}{(2\pi f_o)^2 C} \qquad (6\text{-}76)$$

and an ac open at f_i by making C_2 equal to

$$C_2 = \frac{C_1}{L_1 C_1 (2\pi f_i)^2 - 1}. \qquad (6\text{-}77)$$

Similarly, the elements of the output isolation network, L_2, L_3, and C_3, are chosen to resonate in parallel at f_o and in series at f_i by making

$$L_2 = \frac{1}{(2\pi f_o)^2 C_3} \qquad (6\text{-}78)$$

and

$$L_3 = \frac{L_2}{L_2 C_3 (2\pi f_i)^2 - 1}, \qquad (6\text{-}79)$$

respectively. In order for these networks to introduce small losses and function properly, L_1, L_2, L_3, C_1, C_2, and C_3 should be high-quality-factor components at the highest frequency of interest, that is, at f_i, with self-resonance well above f_i.

The following considerations govern the selection of a charge-controlled switch components. The step recovery diode is chosen with a transition time one tenth (or less) of the input wave period. The inductor L resonates in series with the total diode capacity, C_T, at the medium frequency, f_m:

$$L = \frac{1}{(2\pi f_m)^2 C_T}, \qquad (6\text{-}80)$$

where

$$f_m = \frac{f_i + f_o}{2} \qquad (6\text{-}81)$$

and

$$C_T = C_p + C_V, \qquad (6\text{-}82)$$

with C_p = diode package capacitance,
 C_V = diode capacitance at a reverse bias voltage, V.
To realize L at uhf, it is important to select a diode with the lowest package capacitance.

The capacitor C_B serves two functions: (a) the capacitor ac-couples the signal source to the diode, and hence the value of C_B should be such that

$$\frac{1}{2\pi f_i C_B} \leqslant 0.1\Omega \qquad (6\text{-}83)$$

in order to present a negligible impedance at f_i; and (b) C_B provides, in conjunction with R_B, a dc bias voltage for the diode by charging to the peak of the input signal, E_g, during the positive portion of the cycle of the input wave. When the amplitude of the input wave is negative, the diode is cut off and C_B is not allowed to discharge through R_B by making the resistor such that

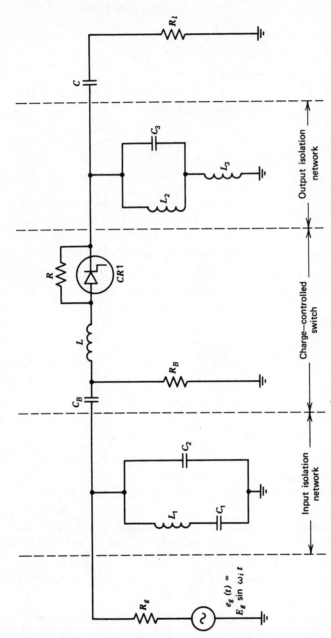

Figure 6-33. Carrier-storage divider; schematic diagram. Reprinted from W. J. Goldwasser "Design Shortcuts for Microwave Frequency Dividers," *The Electronic Engineer*, **May 1970, pp. 61-65.**

$$C_B R_B \cong \frac{100}{2\pi f_i}. \tag{6-84}$$

The value of the resistor R is selected experimentally in the event that parasitic oscillations of the diode take place. A value of 10 kΩ may be quite adequate.

Finally, the capacitor C ac-couples the divider to the load; hence

$$\frac{1}{2\pi f_o C} \leqslant 0.1\Omega. \tag{6-85}$$

At microwave frequencies input and output matching can be implemented with open- and short-ended transmission lines as shown in Fig. 6-34 (Ref. 25). By designing the input transmission line to be a half-wavelength at f_i (i.e., a quarter-wavelength long at f_o) and open-ended, one achieves the same effect as with L_1, C_1, and C_2 in Fig. 6-33. Similarly, by making the output transmission line a half-wavelength long at f_i and short-ended, one reproduces the performance achieved with L_2, L_3, and C_3.

The major advantage of the carrier-storage frequency divider is high frequency of operation. The upper frequency limit is mainly determined by the diode transition time. Relatively inexpensive circuits can be made to

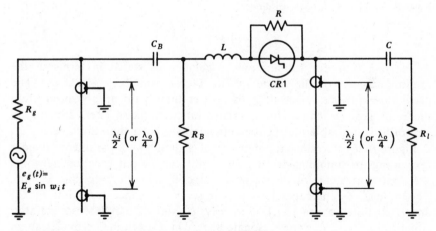

Figure 6-34. Microwave carrier-storage frequency divider.

operate well into the GHz region. Lack of rf output in the absence of an input signal is another advantage. Also, the divider can be made to operate over an octave band (Ref. 25). However, division ratios higher than 2 are attainable at the cost of a narrow band of operation and poor phase stability. Under these conditions, it is not practical to make the circuit a variable-ratio divider.

Divide-by-2 Parametric Frequency Divider

The divide-by-2 parametric divider can be described by a negative conductance generator and input/output isolation networks, as shown in Fig. 6-35. An amplifier-driver provides the required rf power.

The principal element of the negative conductance generator is a voltage-variable capacitance, a varactor diode, which presents a negative conductance to an external circuit tuned to f_o whenever a rf voltage at twice the frequency is applied across the diode. If the losses in the diode are less than this negative conductance, subharmonic oscillations take place and rf power at f_o is delivered to the load.

The reasons for using input and output isolation networks are given in the preceding section on the carrier-storage divider.

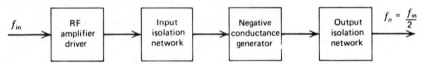

Figure 6-35.   Parametric frequency divider.

The condition for subharmonic oscillation is stated in terms of the minimum value of the diode quality factor at a bias point, Q_B. The oscillation will take place if Q_B is greater than 4 for an abrupt-junction diode or greater than 6 for a graded-junction diode (Ref. 26), but in general it is desirable that Q_B be appreciably greater than these values.

Figure 6-36 is a schematic diagram of a divide-by-2 parametric divider (excluding the amplifier-driver). The input and output isolation networks are dual equivalents of the networks described in the preceding section. The inductor L_1 and capacitor C_1 are in series resonance at f_i, while L_1, C_1, and C_2 parallel-resonate at f_o. Similarly, L_2 and C_3 parallel-resonate at f_i, while L_2, L_3, and C_3 series-resonate at f_o. The capacitor C provides an ac short at f_o. The expressions defining these quantities are as follows:

$$L_1 = \frac{1}{(2\pi f_i)^2 C_1}, \tag{6-86}$$

$$C_2 = \frac{C_1}{L_1 C_1 (2\pi f_o)^2 - 1}, \tag{6-87}$$

$$L_2 = \frac{1}{(2\pi f_i)^2 C_3} , \tag{6-88}$$

$$L_3 = \frac{L_2}{L_2 C_3 (2\pi f_o)^2 - 1} , \tag{6-89}$$

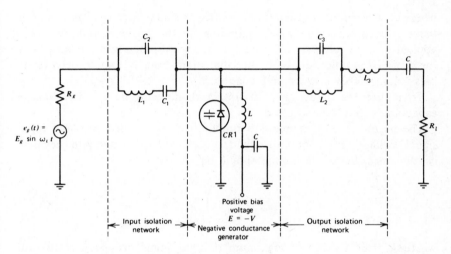

Figure 6-36.   Parametric divider; schematic diagram.

and

$$\frac{1}{2\pi f_0 C} \leqslant 0.1\Omega. \tag{6-90}$$

As in the case of the carrier-storage divider, for these networks to function properly L_1, L_2, L_3, C_1, C_2, and C_3 should be high-quality-factor components with self-resonance well above f_i.

The discussion that follows is based on the material presented in Ref. 26. The choice of the varactor diode used in the negative conductance generator is based on the Q_B requirements stated above. The inductor L parallel-resonates with the total diode capacitance, C_T, at the frequency of the output wave, f_o, that is,

$$L = \frac{1}{(2\pi f_o)^2 C_T} , \tag{6-91}$$

where C_T is defined in the preceding section.

The diode capacitance, $C(V)$, can be expressed in terms of a known capacitance as

$$C(V)=C_s\left(\frac{\phi-V_s}{\phi-V}\right)^n,\qquad(6\text{-}92)$$

where C_s is the diode capacitance specified at some dc bias voltage, V_s, ϕ is the diode contact potential (approximately 0.7 V), V is an externally applied dc bias voltage, and $n=\frac{1}{2}$ for an abrupt-junction diode and $n=\frac{1}{3}$ for a graded-junction diode. Notice that, since both V_s and V are reverse-bias voltages, they are negative quantities.

To realize the value of L at uhf, a diode with the lowest package capacitance should be selected.

The negative conductance is a function of the applied bias and input signal voltage. For optimum divider performance (maximum negative conductance) the diode dc bias should be

$$-V=\left(\frac{n+1}{n}\right)V_r-\phi,\qquad(6\text{-}93)$$

where V_r = the voltage of maximum rf input signal excursion relative to the contact potential (see Fig. 6-37).

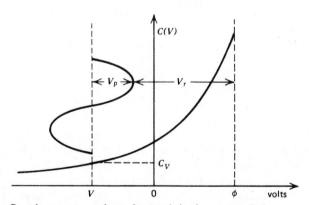

Figure 6-37.   Capacitance versus voltage characteristic of a varactor diode.

This value of bias voltage is not critical and can be varied over a 3-to-1 range with less than a 10% loss in negative conductance. Under this optimum condition the negative conductance is

$$G_- = \frac{\pi f_o n C_V}{n+1}.$$ \hfill (6-94)

Hence, by increasing C_V (i.e., by operating at a lower bias voltage), one improves the divider performance. Operating at a very low bias voltage is, however, undesirable, because it is likely that the diode will be driven into the forward conduction region, which is characteristic of high losses. The final value of bias voltage is determined experimentally and usually is a compromise between the highest Q_B, the highest C_V, and the available input rf power.

To achieve the maximum negative conductance, the required rf power at f_i is

$$P_i = \frac{4\pi f_o C_V}{Q_B}(V_p)^2,$$ \hfill (6-95)

where V_p = peak value of the voltage at f_i appearing across the diode.

If one assumes that the input and output isolation networks present negligible losses to the waves at f_i and f_o, respectively, the divider efficiency is

$$\eta = 1 - \frac{2}{nQ_B}\left(\frac{\phi - V}{V_p}\right).$$ \hfill (6-96)

This circuit is very similar to the carrier-storage divider described in the preceding section and exhibits the same advantages and limitations.

Phase-Locked Loop Frequency Divider

The functional block diagram of a PLL divider is shown in Fig. 6-38. The PLL utilizes a frequency multiplier in the feedback path. This circuit does not have any outstanding advantages as compared to other dividers described above. Moreover, it is relatively expensive and is presented here only to point out that, in combination with other circuits added in series with a times-N multiplier such as a mixer, the PLL may be designed to perform a number of functions simultaneously, a convenient feature. The analog PLL design is considered in great detail in Chapter 4.

6-5 Voltage-Controlled Oscillators

In this section of Chapter 6 we present the design of voltage-controlled oscillators (VCOs) used in phase-locked loops. The Colpitts oscillator is chosen for this purpose; however, the approach can be applied to the

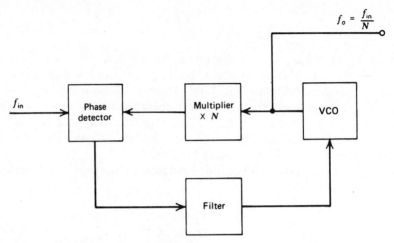

Figure 6-38.   Phase-locked loop frequency divider.

design of any other oscillator type with equal success. It is assumed that the reader is familiar with PLL terminology and with the material of Chapters 2, 4, and 5.

Colpitts Oscillator

Figure 6-39 is a block diagram of an oscillator. The amplifier rf gain is A, and the network feedback factor is β. The oscillation is sustained when the

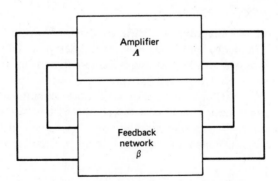

Figure 6-39.   Functional block diagram of an oscillator.

loop gain at the fundamental frequency is unity and the phase shift is zero, that is, when

$$A\beta = 1 + j0. \tag{6-97}$$

For the purpose of analysis, we select a feedback network consisting of two capacitors and an inductor, which are arranged as shown in Fig. 6-40.

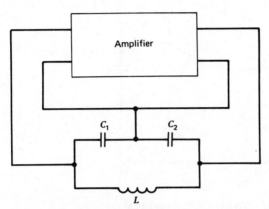

Figure 6-40.   The Colpitts oscillator.

The schematic diagrams of a transistor and a FET oscillator based on this design are shown in Figs. 6-41 and 6-42, respectively. Grounded-base and grounded-gate configurations are chosen for good temperature stability. The dc bias of the circuits is determined by the rf output power requirement. For the transistor oscillator it is determined in the same manner as the bias of an rf amplifier, previously described. (The reader is referred to Refs. 27 and 28 for details of the FET dc biasing procedures.) The combination of R_F and C_F forms a low-pass filter for $B+$ isolation. The capacitor C in Fig. 6-41 is an ac bypass capacitor. The frequency of oscillations can be expressed in terms of the feedback network parameters as

$$f_0 \cong \frac{1}{2\pi \sqrt{LC_{\text{tot}}}}, \tag{6-98}$$

where

$$C_{\text{tot}} \cong C_{\text{tune}} + \frac{C_1 C_2}{C_1 + C_2}, \tag{6-99}$$

and where C_{tune} is a trimmer capacitor. The necessary condition for oscillation is (Ref. 29)

$$h_{fe} \geqslant \frac{C_2}{C_1} \tag{6-100}$$

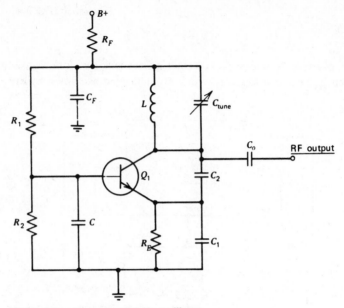

Figure 6-41.   Transistor Colpitts oscillator.

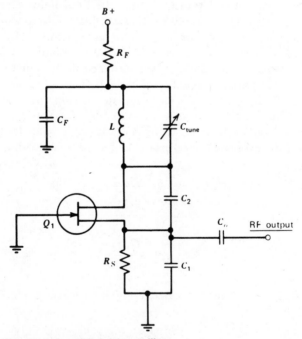

Figure 6-42.   FET Colpitts oscillator.

Compared to a transistor oscillator, FET oscillator biased for zero dc drift displays superior performance with respect to frequency drift with temperature. For this reason we use the FET oscillator in this discussion. Figure 6-43 is a schematic diagram of the oscillator with error voltage and tuning voltage networks added to the circuit. The networks consist of two voltage-variable-capacitance diodes, varicaps, and associated bias circuitry. The error voltage varicap, $CR1$, is reverse-biased at a dc voltage that is generated by the phase detector (or phase comparator) of the phase-locked loop; each capacitor C is used either for ac coupling or for ac bypassing (i.e., it presents a negligibly small impedance at the frequency of oscillation); and the resistor R_1 provides a high impedance at that frequency. The tuning voltage varicap, CR2, is biased in a similar manner at a dc voltage generated by a D/A converter or some other tuning device. The tuning voltage varicap coarse-tunes the VCO to within the PLL capture range, while the error voltage varicap provides the fine-tuning control necessary to keep the VCO frequency-locked to a reference source. Two separate controls are used to prevent a significant change in the tuning-voltage-varicap conversion gain, as the VCO is coarse tuned, from affecting the PLL gain. When a VCO is tuned over a narrow frequency band, this effect is small, and only one control circuit is employed.

The varicap capacitance can be expressed in terms of the bias voltage as given by Eq. 6-92. The expression for the frequency of oscillation, f_0 is

$$f_0 = \frac{1}{2\pi\sqrt{LC_T}}. \tag{6-101}$$

where

$$C_T = C_{\text{osc}} + C_V, \tag{6-102}$$

and C_{osc} = capacitance that tunes with the inductor L at f_0, excluding the capacitance of the varicap,
C_V = varicap capacitance at a reverse bias voltage V.

It can be shown that for $n = \frac{1}{2}$ and for a small frequency change

$$\Delta C_V = -\left(\frac{\Delta f_0}{2L\pi^2 f_0^3}\right) \tag{6-103}$$

and

$$\frac{\Delta f_0}{\Delta V} = \frac{C_s\sqrt{\phi - V_s}}{8\pi\sqrt{L}\left[C_T(\phi - V)\right]^{3/2}}. \tag{6-104}$$

The usefulness of Eqs. 6-101 to 6-104 will become evident as the material of this section is presented.

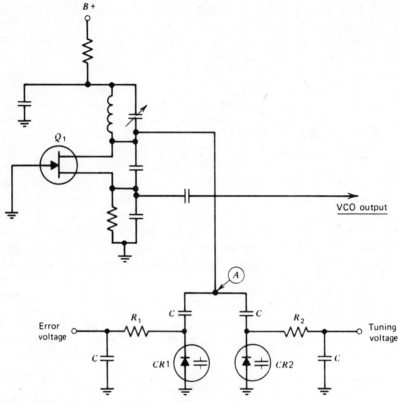

Figure 6-43.   LC voltage-controlled oscillator.

Consider a numerical example. A tuning-voltage and an error-voltage-varicap circuits are to be designed for a VCO meeting the following requirements:

1. Frequency range: 100 to 120 MHz.
2. VCO gain constant, K_{VCO}:   400 kHz/V minimum at the
                                      nominal error voltage of 10 V.

3. Discrete FM spurious outputs: −60 dB for offset frequencies
                                      between 60 Hz and 10 kHz;
                                      −80 dB for offset frequencies
                                      above 10 kHz.

4. Phase-noise level due to   −100 dB/Hz bandwidth at 10
   both tuning networks: kHz offset frequency.

Assume that the total capacitance of the VCO tuned circuit is 30 pF at 120 MHz, that is, $C_{T,120} = 30$ pF. Then, according to Eq. 6-101, $L = 0.059$ μH. At 100 MHz the total capacitance that tunes with 0.059 μH is 43 pF. Hence the tuning voltage varicap has to provide a change in capacitance equal to

$$\Delta C_T = C_{T,100} - C_{T,120} = 13 \text{ pF}. \tag{6-105}$$

The variables to be considered in selecting a varicap are the available tuning voltage, the value of the varicap capacitance at the highest tuning voltage, the varicap quality factor, and the diode temperature stability. In low-phase-noise applications tuning voltage selection is governed by the phase-noise requirement, and the tuning voltage varicap is operated at the highest possible voltage. Such an approach results in low varicap conversion gain, high quality factor, and good temperature stability. For illustrative purposes assume that the varicap is operated between approximately 5 and 25 V.

We select varicap type SQ1738 from Table 6-3 as our first try. The capacitance of this diode at $V_s = -4$ V bias voltage is $C_s = 33$ pF. At the tuning voltage of -25 V the varicap capacitance, Eq 6-92, is

$$C_{25} = 33 \sqrt{\frac{0.7 + 4}{0.7 + 25}} \cong 14.1 \text{ pF},$$

and, for the required capacitance change of 13 pF, the diode capacitance at the bias voltage, which has yet to be determined, is

$$C_V = C_{25} + \Delta C_T = 27.1 \text{ pF}.$$

Solving for V in Eq. 6-92 results in

$$-V = (\phi - V_s)\left(\frac{C_s}{C_V}\right)^2 - \phi. \tag{6-106}$$

Hence

$$-V = (0.7 + 4)\left(\frac{33}{27.1}\right)^2 - 0.7,$$

$$V \cong -6.3 \text{ V dc}.$$

We select the error voltage varicap in a similar manner, but in this case Eq. 6-104 is used because of a small change in capacitance required to

Table 6-3. Capacitance, Tuning Ratio, and Figure of Merit of Series SQ1716 to SQ1750 Voltage-Variable Capacitance Diodes Manufactured by MSI Electronics, Inc.[a]

	C_s, Diode capacitance[b] (pF) $V_s = 4$ V dc, $f = 1$ MHz			TR, Tuning ratio C_2/C_{30} $f = 1$ MHz	Q, Figure of merit $V_s = 4$ V dc, $f = 50$ MHz
Device	Minimum	Nominal	Maximum	Minimum	Minimum
SQ1716	2.7	3.0	3.3	2.40	1200
SQ1717	3.2	3.6	4.1	2.40	1200
SQ1718	4.0	4.5	5.1	2.50	1000
SQ1719	5.0	5.6	6.2	2.50	1000
SQ1720	6.1	6.8	7.5	2.70	1000
SQ1722	7.4	8.2	9.0	2.80	1000
SQ1724	9.0	10.0	11.0	2.80	1000
SQ1726	10.8	12.0	13.2	2.80	900
SQ1728	13.5	15.0	16.5	2.80	900
SQ1730	16.2	18.0	19.8	2.90	900
SQ1732	18.0	20.0	22.0	2.90	800
SQ1734	19.8	22.0	24.2	2.90	800
SQ1736	24.3	27.0	29.7	2.90	800
SQ1738	29.7	33.0	36.3	2.90	700
SQ1740	35.1	39.0	42.9	2.90	600
SQ1742	42.3	47.0	51.7	2.90	500
SQ1744	50.4	56.0	61.6	2.90	450
SQ1746	61.2	68.0	74.8	2.90	300
SQ1748	73.8	92.0	90.2	2.90	300
SQ1750	90.0	100.0	110.0	2.90	300

[a]Courtesy of MSI Electronics, Inc.
[b]To order devices with C_s nominal values of $\pm 5.0\%$ or $\pm 2.0\%$, add suffix B or C, respectively.

obtain a VCO gain of 400 kHz/V. The computations are performed at 100 MHz because K_{VCO} is smaller at the low than at the high end of the frequency band. The type SQ1726 diode is selected. The capacitance of the diode at -4 V bias is 12 pF. The parameters used are as follows:

$$L = 0.059 \mu H, \qquad C_s = 12 \text{ pF},$$

$$C_T = 43 \text{ pF}, \qquad V_s = -4 \text{ V},$$

$$\phi = 0.7 \text{ V}, \qquad V = -10 \text{ V}.$$

Substituting these values in Eq. 6-104, we obtain

$$K_{VCO} = \frac{\Delta f_0}{\Delta V} = \frac{12 \times 10^{-12} \sqrt{0.7 + 4}}{8 \times 3.14 \sqrt{0.059 \times 10^{-6}} \left[43 \times 10^{-12} (0.7 + 10) \right]^{3/2}},$$

$K_{VCO, 100} = 420 \text{ kHz/V}.$

At 120 MHz, $C_T = 30 \text{ pF}$. Hence

$$K_{VCO, 120} = 730 \text{ kHz/V}.$$

The error voltage varicap seldom presents any discrete FM or phase-noise problems. It is usually the tuning-voltage-varicap circuit that has to be analyzed for the feasibility of allowable phase-noise and spurious-signal levels. Such analysis is performed next.

First we compute the tuning-voltage-varicap conversion gain at the low and high ends of the frequency band. The values of the parameters used in Eq. 6-104 at 100 MHz are as follows:

$$L = 0.059 \ \mu H, \qquad C_s = 33 \text{ pF},$$
$$C_T = 43 \text{ pF}, \qquad V_s = -4 \text{ V},$$
$$\phi = 0.7 \text{ V}, \qquad V = -6.3 \text{ V}$$

Hence

$$\frac{\Delta f_o}{\Delta V}\bigg/_{100 \text{ MHz}} = \frac{33 \times 10^{-12}\sqrt{0.7+4}}{8 \times 3.14\sqrt{0.059 \times 10^{-6}}\left[43 \times 10^{-12}(0.7+6.3)\right]^{3/2}}$$

$$= 3.91 \text{ MHz/V}.$$

The parameters used at 120 MHz are the same as those employed at 100 MHz with the exception of

$$C_T = 30 \text{ pF}, \qquad V = -25 \text{ V}.$$

This results in

$$\frac{\Delta f_o}{\Delta V}\bigg|_{120 \text{ MHz}} = 550 \text{ kHz/V}.$$

It is worthwhile to compare the tuning-voltage-varicap conversion gain at 100 MHz to that at 120 MHz. The difference in decibels is

$$\text{change in varicap conversion gain (dB)} = 20\log_{10}\left(\frac{3.91}{0.55}\right) \quad = 17.$$

It is expected, therefore, that at the output of the VCO the phase noise and the levels of discrete FM components due to the tuning-voltage-varicap circuit will increase by 17 dB as the VCO is tuned from 120 to 100 MHz, even though the noise voltage and the levels of spurious signals at the input of the varicap circuit remain unchanged. The preliminary analysis that follows is carried out at the frequency of the highest $\Delta f_o / \Delta V$, that is, at 100 MHz.

Next we compute the allowable voltage level of discrete spurious signals that may appear at the input of the tuning-voltage-varicap circuit, V_{spur}. The requirement for spurious outputs is -60 dB at offset frequencies between 60 Hz and 10 kHz. Using Eq. 2-19, we arrive at the allowable voltage level as follows:

$$-60 \text{ dB} = 20 \log_{10} \left(\frac{\Delta f_{\text{rms}}}{\sqrt{2} \, f_m} \right)$$

Hence

$$\frac{\Delta f_{\text{rms}}}{\sqrt{2} \, f_m} = 10^{-3}$$

or

$$\Delta f_{\text{rms}} = (1.414 \times 10^{-3}) f_m.$$

Hence the allowable level at 100 MHz is

$$V_{\text{spur},100} = \frac{\Delta f_{\text{rms}}}{\Delta f_o / \Delta V |_{100 \text{ MHz}}} = \frac{1.414 \times 10^{-3}}{3.91 \times 10^6} f_m$$

$$= (3.62 \times 10^{-10}) f_m \text{ V rms.} \qquad (6\text{-}107)$$

At $f_m = 60$ Hz, $V_{\text{spur},100} = 0.02$ μV rms, a value that is not likely to be achieved with practical shielding and circuit layout techniques in the presence of strong magnetic fields at the power line frequency encountered in normal operating conditions.

The FM spurious-signal requirement is -80 dB for offset frequencies above 10 kHz. Hence

$$-80 \text{ dB} = 20 \log_{10} \left(\frac{\Delta f'_{\text{rms}}}{\sqrt{2} \, f_m} \right),$$

and therefore

$$\frac{\Delta f'_{\text{rms}}}{\sqrt{2}\, f_m} = 10^{-4}$$

or

$$\Delta f'_{\text{rms}} = (1.414 \times 10^{-4}) f_m,$$

so that, at 100 MHz and for $f_m \geqslant 10$ kHz, the allowable level of FM spurious signals is

$$V'_{\text{spur,100}} = \frac{\Delta f'_{\text{rms}}}{\Delta f_o / \Delta V|_{100\,\text{MHz}}} = \left(\frac{1.414 \times 10^{-4}}{3.91 \times 10^6} \right) f_m$$

or

$$V'_{\text{spur,100}} = (3.62 \times 10^{-11}) f_m \ \text{V rms.}$$

At $f_m = 10$ kHz, $V_{\text{spur,100}} = 0.362$ μV rms, which is a relaxed but still quite stringent requirement in comparison to the figure arrived at above for 60 Hz offset frequency. It can be met only if precautions are taken in laying out and shielding the VCO and associated circuitry.

Before discussing possible solutions of the problem, let us evaluate the phase-noise requirement. Using Eq. 2-59, we proceed as

$$-100 \ \text{dB/Hz bandwidth} = 20 \log_{10} \left(\frac{\Delta f_{\text{rms},n}}{\sqrt{2}\, f_m} \right),$$

and therefore

$$\frac{\Delta f_{\text{rms},n}}{\sqrt{2}\, f_m} = 10^{-5}$$

or

$$\Delta f_{\text{rms},n} = (1.414 \times 10^{-5}) f_m \ \text{Hz/Hz bandwidth.}$$

At 100 MHz the allowable level of noise voltage at the input of the tuning-voltage-varicap circuit, V_n, is

$$V_{n,100} = \frac{\Delta f_{\text{rms},n}}{\Delta f_o / \Delta V|_{100\,\text{MHz}}} = \left(\frac{1.414 \times 10^{-5}}{3.91 \times 10^6} \right) f_m \qquad (6\text{-}108)$$

or

$$V_{n,100} = (3.61 \times 10^{-12}) f_m \text{ V rms/Hz bandwidth.}$$

At $f_m = 10$ kHz, $V_{n,100} = 3.61 \times 10^{-8}$ V rms/Hz bandwidth. On the assumption that the equivalent input impedance of the varicap circuit is 150 kΩ and is mainly resistive at low frequencies (see Chapter 2 on noise in oscillators), the allowable noise power is

$$P_n = \frac{(V_{n,100})^2}{R_{eq}} = \frac{(3.61 \times 10^{-8})^2}{0.15 \times 10^6} \cong 8.67 \times 10^{-21} \text{ W/Hz,} \qquad (6\text{-}109)$$

or, expressed in dBm, it is

$$10 \log_{10} \left(\frac{8.67 \times 10^{-21}}{10^{-3}} \right) = -171 \text{ dBm/Hz.}$$

With a practical limit of -164 dBm/Hz, the phase-noise requirement will be exceeded by the circuit by 7 dB at the low end of the band. (At the high end of the band one may expect a 17 dB improvement.) The tendency may be to reduce R_{eq}; see Eq. 6-109. This, however, may result in lowering the varicap quality factor and in noise enhancement. Reducing R_{eq} may also have an effect on the acquisition time of the phase-locked loop in which this VCO is used.

This problem of VCO tuning is solved by reducing the frequency range that is covered by the tuning voltage varicap and by electronically switching in fixed values of tuning capacitors as shown in Fig. 6-44. In band 1, all diodes $CR1$ to $CR(n-1)$ are off and the circuit is effectively the same as in Fig. 6-43. When band 2 operation is desired, diode $CR1$ is switched on, which places $C_{tune,2}$ across the oscillator tuned circuit. Band 3 operation requires $CR2$ to be switched on and $CR1$ to be switched off, or left on if desired, and so forth. Each resistor R_c, which is bypassed by the corresponding capacitor C, is used to limit the current supplied to the conducting diode by the power supply. Each inductor L is a rf choke.

The device that performs switching is a PIN diode. The rf resistance of a PIN diode is current controlled. Figure 6-45 shows a typical rf resistance versus dc bias current characteristic of a Hewlett-Packard PIN diode. When no dc current flows in the diode, the diode rf resistance is high. This resistance is very low, however, when a 100 mA dc current is supplied to the device, biasing the diode in the forward direction. Other manufacturers, such as Eartech and Motorola Semiconductor, offer diodes that operate at 20 and 10 mA and exhibit resistance as low as a fraction of an ohm.

Having such a powerful technique of noise reduction at our disposal, we

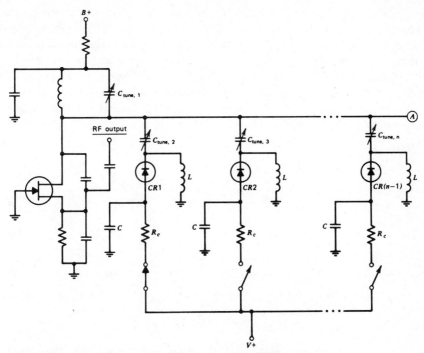

Figure 6-44. Low-phase-noise LC voltage-controlled oscillator.

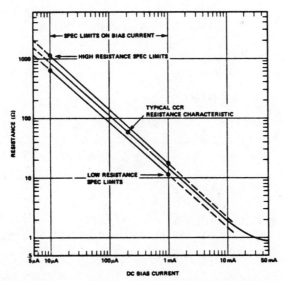

Figure 6-45. Radio-frequency resistance characteristic of the PIN diode, HP type 5082-3004. Courtesy of Hewlett-Packard Company.

design the tuning-voltage-varicap circuit so as to result in the maximum varicap conversion gain being at least 7 dB lower than the gain computed above, that is, at any frequency between 100 and 120 MHz the maximum conversion gain of the tuning voltage varicap should not exceed $3.91 \times 10^6 \times 0.4467 = 1.75$ MHz/V. Let $(\Delta f_o / \Delta V)_{max} = 1.0$ MHz/V, giving us a margin of approximately 5 dB. Using the same tuning voltage varicap, type SQ1738 with $C_s = 33$ pF and $V_s = -4$ V, and working at the low end of each newly formed frequency band, we proceed as follows.

Band 1:

$$\left.\frac{\Delta f_o}{\Delta V}\right|_{100 \text{ MHz}} = 1.0 \text{ MHz/V} = \frac{33 \times 10^{-12}\sqrt{4.7}}{8 \times 3.14\sqrt{5.9 \times 10^{-8}}\left[43 \times 10^{-12}(0.7 - V)\right]^{3/2}},$$

$$V = -11.3 \text{ V}$$

At $V = -11.3$ V, $C_{11.3} = 33\sqrt{4.7/12} = 20.7$ pF. Similarly, at $V = -25$ V, $C_{25} = 33\sqrt{4.7/25.7} = 14.1$ pF. Hence

$$\Delta C_{B1} = 20.7 - 14.1 = 6.6 \text{ pF},$$

so that at the high end of band 1 the total capacitance that tunes with $0.059 \ \mu\text{H}$ is $43 - 6.6 = 36.4$ pF, and the upper frequency of oscillations is

$$f_{H,B1} = \frac{1}{2\pi\sqrt{5.9 \times 10^{-8} \times 36.4 \times 10^{-12}}} = 108.5 \text{ MHz}.$$

Band 2:

$$\left.\frac{\Delta f_o}{\Delta V}\right|_{108.5 \text{ MHz}} = 1.0 \text{ MHz/V}$$

$$= \frac{33 \times 10^{-12}\sqrt{4.7}}{8 \times 3.14\sqrt{5.9 \times 10^{-8}}\left[36.4 \times 10^{-12}(0.7 - V)\right]^{3/2}},$$

$$V = -13.5 \text{ V}$$

At $V = -13.5$ V, $C_{13.5} = 33\sqrt{4.7/14.2} = 19$ pF, and since $C_{25} = 14.1$ pF, $\Delta C_{B2} = 19.0 - 14.1 = 4.9$ pF. Hence, at the high end of band 2, the total capacitance that tunes with $0.059 \ \mu\text{H}$ is $36.4 - 4.9 = 31.5$ pF, and the upper frequency is

$$f_{H,B2} = \frac{1}{2\pi\sqrt{5.9 \times 10^{-8} \times 31.5 \times 10^{-12}}} = 116 \text{ MHz}.$$

Band 3:

$$\frac{\Delta f_o}{\Delta V}\bigg|_{116\,\text{MHz}} = 1.0 \text{ MHz/V}$$

$$= \frac{33 \times 10^{-12}\sqrt{4.7}}{8 \times 3.14\sqrt{5.9 \times 10^{-8}}\left[31.5 \times 10^{-12}(0.7 - V)\right]^{3/2}},$$

$$V = -15.7 \text{ V}$$

We can deduce from the computations performed above that, to meet the phase-noise requirement, the VCO would have to be tuned in three bands. In a practical design one would select bands so as to result in approximately the same tuning voltage range for all bands. This task is left to the interested student.

Having completed the analysis of the tuning-voltage-varicap circuit based on the phase-noise requirement, we shall examine the improvement in the allowable level of discrete spurious signals resulting from this analysis. Originally, we arrived at a level of 0.02 μV rms at 60 Hz offset frequency, based on a varicap conversion gain of 3.91 MHz/V. By reducing this gain to 1 MHz/V, we also relaxed the spurious-signal requirement to $0.02 \times 3.91 = 0.0782$ μV rms. This is still a very stringent requirement. The varicap conversion gain should be further reduced to 320 kHz/V, or by an additional 10 dB.

Looking at the figures of the VCO gain, K_{VCO}, we conclude that the value of 400 kHz/V minimum is not acceptable either; that is, the VCO gain requirement conflicts with the discrete FM spurious-output requirement, and a different error voltage varicap has to be selected. We decide on the SQ1718 diode, which gives us $K_{\text{VCO}, 100} = 123.5$ kHz/V and $K_{\text{VCO}, 120} = 345$ kHz/V. If for reasons of system performance it is necessary to keep the PLL gain as large as it is with $K_{\text{VCO}} = 400$ kHz/V, the phase-detector gain should be increased by a ratio of 400 to 123.5, or by 3.25.

Transmission Line VCO

It was demonstrated in Chapter 2 that the lower the loaded quality factor of the oscillator tuned circuit, Q_l, the poorer is the phase-noise spectrum of that oscillator. It is because of a relationship between the phase noise and Q_l of an oscillator that implementation of paper design of low-phase-noise VCOs operating at frequencies up to 1 GHz is associated with a problem: a poor quality factor of the VCO tuned circuit inductor. This problem can be solved at frequencies above approximately 80 MHz by utilizing a quarter-wave transmission line technique of producing inductance. The principle involved is not, indeed, a novel one; it has been successfully applied to frequency pulling of crystal-controlled oscillators (see Section 6-6). A quarter-wave network is used as a reactance transformer which converts, with little effect on reactance losses, a capacitive reactance connected to one end of the network to an inductive reactance at the other end. This permits the making of a high-Q inductance by terminating a quarter-wave transmission line with a high-Q capacitance. The approach is simple to deal with analytically, inexpensive to implement, and easily adaptable to miniaturization.

The expressions used in the design of a quarter-wave transmission line tuning-voltage-varicap circuit are

$$\lambda = \frac{\lambda_0}{\sqrt{\epsilon_r}} , \qquad (6\text{-}110)$$

where λ = the wavelength in a medium (ft),
λ_0 = the wavelength in free air (ft),
ϵ_r = the dielectric constant of the medium relative to air (2 to 2.05 for teflon at frequencies between 60 Hz and 30 GHz),
and

$$\lambda_0 = \frac{984}{f_0} , \qquad (6\text{-}111)$$

where λ_0 is in feet if f_o is in megahertz. And, finally, for a lossless quarter-wave transmission line

$$Z_0 = \sqrt{X_C X_L} , \qquad (6\text{-}112)$$

where Z_0 = the characteristic impedance of the transmission line,
X_C = the capacitive impedance terminating the line, $1/2\pi f_o C_{ter}$,
X_L = the effective inductive impedance seen at the unterminated end of the line, $2\pi f_o L$.
Hence

$$L = C_{ter} Z_0^2 . \qquad (6\text{-}113)$$

Equations 6-110 and 6-111 are used in determining the length of the transmission line, and Eq. 6-113 gives the approximate change in capacitance required to produce a desired change in inductance. Although Eq. 6-113 is valid for a lossless line at one frequency only, a preliminary paper design and a noise analysis of an oscillator operating over a frequency range as wide as 30% can be carried out to serve as a guide in the development and optimization of the oscillator.

Transmission line VCOs display temperature-stability and phase-noise performance superior to that of LC oscillators.

As an example, consider the VCO requirements stated above. The line length at any frequency between 100 and 120 MHz should be a quarter-wavelength or shorter, so that λ is computed at the upper end of the band, that is, at 120 MHz. Having selected a 50 Ω semirigid coaxial cable, such as type UT-85C manufactured by Uniform Tubes Company, as our transmission line, we proceed as follows:

$$\lambda_0 = \frac{984}{120} = 8.2 \text{ ft,}$$

and for $\epsilon_r = 2$

$$\lambda = \frac{8.2}{\sqrt{2}} = 5.8 \text{ ft,}$$

$$\frac{\lambda}{4} = \frac{5.8}{4} \cong 17.5 \text{ in.}$$

We have selected $C_T = 30$ pF and $L = 0.059$ μH to resonate at 120 MHz. At the upper end of the band, the transmission line terminating capacitance, therefore, should be

$$C_{ter, 120} = \frac{0.059 \times 10^{-6}}{(50)^2} = 23.6 \text{ pF.}$$

At the low end of the band, the required inductance which resonates with 30 pF is

$$L = \frac{1}{(6.28 \times 10^8)^2 \times 30 \times 10^{-12}} = 0.0845 \mu H,$$

so that the terminating capacitance is

$$C_{ter, 100} = \frac{0.0845 \times 10^{-6}}{(50)^2} = 33.8 \text{ pF.}$$

The tuning-voltage varicap has to provide a capacitance change equal to

$$\Delta C_{ter} = C_{ter, 100} - C_{ter, 120} = 10.2 \text{ pF.}$$

From this step on, the analysis of the tuning-voltage-varicap circuit is carried out in the same manner as previously described, except that Eqs. 6-103 and 6-104 have to be expressed in terms of the transmission line characteristic impedance as

$$\Delta C_V = -\left(\frac{\Delta f_o}{2C_T(\pi Z_0)^2 f_o^3}\right),\qquad(6\text{-}114)$$

where C_T is the total capacitance of the VCO tuned circuit, and

$$\frac{\Delta f_o}{\Delta V} = \frac{1}{8\pi Z_0\left[(\phi - V_s)(C_T C_s)^2(\phi - V)^3\right]^{1/4}}.\qquad(6\text{-}115)$$

Figure 6-46 is a schematic diagram of a transmission line VCO. The section of the circuit to the left of point B is the same as the section of the circuit to the left of point A in Fig. 6-43, except that the dc source resistance, R_S, is rf bypassed, a low-loss rf choke, L_s, is added to provide a high source impedance, and the error-voltage-varicap decoupling resistor, R_1, is replaced by a low-loss rf choke, L_1. To the right of point B we have the transmission line tuned-inductance circuit. The inductors L_2 and L_3 are low-loss rf chokes. For very low phase-noise operation we divide the specified frequency band into a number of subbands as is shown in Fig. 6-47. Notice that the bands-switching network could be connected directly

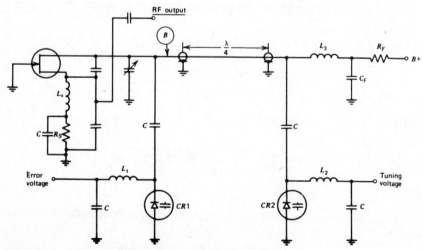

Figure 6-46.   Transmission line voltage-controlled oscillator.

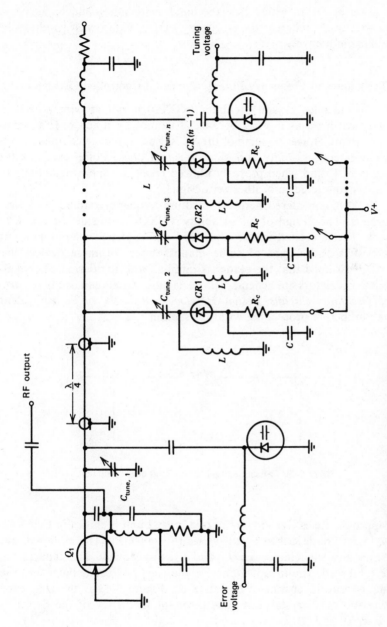

Figure 6-47. Low-phase-noise transmission line oscillator.

397

to the drain of Q_1, in which case the band switching would be accomplished capacitively, whereas the frequency tuning within each band would still be done inductively.

6-6 Techniques of Frequency Pulling of Crystal-Controlled Oscillators

A voltage-controlled crystal oscillator (VCXO) is not as widely used in frequency synthesis as a VCO because it cannot be tuned over a broad frequency band. However, some of the VCXO performance characteristics, such as frequency stability and close-to-signal phase noise, excel by far those of a VCO and when properly utilized result in superior solutions to problems encountered in synthesizer design.

The VCXOs considered here are medium-frequency-stability and low-cost circuits. The design of high-stability VCXOs is outside the scope of this book. The interested reader is referred to Chapter 8, where important characteristics and names of some manufacturers of these devices are given. It is assumed that the reader is familiar with the design of crystal-controlled oscillators in general. (If he is not, he should study *Quartz Crystal Oscillator Circuits Design Handbook*, Ref. 30, or an equivalent publication, before proceeding with this section.)

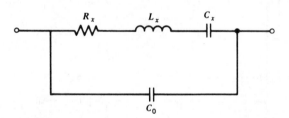

Figure 6-48.   Simplified equivalent circuit of a crystal unit.

An approximate equivalent circuit of a crystal unit is shown in Fig. 6-48. Crystal losses are described by the series resistance R_x. The motional-arm inductance and capacitance are L_x and C_x, respectively. The capacitance C_0, the total static shunt capacitance, is the electrode capacitance and the capacitance due to crystal case and leads. Figure 6-49 is the frequency characteristic of a crystal unit. The resonant frequency of the device is given, Ref. 31, as

$$f_r = f_s - \frac{R_x^2}{4\pi L_x X_{C_0}} \qquad (6\text{-}116)$$

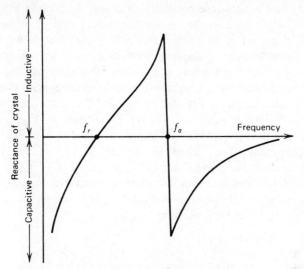

Figure 6-49.   Frequency characteristic of a crystal.

where

$$f_s = \frac{1}{2\pi\sqrt{L_x C_x}} \tag{6-117}$$

and

$$X_{C_0} = -\frac{1}{2\pi f_r C_0}. \tag{6-118}$$

The antiresonant (i.e., the parallel resonant) frequency occurs at

$$f_a = f_s + \frac{|X_{C_0}|}{4\pi L_x}. \tag{6-119}$$

The quality factor of the series-resonant circuit is

$$Q_x = \frac{2\pi f_r L_x}{R_x}. \tag{6-120}$$

The region of interest in the VCXO design lies below f_a. The technique of crystal frequency pulling depends on the oscillator circuit configuration. For example, if the crystal is used to replace the inductor in the feedback network of the Colpitts-type oscillator discussed in Section 6-5, frequency pulling can occur only in the region between f_r and f_a. A better circuit configuration is that in which the crystal is utilized at f_r. By shifting the crystal resonant frequency above and below f_r, one can pull the oscillator frequency by a significantly larger amount. Frequency pulling is achieved by placing an inductor or a capacitor (or both) in series with the crysal unit with the results shown in Fig. 6-50a and b. This effect, for example, is achieved in the transformer-coupled VCXO shown in Fig. 6-51. In this circuit a crystal unit provides a low-impedance feedback path at the crystal resonant frequency. The values of the inductance of the transformer secondary, L, and the voltage-variable capacitance, $C(V)$, are selected so that the $L - C(V)$ combination is slightly inductive or capacitive, depending on the value of the tuning voltage, V. This results in a positive or negative shift of the crystal resonant frequency.

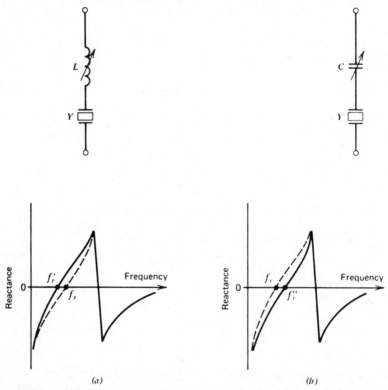

Figure 6-50. Effects of external reactance on crystal resonant frequency. (*a*) Inductive pulling; (*b*) capacitive pulling.

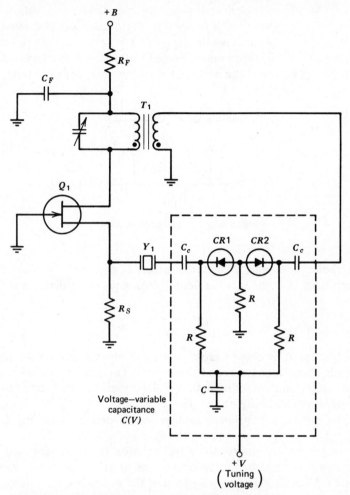

Figure 6-51.   Transformer-coupled wide-pulling-range VCXO.

The variable-voltage capacitance circuit consists of two varicaps, $CR1$ and $CR2$, in back-to-back configuration, two coupling capacitors C_c, which block the tuning voltage and limit the capacitance variation of the varicaps in series with the crystal, decoupling resistors R, and rf bypass capacitor C. At very high frequencies of oscillation each R is replaced by a rf choke.

A somewhat different technique of crystal frequency pulling consists of transforming the series-resonant circuit of a crystal unit into a parallel-resonant circuit used as a collector (or drain) load in an oscillator (Refs. 32

and 33). The network performing the transformation is a quarter-wave transmission line (at vhf) or a lumped-constant quarter-wave network (at hf), such as the π-circuit shown in Fig. 6-52. For simplicity of presentation, the description of this technique is limited to the lumped-constant quarter-wave network; however, the pertinent equations are given for both design cases.

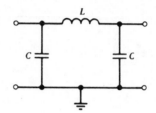

Figure 6-52.   Pi quarter-wave network.

The choice of L and C in the π-circuit is governed by the following considerations. Assume that the desired frequency of oscillations is f_0; then

$$f_0 = \frac{1}{2\pi\sqrt{LC}} . \tag{6-121}$$

Each capacitor C is chosen large enough to absorb any variation of the total static shunt capacitance of the crystal. The inductance L is selected so as to result in minimum reduction of the crystal quality factor caused by the impedance transformation. The effect of L on Q of the parallel equivalent circuit will become evident as the design equations are presented.

The length of the quarter-wave transmission line is determined in the same manner as described in Section 6-5 on voltage-controlled oscillators.

Figure 6-53 is the schematic diagram of the π-network terminated into a crystal unit and an equivalent parallel-resonance circuit of the network. In terms of the design constants, the parameters of the parallel-resonance circuit are (Ref. 32) as follows:

$$L' = \frac{LC_x}{C} = Z_0^2 C_x, \tag{6-122}$$

$$C' = \frac{CL_x}{L} = \frac{L_x}{Z_0^2}, \tag{6-123}$$

$$R' = \frac{Z_0^2 R_p}{Z_0^2 + R_p(R_x + R)}, \tag{6-124}$$

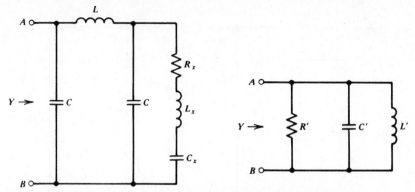

Figure 6-53.   Series resonant circuit to parallel resonant circuit transformation.

where Z_0 = the characteristic impedance of the transmission line or π-network,

  R = the series equivalent resistance of L (or an equivalent transmission line loss).

The characteristic impedance of the π-network is

$$Z_0^2 = \frac{L}{C}.$$

$$(6\text{-}125)$$

The considerations governing a choice of R_p are given later in this section. The quality factor of the equivalent parallel-resonant circuit is

$$Q' = \frac{R'}{2\pi f_0 L'},$$

$$(6\text{-}126)$$

or, in terms of the $L - C$ ratio, it is

$$Q' = \frac{R_p}{2\pi f_0 C_x \left[(L/C) + R_p (R_x + R) \right]}.$$

$$(6\text{-}127)$$

The Q-reduction factor is Q_x / Q', where Q_x is given by Eq. 6-120.

This circuit would present a problem if used as shown in Fig. 6-53. The π-network has a parallel resonance at

$$f_\pi = \sqrt{2}\; f_0.$$

$$(6\text{-}128)$$

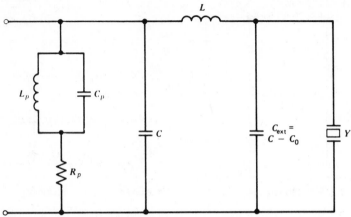

Figure 6-54.   Frequency pulling of a crystal by way of pi quarter-wave network. W. S. Motley, "Frequency-Modulated Quartz Oscillators for Broadcasting Equipment," *IEE Proceedings*, Part b, May 1957 (Vol. 104, No. 15), pp. 239–249.

At this frequency R' is higher than at f_0, because crystal impendance is high and crystal losses can be neglected, so that the circuit will oscillate at f_π rather than at f_r. A lightly coupled tuned circuit that is parallel resonant at f_r is added to the network as shown in Fig. 6-54. The tuned circuit prevents the oscillator from oscillating at f_π. The values of L_p and C_p are determined from

$$f_0 = \frac{1}{2\pi\sqrt{L_pC_p}} \, . \qquad (6\text{-}129)$$

The selection of R_p is based on the requirement for oscillations at f_r and the desired Q-reduction factor. To ensure the oscillations, R_p should be made as small as possible. However, to achieve the desired Q-reduction factor, the value of R_p should be as high as possible. The final value of R_p is arrived at experimentally and is a compromise between a stable operation at f_r and VCXO long- and short-term stability. Frequency pulling is achieved by varying C at the drain side of the π-network. By doing so, we change the susceptance of the equivalent parallel circuit, · that is, the frequency of oscillations. A schematic diagram of a VCXO utilizing a quarter-wave π-network is shown in Fig. 6-55.

The design procedure is best illustrated with a numerical example. Assume that the frequency of oscillations is 6 MHz. The crystal unit

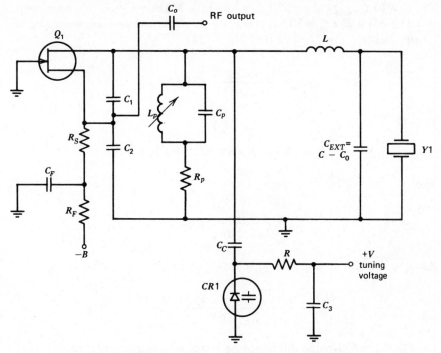

Figure 6-55.   A VCXO utilizing a quarter-wave network for frequency pulling.

parameters are as follows:

$$R_x = 30 \ \Omega,$$

$$L_x = 88 \ \text{mH},$$

$$C_x = 8 \times 10^{-15} \ \text{F},$$

$$C_0 = 1.5 \text{ to } 3.8 \ \text{pF}.$$

Hence

$$Q_x = \frac{6.28 \times 6 \times 10^6 \times 88 \times 10^{-3}}{30} = 110{,}000$$

and

$$LC = \frac{1}{\left(6.28 \times 10^6 \times 6\right)^2} = 7.05 \times 10^{-16}.$$

Let $C \geqslant 10 \, C_{0,\text{nom}}$ (where $C_{0,\text{nom}} = 2.65$ pF). Making $C = 30$ pF, we obtain $L = 23.5 \, \mu$H and $C_{\text{ext}} = 30 - 2.65 = 27.35$ pF. The nearest standard value is 27 pF; hence $C_{\text{ext}} = 27$ pF. From Eq. 6-125

$$Z_0^2 = \frac{23.5 \times 10^{-6}}{30 \times 10^{-12}} = 7.84 \times 10^5,$$

so that

$$L' = 7.84 \times 10^5 \times 8 \times 10^{-15} = 6.26 \, \mu\text{H}$$

and

$$C' = \frac{88 \times 10^{-3}}{7.84 \times 10^5} = 0.112 \, \mu\text{F}.$$

Assume that the measured quality factor of L at 6 MHz is 60. Then

$$R = \frac{2\pi f_0 L}{Q} = \frac{6.28 \times 6 \times 10^6 \times 23.5 \times 10^{-6}}{60} = 14.73 \, \Omega.$$

Let the desired Q-reduction factor be equal to 5 or greater; then

$$Q' = \frac{110{,}000}{5} = 22{,}000$$

and

$$R' = 6.28 \times 6 \times 10^6 \times 22{,}000 \times 6.26 \times 10^{-9} = 5.175 \text{ k}\Omega.$$

Solving for R_p in Eq. 6-124, we have

$$R_p \geqslant \frac{Z_0^2 R'}{Z_0^2 - R'(R_x + R)} \tag{6-130}$$

or

$$R_p \geqslant \frac{7.84 \times 10^5 \times 5.175 \times 10^3}{7.84 \times 10^5 - 5.175 \times 10^3(30 + 14.73)} = 7.35 \text{ k}\Omega.$$

The nearest 5% tolerance standard value of resistance is 7.5 kΩ. Finally, using Eq. 6-129, we compute L_p and C_p as

$$L_p C_p = \frac{1}{\left(6.28 \times 6 \times 10^6\right)^2} = 7 \times 10^{-16}.$$

Let $L_p = 7$ μH; then $C_p = 100$ pF.

6-7 Phase Detectors

Sinusoidal Phase Detector

A commonly used analog phase detector is shown in Fig. 6-56a. It consists of two transformers, T_1 and T_2, which add VCO and reference signal voltages, and two like detectors. The transformers are connected in such a fashion as to result in zero dc output voltage when E_2 is 90 degrees out of phase with E_1 and in a positive or negative voltage whenever the phase difference between E_2 and E_1 is anywhere between 0 and 180 degrees, excluding 90 degrees.

The operation of the phase detector is best described in terms of vector addition of these voltages, Fig. 6-56b. The upper detector, $CR1$, sees the vector sum of E_1 and $E_2/2$ and develops a voltage across R_1, E_3. The lower detector, $CR2$, detects the vector difference of E_1 and $E_2/2$ and develops a voltage across R_2, E_4, which is in opposition to E_3. A 100% detector efficiency is assumed for both $CR1$ and $CR2$. The output dc voltage, e_o, is the difference between E_3 and E_4. When the phase angle between E_1 and E_2 is 90 degrees, $|E_3|$ is equal to $|E_4|$ and e_o is zero. When this angle is $90 \pm \theta$ degrees, where $0 < \theta \leqslant 90$, e_o is a positive or negative voltage, depending on the sign of θ, as shown in Fig. 6-56c.

An expression for the output voltage in terms of E_1 and E_2 is derived as follows. Assume that the angle of E_1 is constant. Then, using the law of cosines, we have

$$|E_3| = \sqrt{|E_1|^2 + \left|\frac{E_2}{2}\right|^2 - 2|E_1|\left|\frac{E_2}{2}\right|\cos\left(\frac{\pi}{2} + \theta\right)}$$

or

$$|E_3| = \sqrt{|E_1|^2 + \left|\frac{E_2}{2}\right|^2 + |E_1||E_2|\sin\theta} \qquad (6\text{-}131)$$

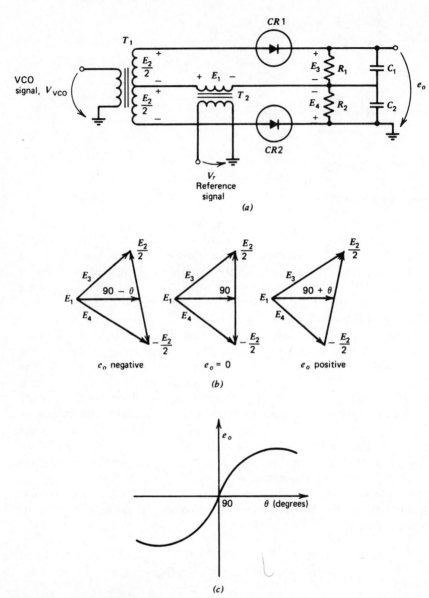

Figure 6-56. Sinusoidal phase detector. (a) Schematic diagram; (b) voltage vector diagram; (c) output voltage versus phase angle.

Similarly,

$$-|E_4| = -\sqrt{|E_1|^2 + \left|\frac{E_2}{2}\right|^2 - 2|E_1|\left|\frac{E_2}{2}\right|\cos\left(\frac{\pi}{2} - \theta\right)}$$

or

$$-|E_4| = -\sqrt{|E_1|^2 + \left|\frac{E_2}{2}\right|^2 - |E_1||E_2|\sin\theta} \qquad (6\text{-}132)$$

The output dc voltage is

$$e_o = |E_3| - |E_4|$$

or

$$e_o = \sqrt{|E_1|^2 + \left|\frac{E_2}{2}\right|^2 + |E_1||E_2|\sin\theta} - \sqrt{|E_1|^2 + \left|\frac{E_2}{2}\right| - |E_1||E_2|\sin\theta}$$

$$(6\text{-}133)$$

The right-hand-side vector diagram of Fig. 6-56b is used in deriving $|E_3|$ and $-|E_4|$.

When $|E_1| = |E_2|$, the expression for the phase-detector output simplifies to

$$e_o = E_1\left(\sqrt{1\tfrac{1}{4} + \sin\theta} - \sqrt{1\tfrac{1}{4} - \sin\theta}\right) \text{ V dc.} \qquad (6\text{-}134)$$

The obvious shortcoming of this device is a direct dependence of e_o on the level of rf input voltage. To overcome it, the amplifier-drivers preceding the phase detector are designed to operate in a limiting mode, or E_2 is made much larger than E_1 so that variations in E_1 have little effect on E_3 and E_4 and variations in E_2 are significantly reduced by the limiting action of $CR1$ and $CR2$, which, under this condition, operate in a saturation mode.

The values of R_1 and R_2 are usually 500 kΩ to 1 MΩ, and capacitors C_1 and C_2 provide an ac short at the frequency of the phase-detector rf input signal. The transformer T_1 should be well balanced, and the diodes should be matched for current and shunt capacitance in order to achieve a small voltage error.

When the leakage of the input rf signal to the phase-detector output results in an excessive phase modilation of the VCO signal, the tendency is to increase C_1 and C_2. Such a solution, however, is often unacceptable because it results in a reduced PLL capture range. The values of C_1 and C_2 should be selected so that the cutoff frequency of the $R - C$ combination is a number of octaves higher than the PLL loop bandwidth, and if more attenuation at the phase-detector frequency is required, a properly designed multipole low-pass filter should be inserted between the phase detector and the VCO.

A poor attenuation characteristic of the sinusoidal phase detector is one of the reasons why the sample-and-hold phase comparator is preferred in digital PLLs to the circuit described above.

Sample-and-Hold Phase Comparator

The sample-and-hold phase comparator finds wide use in digital phase-locked loops. There are two advantages to using this circuit: (*a*) the comparator can be designed to attenuate input signals by more than 80 dB; and (*b*) the circuit operates over a 360 degree phase range as compared to the 180 degree range of the sinusoidal phase detector. It is the high attenuation of the input phase comparator signals, essential in digital PLL design, that accounts for the comparator popularity.

The phase comparator operates on a timing principle rather than by rf vector addition. The reference signal is used to generate a sawtooth voltage by periodically triggering a switch that discharges the voltage across the charge capacitor, C_{ch} (see Fig. 6-57), so that one period of the sawtooth voltage, T_r, is equal to the period of the reference signal. The VCO signal is used to sample the sawtooth voltage by opening a sampling gate and letting the hold capacitor, C_{hold}, charge to the sampled voltage. The phase control action is evident from an examination of the timing diagram, Fig. 6-58. When the VCO mistuning, $\Delta\omega$, is small (i.e., much smaller than the capture range), the sampling pulse samples in the vicinity of the center of the sawtooth wave as shown in Fig. 6-58*a*. In the event that the VCO frequency tends to rise above the nominal value, for example, because of a change in ambient temperature, the phase of the sampling pulse falls back and sampling is done at a lower sawtooth voltage. By lowering the error voltage (the phase-comparator output), one adds capacitance to the VCO tuned circuit, which compensates for the initial drift in the VCO-tuned-circuit reactance and keeps the VCO frequency locked to the reference frequency. Similarly, if the VCO frequency tends to fall below the nominal value, the phase of the sampling pulse advances so as to sample at a higher sawtooth voltage. By increasing the error voltage, one subtracts capacitance from the VCO tuned circuit, keeping the VCO frequency locked.

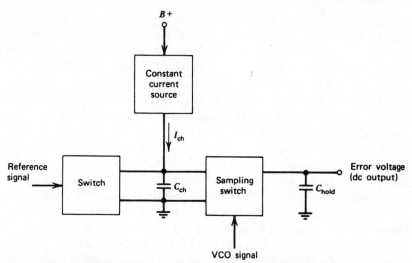

Figure 6-57. Sample-and-hold phase comparator; basic block diagram.

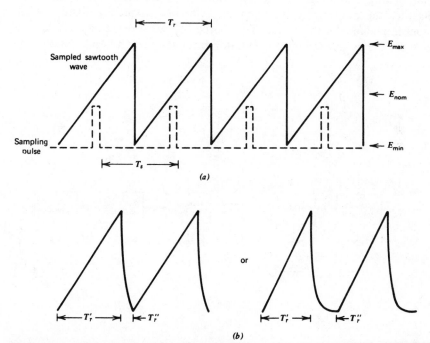

Figure 6-58. Timing diagram of the phase-comparator voltages. (*a*) Theoretical sawtooth and sampling waveforms; (*b*) practical sawtooth waveforms.

The hold-in range in terms of error voltage is that value of VCO mistuning that results in ± 180 degrees of phase variation of the sampling pulse. Theoretically, the error voltage range corresponds to the minimum and maximum values of the sawtooth voltage, E_{min} and E_{max}, respectively. In practical circuits, however, the error voltage range is smaller than the sawtooth voltage amplitude and depends on the width of the sampling pulse. The 360 degree phase range of the comparator is not always attained in practical circuits either. If one assumes a negligible "on" resistance of the switch (a fast discharge time compared to the period of the sawtooth voltage), the time during which the capacitor C_{ch} discharges, T_r'' in Fig. 6-58b, is equal to the width of the reference pulse. Hence the wider the reference pulse, the smaller is the operating phase range of the comparator, or the VCO mistuning that can be compensated for by the PLL.

Figure 6-57 is a simplified block diagram of the phase comparator. To implement this approach, load matching of various constituent parts of the diagram should be done. This is shown in the functional block diagram of the comparator, Fig. 6-59, where a number of source followers (or emitter followers, if desired) are added to indicate the loading points in the circuit that require matching. The source follower at the reference signal input matches the output impedance of the reference frequency divider to the input impedance of the switch. The source follower preceding the sampling switch isolates the sawtooth generator from the switch. And, finally, the

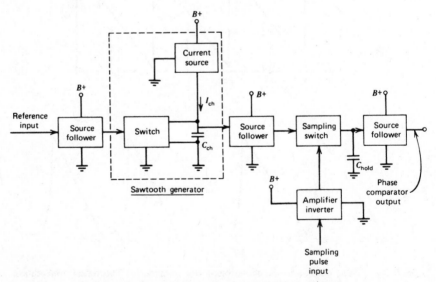

Figure 6-59.   Sample-and-hold phase comparator; functional block diagram.

source follower at the output of the phase comparator presents a high load to the hold capacitor and a low driving source impedance to the Twin-T filter that usually follows the comparator. The function of the amplifier-inverter is described below.

The functional block diagram is implemented by a circuit in Fig. 6-60a. Since the designs of a source follower, FET switch, transistor current source, and transistor inverter are common knowledge, this discussion is limited to a few design elements that are characteristic of this particular circuit.

To keep each FET switch in the "off" state in the absence of an input signal (usually a pulse), Q_1 and Q_2 are negatively biased by a network consisting of a diode and two resistors (Ref. 34). The bias voltage in the case of Q_1 is equal to

$$V_B = -(B_2 - V_1)\left(\frac{R_2}{R_1 + R_2}\right). \qquad (6\text{-}135)$$

Similarly, for Q_2

$$V_B' = -(B_2 - V_2)\left(\frac{R_2'}{R_1' + R_2'}\right), \qquad (6\text{-}136)$$

where $V_1 = V_2 \cong 0.7$ V for a silicon diode.

When a positively going pulse is applied to the diode ($CR1$ or $CR2$), the diode is cut off for the duration of the pulse. A positive voltage is, thus, applied to the gate of Q_1 (or Q_2) through the speed-up capacitor, C_s, and turns the switch on. The function of C_s is to remove the negative charge on the gate so that the switch can be turned on even though the diode is already turned off. An inverter, Q_3, should be used to drive the sampling switch, Q_2, only when the output of the variable-ratio divider is a negatively going pulse.

During the period when Q_1 is in the "off" state, T_r', the capacitor C_{ch} charges to E_{max} (see Fig. 6-61). When Q_1 is turned on, C_{ch} discharges to E_{min} through the "on" resistance of Q_1, R_{on}. The voltage waveform across C_{ch} and the corresponding approximate equivalent circuit of the sawtooth generator during the charge and discharge modes are shown in Fig. 6-61.

There are several ways of computing C_{ch}. When $T_r' \gg T_r''$, that is, when a high-ratio reference divider precedes the phase comparator, one has to make sure that C_{ch} is large enough to discharge to the minimum sawtooth voltage, E_{min}, during T_r''. To determine the required value of C_{ch} under this condition, the equivalent circuit of the comparator during the discharge mode, Fig. 6-61c, is used. During this period the voltage across the

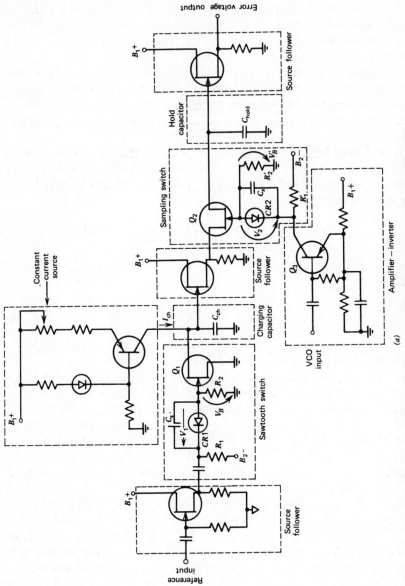

Figure 6-60. Phase comparator; schematic diagram. (*a*) Basic circuit; (*b*) high-speed comparator. By permission of Watkins-Johnson Company.

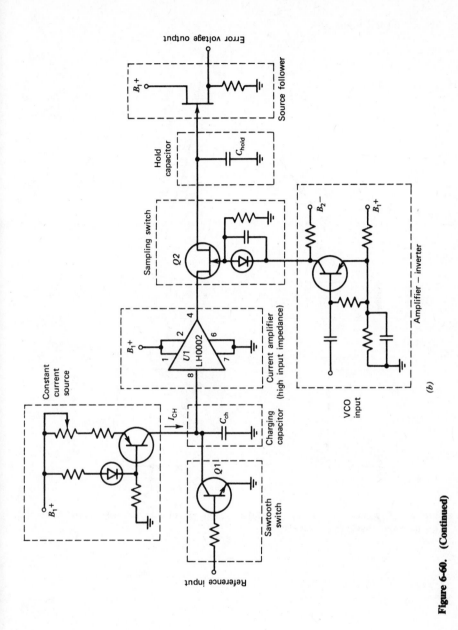

Figure 6-60. (Continued)

415

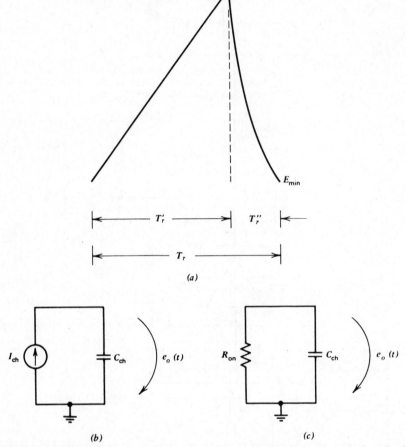

Figure 6-61. Sawtooth generator. (*a*) Output waveform; (*b*) charging mode equivalent circuit; (*c*) discharging mode equivalent circuit.

capacitor is equal to

$$e_o(t) = E_{\max} e^{-t/R_{on}C_{ch}}. \tag{6-137}$$

At the end of the discharge $e_o(t) = E_{\min}$ and $t = T_r''$; hence

$$\frac{E_{\min}}{E_{\max}} = e^{-T_r''/R_{on}C_{ch}}$$

or

$$C_{ch} = - \frac{T_r''}{R_{on} \log_e (E_{min}/E_{max})} \text{ F,} \qquad (6\text{-}138)$$

where $T_r'' =$ the width of the reference pulse triggering Q_1 (sec). Once the value of C_{ch} is determined, an appropriate current to be supplied by the constant current source is computed on the basis of the following considerations.

During the charging mode

$$e_o(t) \cong E_{min} + \frac{I_{ch}}{C_{ch}} \int_0^t dt. \qquad (6\text{-}139)$$

At the end of this period $e_o(t) = E_{max}$ and $t = T_r'$; hence

$$E_{max} \cong E_{min} + \frac{I_{ch} T_r'}{C_{ch}}$$

or

$$I_{ch} = \frac{C_{ch}(E_{max} - E_{min})}{T_r'} \text{ A.} \qquad (6\text{-}140)$$

When the charge and discharge times are equal to each other (i.e., when the reference signal is fed directly into the phase comparator), one is more concerned with C_{ch} charging to the maximum sawtooth voltage, E_{max}, at a rate that can be handled by the current source. Under this condition one first designs the current source to deliver a small current (say, 2 or 3 mA) and then computes the value of C_{ch} from Eq. 6-140.

The considerations governing the selection of C_{hold} were given in Chapter 5 (p. 316) and are not repeated here.

A slight modification of the circuit in Fig. 6-60a makes it a high-speed phase comparator. This characteristic is very useful in the fast-tuning-time PLLs, when it is required that the hold capacitor, C_{hold}, charge to its final voltage in one period of the sampling pulse and the width of the pulse is very narrow—a typical condition at very large division ratios of the variable-ratio frequency divider. This performance is achieved by replacing the source follower, matching the sawtooth generator to the sampling switch by a high-input-impedance current amplifier, such as LH0002 manufactured by National Semiconductor, as shown in Fig. 6-60b, $U1$. The amplifier provides the high current necessary to charge C_{hold} to the required voltage during the short period when $Q2$ is on.

The circuit in Fig. 6-60b differs from that in Fig. 6-60a in another respect: it presents a simplified version of the sawtooth switch.

6-8   Frequency Discriminators

Foster-Seely Discriminator

In a double-tuned loosely coupled circuit the relationship between the primary and secondary voltages at any frequency, f, is

$$\frac{E_2}{E_1} = j\sqrt{\frac{L_2}{L_1}} \left\{ \frac{kQ_2}{1+j2Q_2[(f/f_0)-1]} \right\}, \qquad (6\text{-}141)$$

where k = the coefficient of coupling between the primary and secondary windings,

Q_2 = the quality factor of the tuned circuit of the secondary winding,

f_0 = the center frequency of each tuned circuit,

and where

$$k \triangleq \frac{M}{\sqrt{L_1 L_2}}, \qquad (6\text{-}142)$$

$$Q_2 = \frac{R_2}{2\pi f_0 L_2}, \qquad (6\text{-}143)$$

and M is the mutual inductance between the two windings. (See Fig. 6-62 for definitions of other quantities.) As the frequency of the input signal, E_1, is varied, so is the angle between the voltages E_1 and E_2. At resonance (i.e., when $f = f_0$),

$$\frac{E_2}{E_1} = j\sqrt{\frac{L_2}{L_1}} \, (kQ_2) \qquad (6\text{-}144)$$

and the phase shift between E_1 and E_2 is 90 degrees with E_2 leading. The operation of the Foster-Seeley discriminator is based on the principle of this frequency dependence of the phase angle between the input and output voltages of a double-tuned circuit.

Figure 6-63a is a schematic diagram of the Foster-Seeley discriminator. The primary winding of a loosely coupled transformer is directly connected to the center tap of the secondary winding so that E_3, the voltage detected by the diode $CR1$, is a vector sum of E_1 and $E_2/2$, and E_4, the

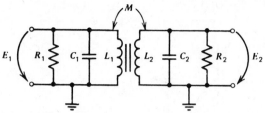

Figure 6-62. Double-tuned magnetically coupled circuit.

voltage detected by $CR2$, is a vector difference of E_1 and $E_2/2$. In this analysis a 100% detector efficiency is assumed for both $CR1$ and $CR2$. The discriminator output voltage, e_o, is the sum of the two detected voltages. As the frequency of the input signal is varied, so is the phase angle between E_1 and E_2 with the result that E_3 and E_4 also vary in magnitude. This action is described in vector form in Fig. 6-63b. At resonance E_2 is 90 degrees out of phase with E_1; hence $|E_3| = |E_4|$ and e_o is zero. At any other frequency $|E_3|$ is either larger or smaller than $|E_4|$ and the discriminator output is either a positive or a negative voltage. A sketch of discriminator output voltage versus frequency of the input wave is shown in Fig. 6-63c.

The expressions for $|E_3|$, $|E_4|$, and e_o derived in Section 6-7, Eqs. 6-131, 6-132, and 6-133, apply in this case as well. Hence

$$e_o = \sqrt{|E_1|^2 + \left|\frac{E_2}{2}\right|^2 + |E_1||E_2|\sin\theta} \ - \sqrt{|E_1|^2 + \left|\frac{E_2}{2}\right|^2 - |E_1||E_2|\sin\theta},$$

$$(6\text{-}145)$$

and for $|E_1| = |E_2|$

$$e_o = E_1\left(\sqrt{1\tfrac{1}{4} + \sin\theta} \ - \sqrt{1\tfrac{1}{4} - \sin\theta}\,\right), \qquad (6\text{-}146)$$

where

$$\theta = 2\pi k (f_0 - f). \qquad (6\text{-}147)$$

It can be shown (Ref. 35) that the discriminator sensitivity at resonance, S, in volts per hertz of frequency deviation is

$$S = 8\pi L_1 Q_1^2 E_1 \frac{\sqrt{(L_2/L_1)}k}{(1+k^2)\sqrt{1 + \left[(L_2/L_1)k^2/4\right]}} \ \ \text{V/Hz.} \qquad (6\text{-}148)$$

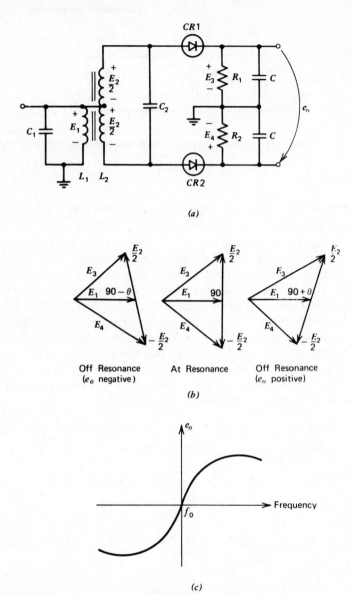

Figure 6-63. Foster-Seely frequency discriminator. (*a*) Schematic diagram; (*b*) voltage vector diagram; (*c*) output voltage versus frequency.

420

Hence the discriminator sensitivity at resonance is proportional to L_1, Q_1^2, and E_1 and is a function of L_2/L_1 and k. A substantially linear performance is obtained over approximately 80% of the 3 dB bandwidth of the primary and secondary tuned circuits.

In terms of the secondary-to-primary inductance ratio the coefficient of coupling is

$$k = \frac{\left[\sqrt{1 + 2(L_2/L_1)} - 1\right]^{1/2}}{L_2/L_1}. \tag{6-149}$$

If square-law detectors are used, the coupling resulting in the optimum sensitivity is equal to 0.578 times the critical coupling.

Given a center frequency, f_0, and a frequency range over which linear operation is required, $\pm \Delta f_l$, the designer may proceed as follows. Compute the 3 dB bandwidth of each tuned circuit from

$$B_{3\,dB} = \frac{2\Delta f_l}{0.8}. \tag{6-150}$$

Determine the quality factors,

$$Q_1 = Q_2 = \frac{f_0}{B_{3\,dB}}. \tag{6-151}$$

Choose the highest possible inductance of the primary winding, L_1, and select the optimum coupling coefficient, $k = 0.578$. Assume that $|E_1| = |E_2|$ and compute L_2 from Eq. 6-144. Compute C_1 and C_2 from

$$f_0 = \frac{1}{2\pi\sqrt{L_1 C_1}} = \frac{1}{2\pi\sqrt{L_2 C_2}}. \tag{6-152}$$

The effective primary and secondary loadings, R_1 and R_2, are computed from

$$R_1 = 2\pi f_0 Q_1 L_1 \tag{6-153}$$

and

$$R_2 = 2\pi f_0 Q_2 L_2. \tag{6-154}$$

In most practical applications the primary of the transformer is the collector load of a transistor amplifier. In such cases an ac coupling capacitor, C_c in Fig. 6-64, is used to connect the primary winding to the center tap of the secondary winding. The capacitor provides an ac short at f_0. If a single-ended output is required, a rf choke that presents a high impedance at f_0 is added to the circuit as shown in Fig. 6-64. If the impedance of this choke is not high enough, the performance of the discriminator will significantly degrade.

The Foster-Seeley discriminator is used in phase-locked loops only as an auxiliary circuit in conjunction with a phase detector because the frequency accuracy of this device is highly dependent on the stability of the center frequency of both primary and secondary tuned circuits, which is temperature dependent. The circuit can be modified to operate as a frequency discriminator and phase detector. We call this modified version of the Foster-Seeley discriminator the balanced discriminator.

Balanced Discriminator

If a reference signal is injected into the Foster-Seeley discriminator as shown in Fig. 6-65, the circuit operates in two modes. When the frequency of the VCO signal is far from the reference signal frequency, the circuit operates in the frequency discriminator mode and generates a dc voltage that depends on the frequency difference between the two signals. When,

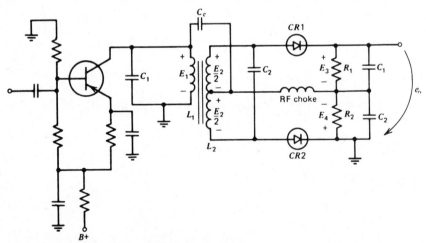

Figure 6-64. Single-ended Foster-Seely discriminator.

because of the PLL action, this difference is made very small, the circuit automatically switches into a phase-detector mode and generates a dc voltage that depends on the phase difference between the two input signals. Indeed, in a practical circuit there is a region of values of the frequency difference for which the circuit operates in both modes simultaneously. To achieve satisfactory operation, the reference signal is injected from a high-impedance source.

The derivation of the output dc voltage in terms of the input rf voltages, the circuit parameters, and the frequency-phase difference between the input voltages is not carried out in this book. However, the techniques of deriving these quantities are demonstrated in this section and in Section 6-7 and can serve as a guide to the interested student if he decides to perform this derivation on his own.

6-9   Sweep Circuits for PLL Acquisition

Frequency acquisition in narrow-bandwidth phase-locked loops may be accomplished by utilizing voltage sweep circuits. The frequency of the VCO of such a loop is set to be within the hold-in but outside the capture range of the loop. (This requires fewer discrete values of VCO tuning voltage and hence saves hardware.) Under such condition the PLL phase detector operates in a mixer mode generating an ac signal at the difference frequency (a beatnote) and a dc voltage. Neither is large enough to pull the VCO in frequency lock. A sawtooth voltage is automatically applied to the input of the VCO error-voltage-varicap circuit from an asynchronism detector external to the loop, and the VCO frequency is swept about its nominal value until locking takes place.

The principle of detection of PLL asynchronous operation is considered in Chapter 4. Here we describe two circuits that generate sawtooth voltage of a desired amplitude and repetition rate upon command from an asynchronism sensor.

Unijunction Relaxation Oscillator

The design equations given in this section were selected from the material presented in Ref. 36, Chapter 13.

The unijunction transistor (UJT) is a three-terminal semiconductor device made of an n-type silicon bar with ohmic contacts at each end (B_1 and B_2 in Fig. 6-66) and a rectifying contact made in between. The most important feature of the UJT is a negative-resistance characteristic (Fig. 6-67) that makes the device suitable for sawtooth voltage generation.

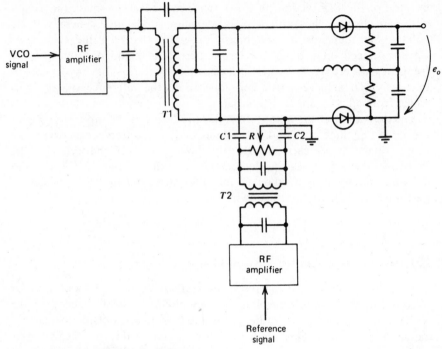

Figure 6-65.   The balanced discriminator.

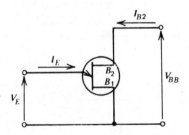

Figure 6-66.   Unijunction transistor.

The UJT parameters of interest in relaxation oscillator design are as follows:

1. Interbase resistance, R_{BB}, which is the resistance between base-1 and base-2 with the emitter open-circuited.
2. Intrinsic standoff ratio, η, defined by means of Eq. 6-156.
3. Peak point emitter-to-base-1 voltage, V_P
4. Peak point emitter current, I_P.

5. Valley emitter-to-base-1 voltage, V_V.
6. Valley emitter current, I_V.
7. Emitter-to-base-1 saturation voltage, $V_{E(sat)}$, which is the forward voltage drop from emitter to base 1 when the UJT is biased in the saturation region.

Some of these parameters are shown graphically in Fig. 6-67.

The schematic diagram and the output waveforms of a unijunction relaxation oscillator are shown in Fig. 6-68. For small values of R the period of oscillation is (Ref. 36)

$$T = R_T C_T \log\left(\frac{1}{1 - \eta}\right)$$ (6-155)

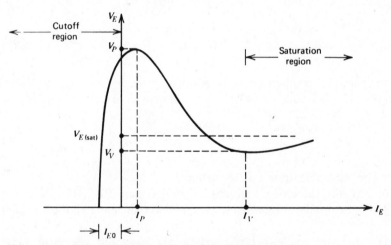

Figure 6-67.   Static emitter characteristic of a unijunction transistor.

with η defined by

$$V_P = \eta V_{BB} + V_D,$$ (6-156)

where V_D = the voltage drop across the emitter diode (internal to the UJT).

At the onset of a sweep, the emitter is reverse biased and the capacitor C_T is charging through the resistor R_T toward the supply voltage $+ V$. As the emitter-to-base-1 voltage reaches the peak point value, V_P, the emitter junction becomes forward biased, and the emitter-to-base-1 dynamic resistance drops to a low value, discharging C_T. When the emitter voltage reaches $V_{E(\text{min})}$, which is equal to approximately one half of $V_{E(\text{sat})}$, the emitter ceases to conduct and the cycle is repeated.

In order for the UJT to be turned on,

$$R_T < \frac{V - V_P}{I_P}.$$

(6-157)

Similarly, for the UJT to be switched off,

$$R_T > \frac{V - V_V}{I_V}.$$

(6-158)

Since V_D varies with temperature, V_P is temperature dependent also. This affects the frequency of oscillations. The frequency stability of a unijunction relaxation oscillator can be improved by selecting R so that

$$R \cong \frac{0.7\,R_{BB}}{\eta V}$$

(6-159)

if temperature compensation is required over an extreme temperature range, such as -60 to $+140°C$. For a narrow temperature range,

$$R \cong \frac{0.4\,R_{BB}}{\eta V}$$

(6-160)

gives better temperature compensation.

Finally, the value of C_T should be greater than 0.01 μF in order not to degrade the oscillator frequency stability and reduce the allowable range of R_T.

To minimize the effect of loading, the sawtooth generator may be coupled to the load by using a direct-coupled emitter follower, as shown in Fig. 6-69. The emitter-follower load across C_T is equal to approximately $(h_{FE}+1)R_E$; hence the values of h_{FE} and R_E should be as large as possible, that is, such that $(h_{FE}+1)R_E$ is much greater than R_T. To ensure the oscillations the minimum values of h_{FE} and R_E are determined from the relationship

$$\frac{(h_{FE}+1)R_E}{R_T+(h_{FE}+1)R_E} > \eta_{\text{max}},$$

(6-161)

where $\eta_{\text{max}}=$ the maximum value of η for the selected type of UJT.

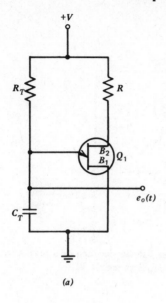

(a)

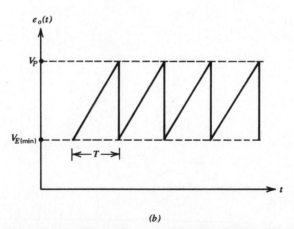

(b)

Figure 6-68. Unijunction relaxation oscillator. (*a*) Schematic diagram; (*b*) output waveform. Courtesy of General Electric Semiconductor Products Department, Syracuse, N. Y.

To turn the sweep voltage off at the time when the PLL acquires lock, a semiconductor switch is connected across C_T as shown in Fig. 6-70.

Figure 6-71 is a block diagram of a complete hunt generator. The reference signal is applied to one port of a phase detector; the VCO signal, to another port via a 90 degree phase shifter. These two circuits constitute an asynchronism sensor. When the loop is locked, the output of the phase

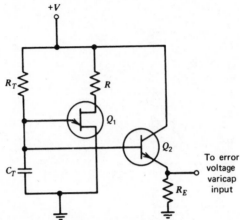

Figure 6-69. Sawtooth generator, emitter-follower coupling to a load. Courtesy of General Electric Semiconductor Products Department, Syracuse, N. Y.

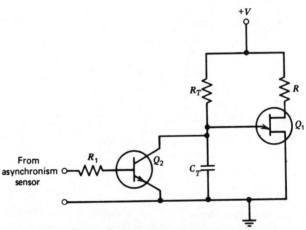

Figure 6-70. Remote control of the sawtooth generator.

detector is a high positive voltage, which keeps the electronic switch closed and prevents the relaxation oscillator from oscillating. When the loop loses synchronism, the phase-detector output drops to a low voltage, which is not adequate to keep the switch closed; the relaxation oscillator generates a periodic sawtooth wave, and the VCO frequency is swept about its

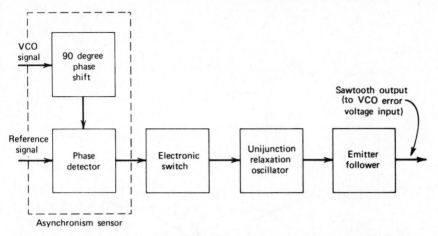

Figure 6-71.   Unijunction relaxation oscillator in a hunt generator, type 2 phase-locked loop.

nominal value until locking takes place. When synchronism is again achieved, the phase-detector output goes high and disengages the relaxation oscillator.

Astable Multivibrator

Another voltage sweep circuit that can operate at significantly faster repetition rates than a unijunction relaxation oscillator is shown in Fig. 6-72*a*. This sawtooth generator is an astable multivibrator, which oscillates whenever a dc voltage is applied to the circuit via an electronic switch. Since the waveform at the collector of Q_2 is a square wave, the signal has to be differentiated to result in a sawtooth wave. Figure 6-72*b* shows the waveforms associated with this circuit. The load resistance, R_l, is selected so that $(R_l + r_{bb'})C_1$ is much smaller than T_2($r_{bb'}$ is the ohmic base resistance of Q_1). The resistors R_1 and R_2 and the capacitors C_1 and C_2 are chosen to give the required period of oscillations (Ref. 15):

$$T = T_1 + T_2, \tag{6-162}$$

where

$$T_1 = (0.69)R_1C_1 \tag{6-163}$$

and

$$T_2 = (0.69)R_2C_2. \tag{6-164}$$

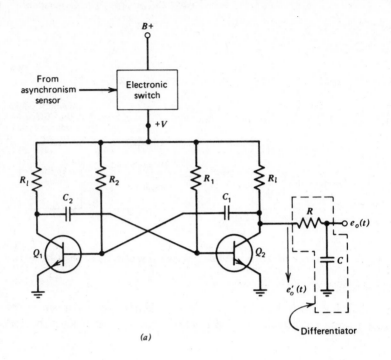

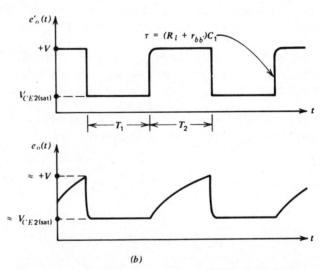

(b)

Figure 6-72. Transistor astable multivibrator. (a) Schematic diagram; (b) relevant waveforms.

430

The period of oscillations may be varied from a second to a fraction of a microsecond by adjusting R and C.

6-10   Voltage-Controlled Oscillator Tuning Circuits

Most broadband VCOs used in phase-locked loops have to be frequency tuned to within PLL capture or hold-in range. The simplest network that provides a set of tuning voltage values upon a digitally executed command is the variable-ratio voltage divider shown in Fig. 6-73. When the resistors R_1 to R_n are connected to ground, one at a time or in groups of two or more, discrete voltages are generated that depend on the value of the resistor R and the resistance from point A to ground. The ground connection is made with transistor (or FET) switches. When remote operation is not required, mechanical switches are employed. For tuning voltages smaller than 25 V, integrated circuits (ICs) may be used. A typical IC is model MEM851P, a four-channel MOS switch manufactured by General Instruments Corporation. In remote operation the control signals are derived from the digital logic, which converts VCO frequency information into discrete values of tuning voltage. The advantages of this network are low cost and ease of selecting any desired voltage, the accuracy of which

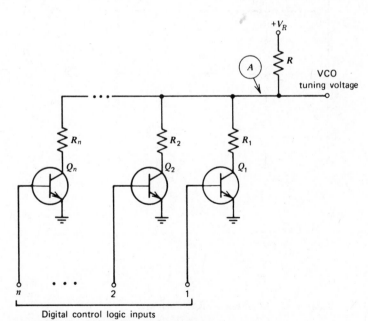

Figure 6-73.   **Variable-ratio voltage divider.**

depends mainly on the accuracy of the reference voltage, V_R, and the tolerance of the voltage divider resistors. Low-noise metal-film resistors should be used in this circuit and the following ones.

When numerous voltage values are to be generated, a more elegant approach may be used. The circuit in question consists of a $2R, R$ resistor ladder and a set of switches interconnected as shown in Fig. 6-74. It has been demonstrated by D. F. Hoeschele (Ref. 37) that the output resistance of the network, R_o, is equal to R and is independent of the position of the switches, S_1 through S_n. The expression for the output voltage derived by Hoeschele is

$$V_o = \left(\tfrac{1}{2}D_1 + \tfrac{1}{4}D_2 + \tfrac{1}{8}D_3 + \cdots + \frac{1}{2^n}D_n \right) \left(\frac{V_R R_l}{R_o + R_l} \right), \qquad (6\text{-}165)$$

where the D represents the state of a particular switch and n is the number of switches used. For the switch S_n making a connection to $+V_R$, D_n is

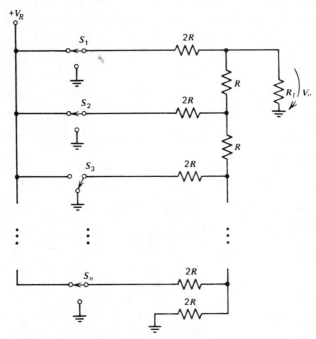

Figure 6-74. $2R, R$ resistor ladder network.

equal to unity; for S_n connected to ground, $D_n = 0$. The resolution of this network, that is, the smallest increment of the output voltage, is

$$\Delta V_o = \frac{1}{2^n}\left(\frac{V_R R_l}{R_o + R_l}\right). \tag{6-166}$$

Hence the accuracy of the output voltage depends not only on the accuracy of the reference voltage and the tolerance of the ladder resistances but also on the resolution of the ladder network. The highest value of the output voltage, $(V_o)_{max}$, occurs when $D_1 = D_2 = D_3 = \cdots = D_n = 1$. This leads to

$$(V_o)_{max} = V_R\left(\frac{R_l}{R_o + R_l}\right)\sum_{k=1}^{n}\frac{1}{2^k}. \tag{6-167}$$

For a given ΔV_o and $(V_o)_{max}$, the number of switches (bits) needed to achieve a specific resolution is determined from

$$\frac{(V_o)_{max}}{\Delta V_o} = \sum_{k=1}^{n} 2^{k-1}, \tag{6-168}$$

which was arrived at by combining Eqs. 6-166 and 6-167.

A practical example demonstrates the use of Eqs. 6-165 to 6-168. Assume that 15 discrete values of VCO tuning voltage, given in Table 6-4, have to be generated by a $2R$, R ladder network with an accuracy of $\pm 5\%$. Assume also that the reference voltage can be held to within $\pm 2\%$ of its nominal value and that the tolerance of the ladder resistors is $\pm 1\%$. The ladder network resolution, therefore, should be better than $\pm 2\%$ of the lowest tuning voltage value or, according to Table 6-4, ± 0.09 V. From this we conclude that $\Delta V_o \leqslant 0.18$ V. We also know that $(V_o)_{max} = 26$ V. Using Eq. 6-168, we arrive at

$$\frac{26}{0.18} = 144.5 = \sum_{k=1}^{n} 2^{k-1}.$$

Since a seven-bit ladder network ($n=7$) does not provide the desired resolution, an eight-bit network is selected. This gives

$$\Delta V_o = \frac{(V_o)_{max}}{\sum_{k=1}^{8} 2^{k-1}} = \frac{26}{255} \approx 0.1 \text{ V},$$

so that the expected error at 4.5 V tuning voltage is less than ± 0.05 V.

Let us assume that R_l is 100 MΩ, and select $R_o = 50$ kΩ. Then

$$\frac{R_l}{R_o + R_l} = 0.9995 \cong 1,$$

Table 6-4.   Required Tuning Voltages and Corresponding 2% Values

Tuning voltage (V dc)	2% accuracy of the nominal value (V)
26	0.52
23	0.46
20	0.40
17	0.34
15	0.30
12	0.24
10	0.20
9	0.18
8	0.16
7	0.14
6.5	0.13
6.0	0.12
5.5	0.11
5.0	0.10
4.5	0.09

Table 6-5.   Generated Tuning Voltages and Corresponding Positions of Ladder Network Switches

Generated tuning voltage (V dc)	Deviation from nominal value (V dc)	Logic inputs							
		S_1	S_2	S_3	S_4	S_5	S_6	S_7	S_8
25.899	-0.10	1	1	1	1	1	1	1	1
22.953	-0.047	1	1	1	0	0	0	1	0
20.007	$+0.007$	1	1	0	0	0	1	0	1
16.9598	-0.04	1	0	1	0	0	1	1	1
14.9290	-0.071	1	0	0	1	0	0	1	1
11.9834	-0.017	0	1	1	1	0	1	1	0
9.9632	-0.0368	0	1	1	0	0	0	1	0
8.9362	-0.0638	0	1	0	1	1	0	0	0
8.0233	$+0.0233$	0	1	0	0	1	1	1	1
7.0070	$+0.0070$	0	1	0	0	0	1	0	1
6.50	0.0	0	1	0	0	0	0	0	0
5.9904	-0.0096	0	0	1	1	1	0	1	1
5.4834	-0.0166	0	0	1	1	0	1	1	0
4.9764	-0.0236	0	0	1	1	0	0	0	1
4.4694	-0.0310	0	0	1	0	1	1	0	0

and Eq. 6-167 becomes

$$V_R = \frac{(V_o)_{\max}}{\displaystyle\sum_{k=1}^{8} (1/2^k)} = \frac{26 \times 256}{255} \cong 26 \text{ V}.$$

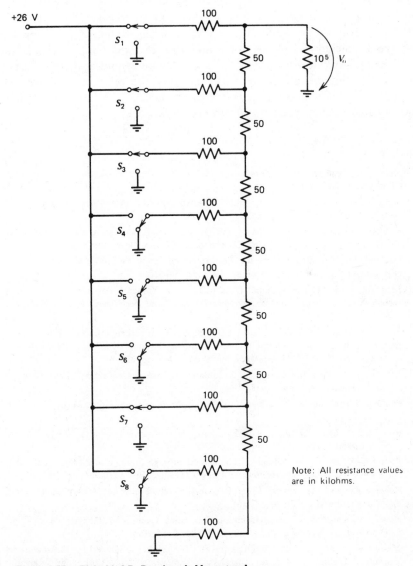

Figure 6-75.   Eight-bit $2R$, R resistor ladder network.

Using Eq. 6-165, we compute all other voltages. Table 6-5 gives a summary of actual tuning voltage values obtained with this network and the positions of the switches corresponding to each voltage. The table provides information about the resolution of the selected tuning network only. To find the total expected error one has to add to each deviation a voltage increment of $\pm 3\%$ of the nominal voltage corresponding to that deviation. For example, a $\pm 3\%$ increment of 26 V is ± 0.78 V; hence the total deviation from the nominal value of 26 V is $+0.68$, -0.88 V.

Figure 6-75 is a schematic diagram of the eight-bit ladder network. The diagram shows the switches in the positions required to generate $+23$ V.

It was previously pointed out that a VCO tuning voltage input is highly sensitive to pickup of phase noise and discrete spurious signals. It is recommended that low-noise resistors such as wire-wound or metal-film be used in the tuning voltage network and that the reference voltage be derived from a low-noise voltage regulator such as the Fairchild model $\mu723$, located in close proximity to the network.

References

1. Hakim, S. S. and R. Barrett. *Transistor Circuits in Electronics* (New York: Hayden Book Company, 1964).
2. Stern, A. P. "Stability and Power Gain of Tuned Transistor Amplifiers," *Proceedings of the IRE*, March 1957, pp. 335–343.
3. Von Recklinghausen, D. R. "Theory and Design of FET Converters," *Application Note*, Texas Instruments, Semiconductor-Components Division, November 8, 1965.
4. Cheadle, D. "Selecting Mixers for Best Intermodulation Performance," *Microwaves*, November 1973, pp. 48–52; December 1973, pp. 58–62.
5. Perlman, B. S. "Current-Pumped Abrupt-Junction Varactor Power-Frequency Converters," *IEEE Transactions on Microwave Theory and Techniques*, March 1965, pp. 150–161.
6. Perlow, S. M. and B. S. Perlman. "A Large Signal Analysis Leading to Intermodulation Distortion Prediction in Abrupt-Junction Varactor Up-converters," *IEEE Transactions on Microwave Theory and Techniques*, November 1965, pp. 820–827.
7. Kuh, E. S. "Theory and Design of Wide-Band Parametric Converters," *Proceedings of the IRE*, January 1962, pp. 31–38.
8. Penfield, P., Jr., and R. P. Rafuse. *Varactor Applications* (Cambridge, Mass.: The MIT Press, 1962).
9. Rouche, N. "Steady Oscillations of Parametric Subharmonic Oscillator," *IRE Transactions on Circuit Theory*, March 1962, pp. 7–12.
10. Ryan, W. D. and H. B. Williams. "The Carrier-Storage Frequency Divider: A Steady-State Analysis," *IEEE Transactions on Circuit Theory*, September 1964, pp. 396–403.
11. Hewlett Packard Associates. "Step Recovery Diode Frequency Multiplier Design," *Application Note* 913, Palo Alto, Calif., May 15, 1967.
12. Hewlett Packard Associates. "Harmonic Generation Using Step Recovery Diodes and SRD Modules," *Application Note* 920. Palo Alto, Calif., June 1968.
13. Members of the Staff of the Department of Electrical Engineering, Massachusetts Institute of Technology. *Magnetic Circuits and Transformers* (New York: John Wiley and Sons, 1943).

14. Friis, H. T. "Analysis of Harmonic Generator Circuits for Step Recovery Diodes," *Proceedings of the IEEE*, July 1967, pp. 1192–1194.

15. Millman, J. and H. Taub. *Pulse, Digital, and Switching Waveforms* (New York: McGraw-Hill Book Company, 1965).

16. Tucker, D. G. "The Synchronization of Oscillators," *Electronic Engineer*, Part I, V. 15 (March 1943), pp. 412-418; Part II, V. 15 (April 1943), pp. 457-461; Part III, V. 16 (June 1943), pp. 26-30.

17. Adler, R. "A Study of Locking Phenomena in Oscillators," *Proceedings of the IRE*, June 1946, pp. 351-357.

18. Paciorek, L. J. "Injection Locking of Oscillators," *Proceedings of the IEEE*, November 1965, pp. 1723-1727.

19. Cleary, J. F. and D. V. Jones. "Cascaded UJT Oscillators Form Stable Frequency Dividers," *Electronic Design*, November 8, 1965, p. 52.

20. Cleary, J. F. and D. V. Jones, "A Unijunction Frequency Divider," *EEE*, May 1964, pp. 52, 53.

21. Veth, G. J. "Tunnel Diode Frequency Divider," *Solid State Design*, Febuary 1963, pp. 30-36.

22. Sulzer, P. G. "Modified Locked-Oscillator Frequency Dividers," *Proceedings of the IRE*, December 1951, pp. 1535-1537.

23. Nichols, J. and C. Shinn. "Pulse Swallowing," *EDN*, October 1, 1970, pp. 39-42.

24. Montevaldo, R. and C. Shinn. "Programmable Divider Performs at 140 MHz," *Computer Hardware*, April 15, 1971, pp. 10-14.

25. Goldwasser, W. J. "Design Shortcuts for Microwave Frequency Dividers," *The Electronic Engineer*, May 1970, pp. 61-65.

26. Hilibrand, J. and W. R. Beam. "Semiconductor Diodes in Parametric Subharmonic Oscillators," *RCA Review*, June 1959, pp. 229-253.

27. Delhom, L. "FET Amplifiers, a Graphic-Analysis Approach, " *EEE*, March 1967, pp. 79-85.

28. Watson, J, and W. E. Eder. "Nomograms Pick FET Biasing Values," *Electronics*, April 3, 1967, pp. 93-95.

29. Hakim, S. S. and R. Barrett. *Transistor Circuits in Electronics* (New York: Hayden Book Company, 1964), p. 173.

30. US Army Electronics Laboratory, Fort Monmouth, N. J. *Quartz Crystal Oscillator Circuits Design Handbook*, AD No. 460377, March 15, 1965.

31. Air Research and Development Command, US Air Force, Wright-Patterson Air Force Base, Ohio. *Handbook of Piezoelectric Crystals for Radio Equipment Designers*, AD No. 110448, October 1956.

32. Mortley, W. S. "Frequency-Modulated Quartz Oscillators for Broadcasting Equipment" *IEE Proceedings*, Part b, May 1957, pp. 239-249.

33. Mortley, W. S. "Circuit Giving Linear Frequency Modulation of a Quartz Crystal Oscillator," *Wireless World*, October 1951, pp. 399-403.

34. Cohen, J. M. "Sample-and-Hold Circuits Using FET Analog Gates," *EEE*, January 1971, pp. 34-37.   7

35. Foster, D. E. and S. W. Seeley. "Automatic Tuning, Simplified Circuits, and Design Practice," *Proceedings of the IRE*, March 1937, pp. 289-313.

36. General Electric, Semiconductor Products Department. *Transistor Manual*, 2nd ed., 1964.

37. Hoeschele, D. F., Jr. *Analog-to-Digital / Digital-to-Analog Conversion Techniques* (New York: John Wiley and Sons, 1968).

7 Frequency Synthesizers

The preceding chapters were devoted to the principles associated with the design of frequency synthesizers. Here we briefly describe various types of synthesizers currently in use, supplementing the descriptions with (*a*) a summary of the most important requirements, (*b*) the basic considerations associated with the design of each synthesizer, and (*c*) a set of applications for these systems. In short, we shall summarize the major topics of the material presented in this book. Evaluation and comparison of the synthesizers are left as an exercise to the interested student.

Some of the commercial synthesizers considered in this chapter will no longer be offered for sale when the book is published. These systems are included here, nevertheless, because their designs demonstrate best the implementation of the frequency synthesis approaches considered in Chapter 1 and help to impart an insight into what it is required to design and develop a high-quality synthesizer.

7-1 Incoherent Synthesizers

Polarad two- and single-tone synthesizers, models 1102 and 1102A, respectively, are typical examples of incoherent frequency synthesizers. The units are used in conjunction with a manually operated spectrum analyzer in testing hf receivers and transmitters. A switching time of 1 sec is more than adequate for satisfactory manual operation. Similarly, the phase-noise purity of a crystal oscillator is more than enough for a 60 dB dynamic range of the analyzer. These two requirements are, therefore, excluded from the specification for models 1102 and 1102A, given in Table 7-1, and from the design considerations.

For clarity of presentation, model 1102A is described first. The synthesizer is composed of a set of crystal-controlled oscillators whose frequencies are either directly added to or subtracted from each other in rf mixers or first multiplied by 2 and then added or subtracted (see Fig. 7-1).

The 5.0 to 5.09, 47.0 to 47.9, and 71.0 to 75.5 MHz oscillators operate over a narrow frequency range in 10 discrete steps by having appropriate

438

Table 7-1. Partial Performance Specification for Polarad Two- and Single-Tone Synthesizers, Models 1102 and 1102A[a]

	Model	
	1102	1102A
Frequency range:	1.0–39.999 MHz	1.0–39.999 MHz
Frequency increments:	1 kHz	1 kHz
Frequency accuracy:	0.08% at 10 MHz, 0.02% at 39.999 MHz	0.08% at 10 MHz, 0.02% at 39.999 MHz
Nonharmonically related spurious outputs:	−70 dB	−70 dB
Two-tone separation:	10 kHz minimum; continuous tuning	NA[b]
Intermodulation distortion:	Greater than −60 dB	NA

[a]Printed with permission of Polarad Electronic Instruments, Division of Polarad Electronics Corporation.
[b] NA = not available.

crystal units manually switched into the circuits. The oscillators are a grounded-base Colpitts type with crystals, which resonate in a series mode, providing an ac ground connection to the transistor base through a thumbwheel switch. The 6.0 to 6.009 and 6.30 to 6.309 MHz oscillators are controlled by one crystal each. The crystal is frequency pulled in 10 discrete steps, utilizing the quarter-wave technique described in Chapter 6. The 85 to 100 MHz signal is generated by four fixed-frequency oscillators energized and driving, one at a time, a times-2 transistor multiplier.

Hence a change of 1 kHz at the final synthesizer frequency is produced by varying the frequency of the 6.0 to 6.009 MHz oscillator; 10 and 100 kHz frequency increments are generated by tuning the 5.0 to 5.09 and 47.0 to 47.9 MHz oscillators, respectively; and 1 MHz increments are generated by the 71.0 to 75.5 MHz oscillator, whose frequency is multipled by 2 before the signal is applied to the 200.0 to 209.999 MHz mixer. Finally, the output synthesizer frequency is incremented in 10 MHz steps whenever one of the 85, 90, 95, or 100 MHz oscillators delivers an rf signal to the times-2 multiplier.

The synthesizer operates in two modes. In mode A the synthesizer is used as a stable and accurate signal generator operating between 1.0 and 39.999 MHz. This requires that the 6.0 to 6.009 MHz oscillator deliver an rf signal to the 58.0 to 58.999 MHz mixer. In LO mode the synthesizer is used as the first local oscillator of the 1.0 to 40.0 MHz spectrum analyzer. Since the analyzer first intermediate frequency is 300 kHz, the synthesizer output frequency is offset by 300 kHz by applying +28 V power supply voltage to the 6.3 to 6.309 MHz (instead of 6.0 to 6.009 MHz) oscillator.

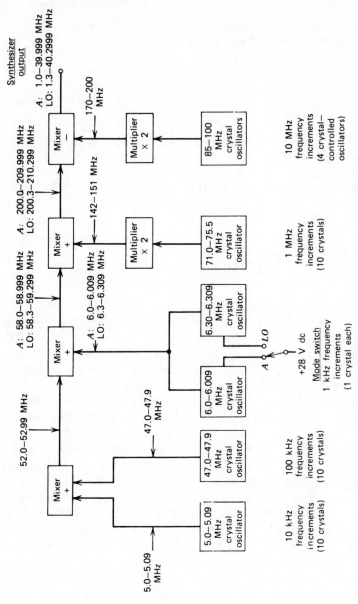

Figure 7-1. 1.0 to 39.999 MHz direct incoherent synthesizer, model 1102A, manufactured by Polarad Electronic Instruments. Courtesy of Polarad Electronic Instruments, Division of Polarad Electronics Corporation.

440

Under these conditions, the frequency to which the spectrum analyzer is tuned can be read directly from the frequency control switches located on the front panel of the synthesizer. The bandwidth of the filters following the 58.0 to 58.999 MHz mixer is large enough to accommodate both modes of operation and to provide adequate suppression of unwanted out-of-band signals. Transistor balanced mixers are used throughout the synthesizer for better isolation between inputs and for some degree of balancing of the higher-frequency input signal applied to each mixer, such as the signal at 47.0 to 47.9 MHz. This relaxes the requirement for the filters that follow the mixers.

The selection of frequencies internal to the synthesizer is governed by three considerations: (1) the level of the in-band and close-to-band spurious signals, which should be lower than -80 dB in relation to the level of the desired signal; (2) the frequency range of each oscillator, which should be sufficiently narrow to permit an operation requiring no retuning of the oscillator tank circuit as various crystal units are switched in; and (3) the upper frequency limit of the crystals, which at the time models 1102 and 1102A were designed was 125 MHz. This is achieved in both models by selecting an appropriate frequency plan, balanced mixers, times-2-frequency multipliers, and separate 85, 90, 95, and 100 MHz oscillators. To keep the cost low, the 6 MHz oscillators are designed to operate with one crystal. The pulling range obtained is approximately 0.15%; however, an identical 6 MHz oscillator used in model 1102 is pulled reliably (i.e., without switching into a spurious mode of oscillations) by more than 0.33%.

The only difference between the two synthesizer models is that the 1102A operates in the single-tone and LO modes only and the 1102 has an additional two modes of operation: mode B, which provides a continuous (instead of a discrete) frequency coverage between two 10 kHz increments anywhere in the 1.0 to 40.0 MHz range, and mode A-B, in which two rf tones with a spacing up to 10 kHz minimum, continuously variable, are simultaneously available at the synthesizer output. Mode A-B is convenient for two-tone intermodulation testing of hf receiver and transmitter mixers and amplifiers.

Figure 7-2 is a functional block diagram of the two-tone synthesizer. The blocks inclosed in dotted lines comprise model 1102A circuitry; the blocks outside the dotted lines are added to generate tone B. In all cases but one, the crystal-controlled oscillators are common to both channels. The variable two-tone frequency separation is obtained by a design calling for separate 6 MHz oscillators in channels A and B, so that the 6.0 to 6.009 MHz oscillator used to generate 1 kHz increments in channel A operates in 10 discrete frequency steps and its counterpart in channel B is continuously tuned over a 20 kHz band between 6.0 and 6.02 MHz with a front

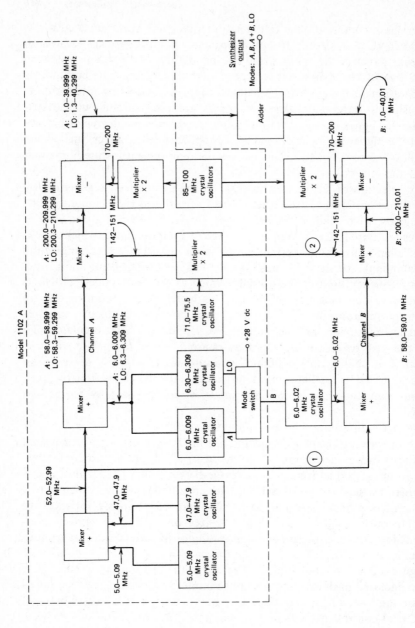

Figure 7-2. 1.0 to 40.01 MHz two-tone synthesizer, model 1102, manufactured by Polarad Electronic Instruments. Courtesy of Polarad Electronic Instruments, Division of Polarad Electronic Corporation.

panel vernier control. The two rf signals at the output of each channel are combined in a passive adder to produce a two-tone output, the levels of third-order intermodulation products being -60 dB below the desired output when the rf power of each tone is set to 0 dBm.

The circuitry in channel A is identical to that in channel B except for the isolation filter-amplifiers at points 1 and 2 (omitted in Fig. 7-2), which prevent spurious signals leaking from the output to the input of the 58 and 200 MHz mixers of one channel from appearing at the inputs of the analogous mixers of the other channel. Since the output mixer in both channels is driven by a broadband signal (85 to 100 MHz), separate times-2 multipliers and associated filters are used in each channel to obtain a channel isolation of better than 80 dB.

In spite of the multiplicity of signals at various frequencies generated within the system, an 80 dB circuit isolation is achieved by (a) using an egg crate type of module construction described in Chapter 3 with partitions separating circuits that operate at different frequencies (also see Fig. 7-3); (b) using a gasket, Polasheet II manufactured by Metex Corporation, in

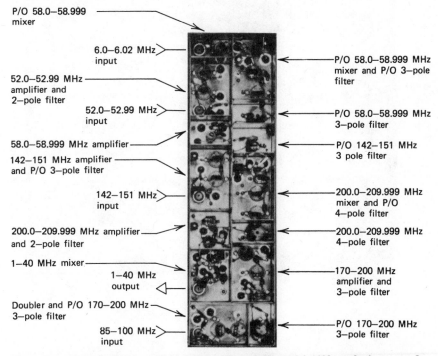

Figure 7-3. Photograph of a typical module (channel B), model 1102 synthesizer, manufactured by Polarad Electronic Instruments. Courtesy of Polarad Electronic Instruments, Division of Polarad Electronics Corporation.

conjunction with each module's covers; (c) selecting most of the coils used in the tuned circuits to be enclosed in shielded coil forms, such as type 1191 manufactured by Cambridge Thermionic Corporation, which provide both electrostatic and magnetic shielding: and (d) placing channels A and B in two separate modules, as Fig. 7-4 indicates.

Figure 7-5 is a front view of a model 1102 frequency synthesizer.

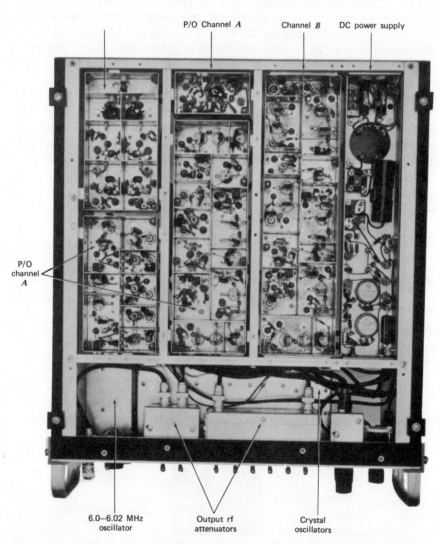

Figure 7-4. Bottom view of model 1102 synthesizer, manufactured by Polarad Electronic Instruments. Courtesy of Polarad Electronic Instruments, Division of Polarad Electronics Corporation.

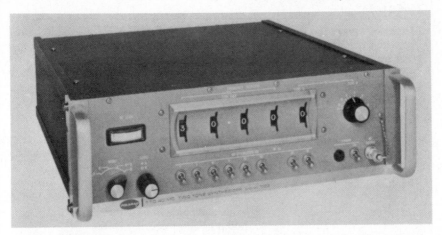

Figure 7-5. Front view of model 1102 frequency synthesizer, manufactured by Polarad Electronic Instruments. Courtesy of Polarad Electronic Instruments, Division of Polarad Electronics Corporation.

7-2   Coherent Synthesizers

The systems discussed here were selected because the approaches used in designing them point out best the problems associated with frequency synthesis and present solutions that result in state-of-the-art performance.

HF Synthesizer, Model 645A, Manufactured by John Fluke MFG. Co., Inc.

Table 7-2 is a partial list of model 645A requirements. The synthesizer operates between dc and 50 MHz in 0.01 Hz increments. This is achieved by direct synthesis utilizing the double-mix-divide approach described in Chapter 1. For clarity of presentation, the synthesizer is divided by the writer into three parts: the fixed-frequency, vhf synthesis, and uhf synthesis sections (see Fig. 7-6). The fixed-frequency section contains a very stable spectrally pure reference oscillator, from which signals at numerous fixed frequencies utilized in the vhf and uhf sections are derived. The vhf section synthesizes a signal at 40 to 41 MHz variable in 0.01 Hz increments. The uhf section produces a dc to 50 MHz output from a set of fixed-frequency signals and the signal synthesized by the vhf section.

The synthesizer is designed so as to result in the lowest possible frequency multiplication factor anywhere along the path of synthesis because of the ultralow phase-noise requirement governing the design of model 645A. The frequency of the reference oscillator is 5 MHz (see Fig. 7-7). A band-pass crystal filter improves the oscillator noise spectrum so as to make the effect of the reference oscillator noise on the synthesizer noise negligible at offset frequencies above 500 Hz. The 5 MHz frequency is

Table 7-2. Partial Performance Specification for the John Fluke Synthesizer, Model 645A[a]

Frequency range:	0.01 Hz to 49.99999999 MHz
Frequency increments:	0.01 Hz
Nonharmonically related spurious outputs:	− 100 dB

Residual phase-noise spectral density (single sideband-to-signal-phase-noise ratio of the synthesizer, excluding the 5 MHz reference oscillator):

Offset from signal (Hz):	20	2×10^2	5×10^3	5×10^4
ratio (dB):	104.6	116.6	129.9	132.0

Switching time: Less than 20 μsec for output amplifier to be within ± 1 dB of final value and phase to be within ± 0.1 rad at its new frequency (without automatic level control)

[a]Material reproduced with permission of John Fluke Mfg. Co., Inc.

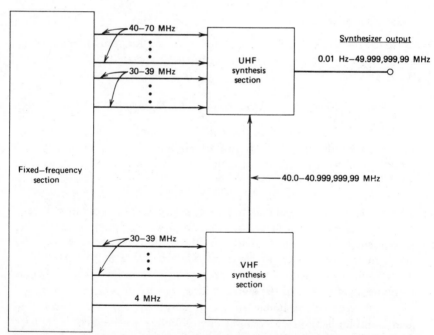

Figure 7-6. Direct current to 50 MHz direct coherent synthesizer, model 645A, manufactured by John Fluke Mfg. Co., Inc. Material reproduced with permission of John Fluke Mfg. Co., Inc. Reprinted from "An Ultra Low Noise Direct Frequency Synthesizer," D. G. Meyer, Proc. 24 AFCS, pp. 209–232, April 1970, available from Electronics Industries Association.

446

multiplied by 2, and the signal is applied to a 10 MHz comb generator, which converts a sine wave into a periodic pulse rich in harmonic content. The rf filter-amplifiers tuned to 30, 40, 50, 60, and 70 MHz select the appropriate harmonics for further frequency synthesis. The signals at 30 to 70 MHz are fed into the uhf section. These signals are also processed in the fixed-frequency section by undergoing frequency division, subtraction, and addition with the result that a set of 10 fixed-frequency signals at 30

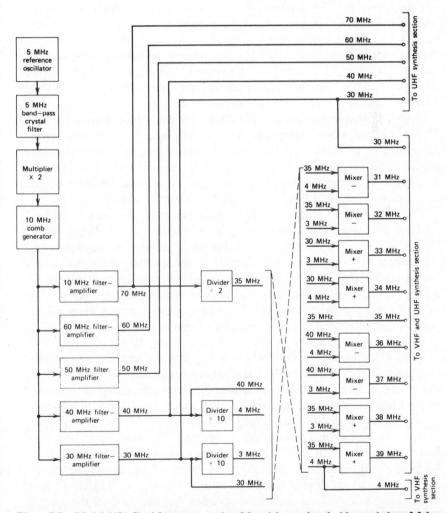

Figure 7-7.   Model 645A fixed-frequency section. Material reproduced with permission of John Fluke Mfg. Co., Inc. Reprinted from "An Ultra Low Noise Direct Frequency Synthesizer," D. G. Meyer, Proc. 24 AFCS, pp. 209–232, April 1970, available from Electronics Industries Association.

through 39 MHz variable in 1 MHz steps is generated. The 30 to 39 MHz frequencies are used in synthesis in both vhf and uhf sections.

A seemingly elaborate method of 30 to 39 MHz generation keeps all frequencies of the signals internal to the synthesizer as high as feasible throughout the frequency synthesis and results in superior phase-noise performance. Other steps taken to keep the noise low consist of (*a*) using negative feedback for $1/f$ noise reduction in rf amplifiers (not shown in Fig. 7-7 or other illustrations that follow) to as great an extent as practical; (*b*) designing for the highest signal power levels compatible with low spurious intermodulation products generation in order to obtain the highest possible signal-to-noise ratio at offset-from-signal frequencies at which thermal noise predominates; and (*c*) optimizing the noise figures of amplifiers and mixers. (The noise floors in various individual circuits throughout the system have been held from approximately -148 to -170 dB/Hz bandwidth relative to the desired signal level.)

Signals at frequencies as low as 3 MHz are used in generating the 30 to 39 MHz set; however, their noise can be neglected because these signals are added to or subtracted from signals whose frequencies (hence noise) are an order of magnitude higher. Figure 7-7 demonstrates how the 30 to 39 MHz signals are synthesized from 3, 4, 30, 35, and 40 MHz without significant phase-noise degradation of the processed signals.

The vhf synthesis section utilizes the signals at 4 MHz and at 30 to 39 MHz to generate 40 to 41 MHz signals variable in 0.01 Hz in the manner shown in Fig. 7-8. The 30 to 39 MHz set is applied to 10 inputs of a digital-gate matrix switch. Eight outputs of the switch carry signals at any frequency of the 30 to 39 MHz set independently of each other. The frequency of each matrix switch output is selected either manually, by using eight front panel switches, or remotely, by applying appropriately coded digital signals to a connector located on the rear panel.

The 30 to 30 MHz signals are divided by 10 to 3.0 to 3.9 MHz at the output of the matrix switch before being applied to the corresponding frequency decades. This relaxes the isolation requirement for the matrix switch digital gates by 20 dB and makes the noise introduced by the switch negligible as compared to the noise of the signal at 33 MHz, to which the signals at 3.0 to 3.9 MHz are added in the process of frequency synthesis.

Every frequency-increment decade operates on three input signals by first adding 4 MHz to 33 MHz, then adding the sum at 37 MHz to 3.0 to 3.9 MHz, and, finally, dividing 40 to 41 MHz by 10, except for the 100 kHz decade, which does not use a divider to provide a signal at 40 to 41 MHz variable in 0.01 Hz increments. Notice that the digits of the synthesized frequency are shifted to an adjacent, less significant decimal position each time the signal at 3.0 to 3.9 MHz is transmitted through a complete

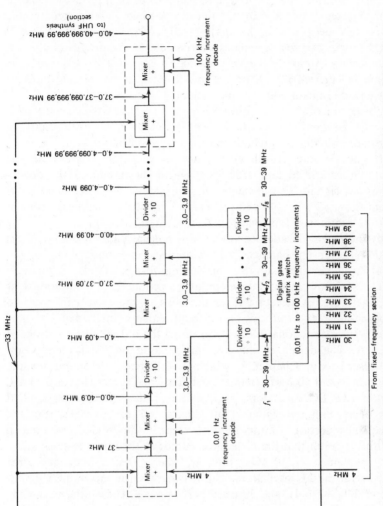

Figure 7-8. Model 645A vhf synthesis section. Material reproduced with permission of John Fluke Mfg. Co., Inc. Reprinted from "An Ultra Low Noise Direct Frequency Synthesizer," D. G. Meyer, Proc. 24 AFCS, pp. 209–232, April 1970, available from Electronics Industries Association.

decade. For example, the 100 kHz frequency steps of the signal at 3.0 to 3.9 MHz injected into the 0.01 Hz decade become 10 kHz steps at the output of this decade and, having passed a total of seven complete decades, become 0.01 Hz frequency increments at the output of the vhf synthesis section. Similarly, having passed no complete decades, the 100 kHz frequency increments of the 3.0 to 3.9 MHz signal injected into the 100 kHz decade remain 100 kHz increments at the vhf output.

The phase noise of the synthesized signal is kept low because of a repetitive use of the divide-by-10 circuits. Hence the noise of the signal at 33 MHz predominates throughout the frequency synthesis.

The frequency plan of the vhf section is prepared to result in the smallest number and highest order possible of the in-band and close-to-band spurious signals. Out-of-band spurious signals are attenuated by rf filters (not shown in Fig. 7-8) and divide-by-10 dividers. The most stringent suppression requirement is placed on the filters used in the 100 kHz frequency-increment decade because it does not have a divider to perform an additional clean-up. Balanced mixers are used throughout.

High rf levels are mandatory for high signal-to-noise ratios. A low level is preferred for low mixer intermodulation products. A compromise is reached, therefore, in selecting rf input levels to the mixers in order to satisfy phase-noise and spurious-output requirements simultaneously.

Figure 7-9 is a functional block diagram of the uhf synthesis section of model 645A. The design of this section is based on the triple-mix approach, described in Chapter 1, which is modified to account for the generation of 1 and 10 MHz frequency increments. A signal at 370 to 410 MHz from five LC oscillators, switched on one at a time, is applied simultaneously with the signal at 30, 40, 50, 60, or 70 MHz from the fixed-frequency section to the input hot-carrier-diode double-balanced mixer with the result that the difference output of the mixer is always at a constant frequency of 340 MHz plus Δf, where Δf is a frequency error associated with the LC oscillators. The two parametric mixers perform an addition of 340, 40 to 41, and 30 to 39 MHz signals. The output frequency of the right-hand parametric mixer is 410 to 420 MHz variable in 0.01 Hz. The final synthesizer output is the difference between the 410 to 420 MHz synthesized signal and the 370 to 420 MHz signal from LC oscillators with the subtraction performed in a low-noise-figure double-balanced mixer. This approach permits signal processing by frequency addition or subtraction to be performed with all in-band intermodulation products approaching a -110 dB level in relation to the desired output and, at the same time, provides cancellation of the phase noise and frequency drift associated with the LC oscillators that are used in the uhf synthesis to generate an auxiliary signal.

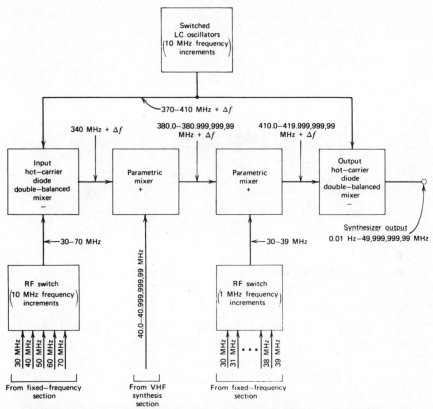

Figure 7-9. Model 645A uhf synthesis section. Material reproduced with permission of John Fluke Mfg. Co., Inc. Reprinted from "An Ultra Low Noise Direct Frequency Synthesizer," D. G. Meyer, Proc. 24 AFCS, pp. 209–232, April 1970, available from Electronics Industries Association.

The noise floor at the output of the broadband hot-carrier-diode mixer is as follows: AM at -150 dB/Hz bandwidth attributed to the power level and noise figure associated with the mixer (the input signal to the mixer is fairly low to achieve a -110 dB spurious-output performance), and PM at -137 dB/Hz due to a summation of noise floors throughout the synthesizer.

A significant effort, which also makes an important contribution to the excellent performance of model 645A, is made to achieve a mechanical design of the synthesizer compatible with a stringent spurious-output requirement. Modular construction avoids placing circuits that are sensitive to pickup near any signal sources operating at such frequencies. All rf inputs and outputs to and from each module are by means of coaxial rf connectors to prevent radiation. Feedthrough capacitors are utilized in all dc connections made between each module and the synthesizer power supply and among modules to prevent rf contamination by way of dc connections. Figure 7-10 is a photograph of a typical module used in a model 645A synthesizer.

Figure 7-10.   Typical module used in model 645A synthesizer, manufactured by John Fluke Mfg. Co., Inc. Photo courtesy of John Fluke Mfg. Co., Inc.

Cable routing is given special attention to prevent the pickup of spurious signals. The equipment is designed for a minimum number of interconnections between the modules to reduce the chance of pickup. Figure 7-11 and 7-12 are a set of photographs of the synthesizer as viewed from the top, bottom, front, and rear of the unit.

This system is a general type of synthesizer which is employed as a stable, spectrally pure signal generator in a variety of applications such as automatic testing of crystal filters and frequency discriminators and generation of local oscillator signals for receivers and spectrum analyzers. Utilization of model 645A in the frequency synthesis of microwave synthesizers is demonstrated further in this chapter.

UHF Synthesizer, Model 5105A / 5110B, Manufactured by Hewlett-Packard Company

The basic approach used in the design of the model 5105A/5110B synthesizer is similar to that for model 645A. The synthesizer may be described in terms of three sections: a fixed-frequency, vhf synthesis, and uhf synthesis (see Fig. 7-13). All frequencies are derived from one reference oscillator. The fixed-frequency section provides signals for vhf and uhf synthesis. The vhf section synthesizes by the double-mix-divide techniques a signal at 30 to 31 MHz, which is upconverted to 500 MHz in the uhf section.

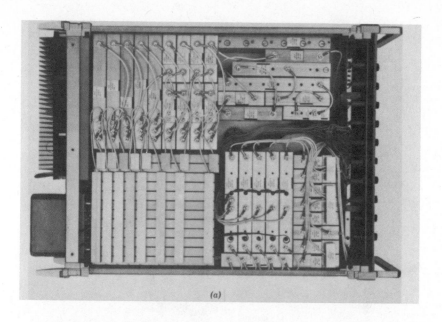

(a)

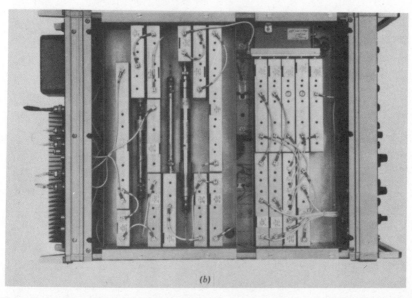

(b)

Figure 7-11. Top and bottom views of model 645A synthesizer, manufactured by John Fluke Mfg. Co., Inc. (a) Fixed-frequency and vhf synthesis sections. (b) UHF synthesis section. Photo courtesy of John Fluke Mfg. Co., Inc. Reprinted from "An Ultra Low Noise Direct Frequency Synthesizer," D. G. Meyer, Proc. 24 AFCS, pp. 209–232, April 1970, available from Electronics Industries Association.

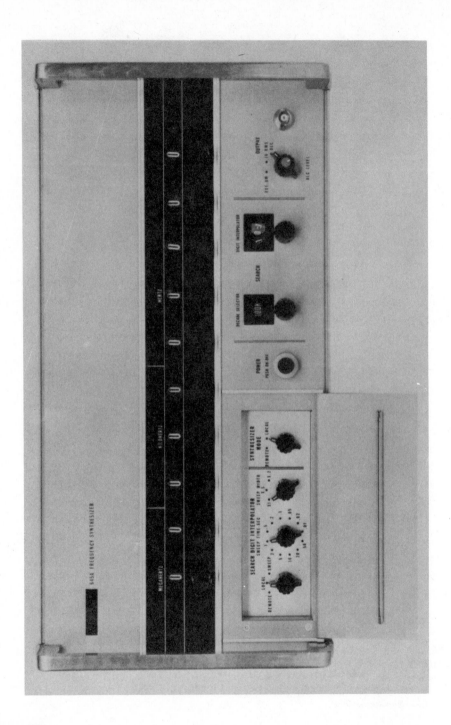

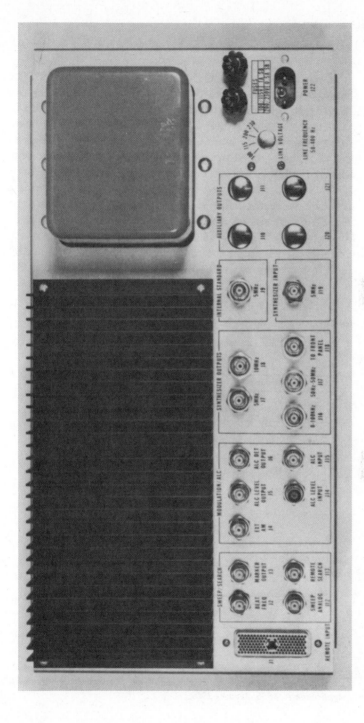

Figure 7-12. Front and rear panels of model 645A synthesizer, manufactured by John Fluke Mfg. Co., Inc. Photocourtesy of John Fluke Mfg. Co., Inc. Reprinted from "An Ultra Low Noise Direct Frequency Synthesizer," D. G. Meyer, Proc. 24 AFCS, pp. 209–232, April 1970, available from Electronics Industries Association.

The differences between models 645A and 5105A/5110B are important enough, however, to justify descriptions of both synthesizers.

Table 7-3 is a partial list of model 5105A/5110B requirements. The synthesizer frequency range is 0.1 to 500 MHz. These frequencies are derived from a 1 MHz reference oscillator. The oscillator, which is followed by a phase-noise clean-up crystal filter (see Fig. 7-15), drives a comb generator the output signal of which, a periodic pulse wave, is used to synthesize a set of signals at 3 through 60 MHz by filtering, addition, multiplication, and division. The rf diode matrix switch of the vhf synthesis section operates at 3.0 to 3.9 MHz (see Fig. 7-16), channeling appropriate frequencies into the double-mix-divide decades. The vhf synthesis is done at 30 to 31 MHz. There are eight frequency decades; hence the signal at 30 to 31 MHz is varied in 0.01 Hz steps at the output of the vhf section. However, because of the nature of the uhf synthesis, the final synthesizer output frequency is incremented in 0.1 Hz. Figure 7-17 shows how this change in the smallest frequency increment comes about. To produce a

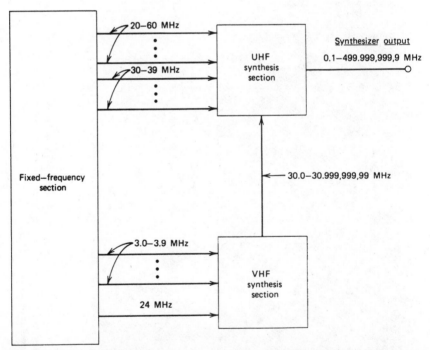

Figure 7-13.   0.1 to 500 MHz direct coherent synthesizer, model 5105A/5110B, manufactured by Hewlett-Packard Company. Reprinted by permission from "Frequency Synthesizers," Hewlett-Packard *Application Note* 96, January 1969.

signal at 0.1 to 500 MHz and keep all spurious products generated in the broadband difference mixer at a level that is lower than −70 dB below the desired signal, both the synthesized and LC oscillator frequencies are first multiplied by 10 (as is the smallest frequency increment), and the signals are then applied to the mixer inputs.

Table 7-3. Partial Performance Specification for Hewlett-Packard Synthesizer, Model 5105A/5110B.[a]

Frequency range:	0.1 to 499.9999999 MHz
Frequency increments:	0.1 Hz
Nonharmonically related spurious outputs:	−70 dB
Phase noise:	See Fig. 7-14
Switching time:	20 μsec typically

[a]Reprinted by permission from Hewlett-Packard *Date Sheet for Frequency Synthesizer*, Model 5105A/5001B, March 15, 1967.

Conceptually, the uhf section of model 5105A/5110B is very much like the uhf section of model 645A and is not discussed in any greater detail here.

Some insight into the sources limiting the frequency switching time may be of value to the reader. In general, the switching time of a direct synthesizer is limited by the bandwidths of the rf filters employed along the path of synthesis and in the control lines of the matrix switch. In model 5105A/5110B the propagation delay through each complete frequency-increment decade is up to 4 μsec. The delay associated with the diode

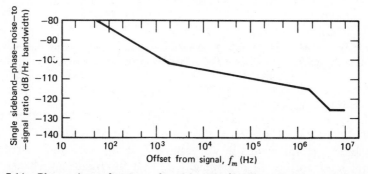

Figure 7-14. Phase-noise performance of model 5105A/5110B synthesizer, manufactured by Hewlett-Packard Company. Reprinted by permission from Hewlett-Packard *Data Sheet for Frequency Synthesizer, Model 5105A/5110B*, March 15, 1967.

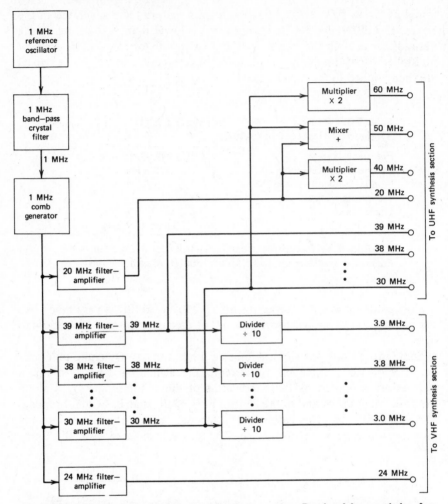

Figure 7-15. Model 5105A/5110B, fixed-frequency section. Reprinted by permission from "Frequency Synthesizers," Hewlett-Packard *Application Note* 96, January 1969.

matrix switch is less than 1 μsec. Frequency injection from vhf to uhf synthesis sections takes place in about 1 μsec. The 350 to 390 MHz LC oscillators in the uhf section are switched on in 2 μsec with a short period of phase drift following. If we assume that the synthesizer frequency control lines are loaded in parallel (i.e., that the frequency change command data are fed into the control lines simultaneously), the major contribution to the synthesizer switching time is made by the propagation delays through the double-mix-divide decades, particularly in cases where

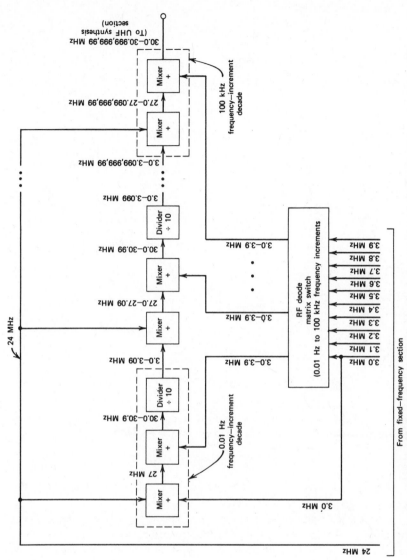

Figure 7-16. Model 5105A/5110B, vhf synthesis section. Reprinted by permission from "Frequency Synthesizers," Hewlett-Packard *Application Note 96*, January 1969.

459

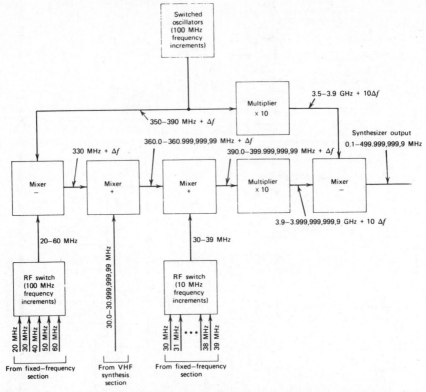

Figure 7-17. Model 5105A/5110B, uhf synthesis section. Reprinted by permission from "Frequency Synthesizers," Hewlett-Packard *Application Note 96,* January 1969.

the synthesizer frequency is incremented in small steps. Changing the frequency in 0.1 Hz increments takes the longest time.

In spite of circuit complexity and numerous sources of time delay, direct synthesizers of the model 5105A/5110B class display switching time performance superior to that of any other type of synthesizers presently in use.

Moreover, almost any desired frequency range can be obtained with the frequency synthesis approach utilized in model 5105A/5110B by modifying the uhf synthesis section appropriately.

Model 5105A/5110B is employed as a stable, spectrally pure signal generator in a variety of applications such as automatic testing of crystal filters and frequency discriminators and generation of local oscillator signals for high-quality communication receivers and transmitters.

A front view of model 5105A/5110B is shown in Fig. 7-18. The vhf and uhf synthesis takes place in the model 5105A synthesizer. The model 5110B driver generates the fixed frequencies utilized in the vhf and uhf synthesis sections and is capable of driving five synthesizers, model 5105A, simultaneously.

GI/ESD Digital Synthesizers

The synthesizers described in this section are intended to serve as the first local oscillators in the hf computer-controlled receivers designed and built at Electronic Systems Division of General Instruments Corporation.

The first synthesizer considered here generates a signal at 92.67 to 122.17 MHz variable in 100 Hz increments. It is composed of three digital phase-locked loops (PLLs), an output PLL operating at 92.67 to 122.17

Figure 7-18. Front view of model 5105A/5110B synthesizer, manufactured by Hewlett-Packard Company. Courtesy of Hewlett-Packard Company.

MHz, and two auxiliary PLLs operating at 15.3 to 16.2 and 137.01 to 147.00 MHz, combined to form the system shown in Fig. 7-19. The synthesizer is to meet the specification given in Table 7-4. Because of stringent phase-noise and switching-time requirements, the synthesizer is designed so that only the output PLL contributes to the synthesizer phase noise, and only the 137 to 147 MHz PLL bandwidth is a governing factor in determining the synthesizer switching time.

Table 7-4. Partial Performance Specification for the First Local Oscillator of Model DCR-30 Computer-Controlled Receiver, Manufactured by GI Electronic Systems Division[a,b]

Frequency range:	92.67 to 122.17 MHz
Frequency increments:	100 Hz
Nonharmonically related	(a) Spurious outputs falling in
spurious outputs:	the passband of the front end
	or first IF frequencies, − 110 dB
	(b) Elsewhere, − 80 dB
Phase noise:	See Fig. 7-19
Switching time:	5 msec

[a] By permission of Electronic Systems Division of General Instrument Corporation.
[b] The front end passband is between 0.5 and 30 MHz. The first IF passband is between 122.662 and 122.678 MHz.

The first condition is satisfied by dividing the frequency of the 137 to 147 MHz PLL output signal by 100 (a 40 dB improvement of the PLL phase noise) and by selecting the frequency of the 15.3 to 16.2 MHz PLL to be significantly lower (− 17.6 dB) than the frequency of the output PLL. A very stable 1 MHz crystal-controlled oscillator is selected for a reference source because the phase-noise requirement, Fig. 7-20, specifies close-to-signal noise only. The oscillator is followed by a single-pole clean-up crystal filter with both the oscillator and the filter located in a proportionally controlled oven. (The phase-noise spectrum of the oscillator-filter combination is shown in Fig. 7-21.) At the same time the bandwidth of the output PLL is made large enough to attenuate the VCO noise so that the noise in the region of interest is mainly the 1 MHz oscillator noise multiplied to the final synthesizer frequency. To relax the loop gain requirement for the output PLL, the 92 to 122 MHz VCO is carefully designed for low noise operation, utilizing the transmission line techniques described in Chapter 6. The circuit is tuned with high-Q varicaps, SQ1730 and SQ1744 by MSI, in three bands. The levels of rf signals throughout the system are kept above − 8 dBm to negate the effect of the noise floors of the amplifiers and mixers on the synthesizer noise.

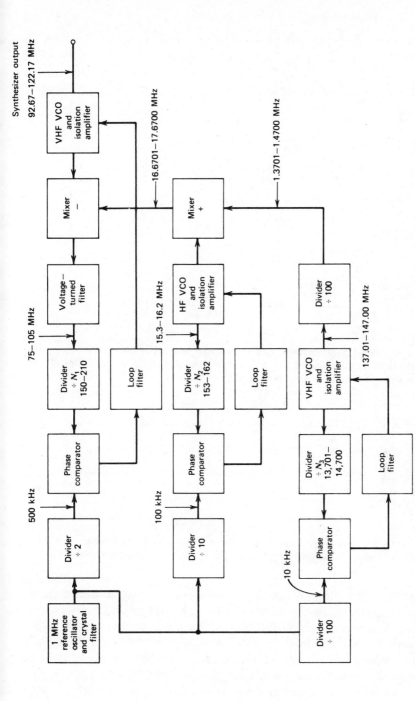

Figure 7-19. First local oscillator for model DCR-30 computer-controlled receiver, manufactured by GI Electronic Systems Division. By permission of Electronic Systems Division of General Instrument Corporation.

463

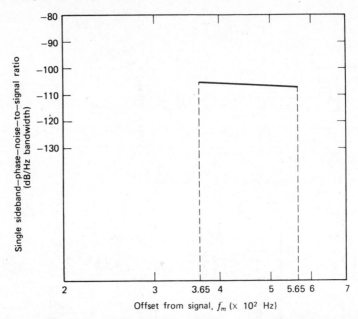

Figure 7-20. Phase-noise requirement for the first local oscillator of model DCR-30 computer-controlled receiver, manufactured by GI Electronic Systems Division. By permission of Electronic Systems Division of General Instrument Corporation.

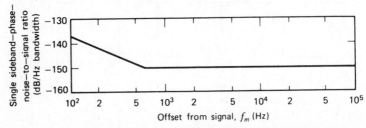

Figure 7-21. Phase-noise specification for 1 MHz reference oscillator followed by a crystal filter. By permission of Electronic Systems Division of General Instrument Corporation.

The second condition is satisfied by selecting the loop bandwidth of the PLLs, that is, the phase-comparator frequencies, so that the narrowest loop bandwidth occurs in the 137 to 147 MHz PLL, and the bandwidths of the other two PLLs are at least an order of magnitude wider. All three PLLs are first-order loops so that frequency acquisition takes place without an overshoot in the phase response. Each VCO is tuned to within the capture range, and no auxiliary capture circuitry, such as a sweep oscillator, is used during the acquisition mode of operation.

In addition to the phase-noise and switching-speed requirements, selection of the frequency plan shown in Fig. 7-19 is affected by considerations of (a) the in-band spurious outputs of both mixers, (b) the 100 Hz frequency-increment requirement, (c) the width of the operating frequency ranges of the VCOs used in the auxiliary PLLs, and (d) the highest reliable frequency of operation of commercially available integrated circuits comprising the variable-ratio frequency dividers, N_1, N_2, and N_3.

Special steps are taken to keep all spurious outputs below the required level. Some of these are typical of all synthesizers; they are described in the discussion of model 645A. Here a few problems peculiar to the selected frequency plan and application of the synthesizer are considered.

The output PLL attenuates all spurious signals injected into the loop with the 16.67 to 17.67 MHz signal by approximately 20 dB inside the loop bandwidth, so that to meet the -80 dB requirement the system is designed for a -65 dB level of in-band and out-of-band spurious signals. Additional filtering of the out-of-band spurious signals is provided by the voltage-tuned filter. The major function of this filter, though, is to attenuate the 92 to 122 MHz signal appearing at the input of the variable-ratio divider N_1 to less than -60 dB in relation to the desired input level at that point. If this signal were allowed to be higher than -60 dB, there would be a higher than -80 dB crossing spurious signal at the output of the synthesizer, which is generated as a second-order product of the signal at 92 to 122 MHz and an appropriate harmonic of the 500 kHz reference pulse every 500 kHz of the synthesizer frequency. For example, if the synthesizer frequency were 91.01 or 90.99 MHz, the spurious signal would be ± 10 kHz away from the synthesizer output signal, manifesting itself as a pair of narrowband FM components. For a frequency separation between the synthesized and spurious signals greater than the loop bandwidth of the output PLL, the PLL provides an additional 6 dB/octave attenuation of such spurious signals.

The -110 dB requirement is satisfied by (a) placing each PLL in a shielded module of its own, with partitions separating circuits that operate at different frequencies (for photographs of a typical module see Figs. 7-22 and 7-23); (b) utilizing the Polasheet II gasket manufactured by Metex Corporation in conjunction with module covers; (c) designing the system with sample-and-hold phase comparators with additional attenuation of the reference signal provided by a Twin-T and low-pass filters following the phase comparators; (d) having separate divide-by-10 and divide-by-100 frequency dividers located in the auxiliary PLLs, in order that strong harmonics of the 10 and 100 kHz fast-rise-time pulses which are in the receiver band do not leak to the front end circuitry of the receiver by way of cables and large ground loops; (e) using semirigid cables for all rf interconnections among the synthesizer modules and between the

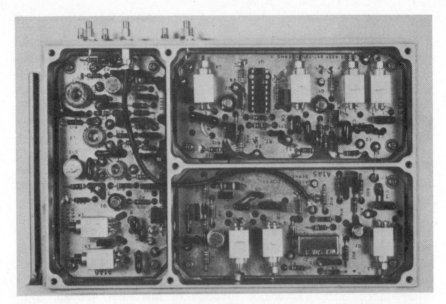

Figure 7-22. Top view of a typical module. First local oscillators of GI/ESD computer-controlled receiver. Courtesy of Electronic Systems Division of General Instrument Corporation.

Figure 7-23. Projection view of the module shown in Fig. 7-22. Courtesy of Electronic Systems Division of General Instrument Corporation.

synthesizer and receiver circuits; (*f*) designing for the synthesizer power supplies (+5 V dc and +28 V dc for the digital and rf circuitry, respectively) separately from the receiver circuitry power supplies; and (*g*) placing a two-stage unity-gain isolation amplifier between the VCO and the mixer in the output PLL in order that the signal at 16.67 to 17.67 MHz leaking to the VCO input of the mixer does not contaminate the synthesizer output. Under these conditions the most important leakage paths for undesired signals are through the control lines for VCO band switching and variable-ratio dividers, which are dealt with by inserting LC and RC low-pass filters in appropriate places.

The 100 Hz frequency-increment requirement is satisfied by designing the 137 to 147 MHz PLL to generate 10 kHz increments without violating the switching-time requirement and by dividing the signal at 137 to 147 MHz by 100 (see Fig. 7-19).

To arrive at a low-cost VCO tuning arrangement, it is desirable to operate the auxiliary PLL VCOs over as narrow a frequency range as possible. This means that the VCO frequency should be as high as feasible. However, two factors limit the VCO frequency; the phase-noise requirement, and the highest frequency at which the variable-ratio dividers operate reliably. Hence in selecting the frequencies of both auxiliary PLLs a compromise is reached that results in simple, low-cost tuning networks, each consisting of a two-step resistive divider.

In selecting an approach that meets a set of requirements, the designer should take into account possible future modifications of the original synthesizer specification and arrive at a system that lends itself to such modifications without major electrical or mechanical redesign. Table 7-5

Table 7-5. Partial Performance Specification for the First Local Oscillator of Model DCR-30B Computer-Controlled Receiver, Manufactured by GI Electronic Systems Division[a,b]

Frequency range:	90.40 to 118.89999 MHz
Frequency increments:	10 Hz
Nonharmonically related spurious outputs:	(*a*) Spurious outputs falling in the passband of the front end or first IF frequencies, −110 dB
	(*b*) elsewhere, −80 dB
Phase noise:	See Fig. 7-24
Switching time:	20 msec

[a]By permission of Electronic Systems Division of General Instrument Corporation.
[b]The front end passband is between 1.5 and 30 MHz. The first IF passband is between 120.392 and 120.408 MHz.

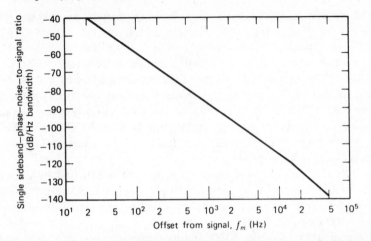

Figure 7-24. **Phase-noise requirement for the first local oscillator of model DCR-30B computer-controlled receiver, manufactured by GI Electronic Systems Division. By permission of Electronic Systems Division of General Instrument Corporation.**

presents the most important requirements for the first local oscillator of the model DCR-30B computer-controlled receiver. Comparing this specification to the requirements of Table 7-4, we notice the similarity between the two sets of requirements in the output frequency range. We also see the major differences: for the model DCR-30B, 10 Hz frequency increments instead of 100 Hz, a much wider range of offset frequencies at which synthesizer phase noise is specified (though it is relaxed by more than 25 dB at offset frequencies between 300 and 600 Hz), and a relaxed switching time (20 msec instead of 5 msec).

The 10 Hz increments are obtained by changing the 137 to 147 MHz PLL so that it operates over an octave band, from 100 to 200 MHz, and by dividing these frequencies by 1000, as shown in Fig. 7-25. This presents a minor problem in the design of the divide-by-N_3 and divide-by-100 circuits, which now operate at 200 MHz. Available on the market are vhf prescalers rated for that frequency.

To meet a relaxed close-in and stringent far-out phase-noise requirement the bandwidth of the main loop is reduced by inserting a low-pass filter and a lag network in the loop. Both circuits are automatically removed from the loop during frequency acquisition by techniques described in Chapter 4, so that their addition does not alter the original VCO tuning arrangement. At the same time the noise of the 90 to 118 MHz VCO is further improved by tuning the VCO in four bands and by replacing all low-cost capacitors used in this circuit with low-loss components.

It is decided during the redesign phase to eliminate the binary-coded-decimal (BCD)-to-9's-complement converter used in controlling the ratios

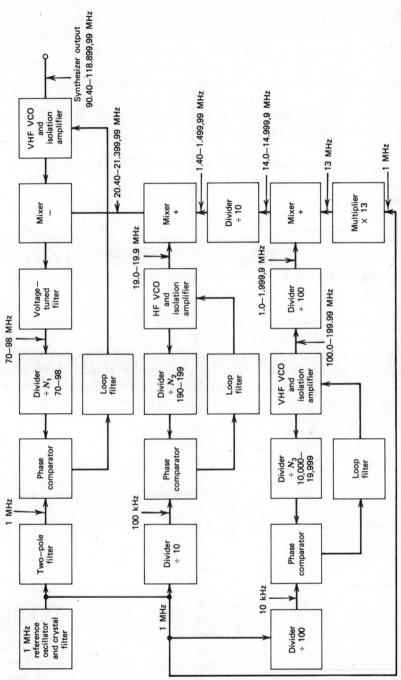

Figure 7-25. First local oscillator for model DCR-30B computer-controlled receiver, manufactured by GI Electronic Systems Division. By permission of Electronic Systems Division of General Instrument Corporation.

469

of the variable-ratio dividers by slightly modifying the original frequency plan. Table 7-6 lists the extreme values of the receiver rf tuning capability and division ratios required. A study of these two sets of values indicates that a direct relationship exists between the 9's complement necessary to set the ratios and the input receiver frequency. By selecting the frequencies of the three PLLs as shown in Fig. 7-25, the need for a BCD-to-9's complement conversion is eliminated.

Table 7-6.   Receiver Frequencies and Ratios of Variable-Ratio Frequency Dividers[a]

Receiver Frequency (MHz)	Synthesizer Output (MHz)	Loop division ratios		
		N_1	N_2	N_3
01.50000	118.89999	98	194	19,999
29.99999	90.40000	70	190	10,000

[a]By permission of Electronic Systems Division of General Instrument Corporation.

Comparison of Fig. 7-19 with Fig. 7-25, reveals that relatively small changes in the original design resulted in a synthesizer with significantly different characteristics. Also, as one would expect, the synthesizer specifications underwent further modifications as some receiver users expressed a desire to tune the receiver in 1 Hz increments. This requirement is satisfied by replacing the 13 MHz fixed-frequency signal with the digital PLL-divider network shown in Fig. 7-26.

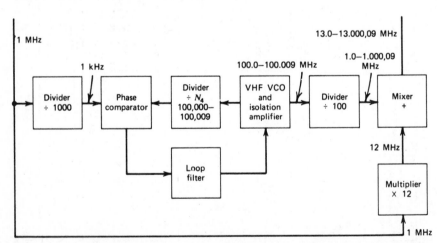

Figure 7-26.   Additional circuitry required to improve the frequency resolution of model DCR-30B computer-controlled receiver, manufactured by GI Electronic Systems Division. By permission of Electronic Systems Division of General Instrument Corporation.

7-3   Microwave Synthesizers

The output frequency range for some applications, particularly in communications, has shifted from hf to shf during the last two decades, whereas other synthesizer requirements, such as those for spurious outputs and phase noise, have remained unchanged or became more stringent. These developments present severe design problems.

Whenever the output-frequency-range, spurious-signal and phase-noise requirements permit, the frequency synthesis approach used to generate signals at microwave frequencies involves a vhf synthesizer followed by a frequency multiplier. A digital synthesizer is well suited for such a purpose because it generates any desired value of frequency increments. For example, if the final synthesized frequencies were specified as between 4 and 5 GHz incremented in 10 kHz, a vhf synthesizer tuned between 250.0 and 312.5 MHz in 625 Hz steps followed by a times-16 multiplier would fulfill the requirement at relatively low cost. This is commonly done whenever phase noise is specified only in the far-from-signal region by designing for a low-noise VCO and narrow-loop-bandwidth PLL so that the noise at the offset frequencies of interest is predominantly the VCO noise. Such techniques are described elsewhere in this book and are not elaborated upon here. An example exhibiting typical problems associated with the design of microwave synthesizers will be of great value to the reader.

We begin by evaluating the specification for a local oscillator of a satellite communication receiver, Table 7-7. It differs from all previously discussed synthesizer specifications in three respects: (a) the operating frequencies are at shf; (b) the FM discrete spurious signals are singled out from the general requirement for nonharmonically related spurious outputs; and (c) a long-term frequency stability requirement is added to the specification.

Table 7-7.   An SHF Synthesizer Specification[a]

Frequency range:	6550 to 7050 MHz
Frequency increments:	10 Hz
Nonharmonically related spurious outputs:	(a) From 6550 to 7250 MHz, -80 dB
	(b) From 5850 to 6350 MHz, -113 dB
	(c) From 7250 to 7750 MHz, -113 dB
	(d) Discrete FM see Fig. 7-27
Phase noise	See Fig. 7-27
Switching time	Not applicable (manual operation)
Long-term frequency stability	± 2 parts in 10^{11} per month

[a]Courtesy of ITT Defense Communications Division.

To demonstrate how difficult it is to meet these requirements, let us attempt to use either the Fluke synthesizer, model 645A, or the Hewlett-Packard synthesizer, model 5105A/5110B, the finest systems of their kind presently available on the market, in conjunction with appropriate frequency multipliers. Since the requirement calls for 10 Hz increments, the smallest multiplication factor that can be used to generate frequencies in the shf band is times-200 for model 645A (which is equivalent of a 46 dB enhancement of FM sidebands and phase noise) and times-20 for model 5105A/5110B, a 26 dB enhancement. The specified LO phase noise and FM sidebands are plotted in Fig. 7-27. The noise and sideband performances of the two synthesizer-multiplier systems are compared to the requirements in Fig. 7-28. Neither of the systems meets the phase-noise or discrete FM requirements. (It is assumed in preparing Fig. 7-28 that some of the nonharmonically related spurious outputs of each synthesizer are in the form of crossing FM sidebands that cannot be filtered out and that are, therefore, enhanced by the multipliers—a valid assumption.) For this reason no attempt is made to analyze these approaches with respect to other requirements of the specification.

A significantly more elaborate approach that meets all requirements of the specification given in Table 7-7 is shown in Fig. 7-29. A reference source supplies a spectrally pure ultrastable signal to a low-noise distribution amplifier that drives a hf synthesizer. The synthesizer frequency is upconverted to uhf and multiplied to shf. Figure 7-30 is a functional block diagram of the system. The reference source is a rubidium cell atomic standard operating at 5 MHz, which is followed by a narrowband crystal filter. The phase-noise performance of the standard with and without the filter is shown in Fig. 7-31. The signal at 5 MHz is distributed with some

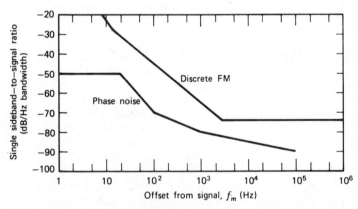

Figure 7-27. Phase-noise and discrete-spurious-outputs requirement for shf synthesizer. Courtesy of ITT Defense Communications Division.

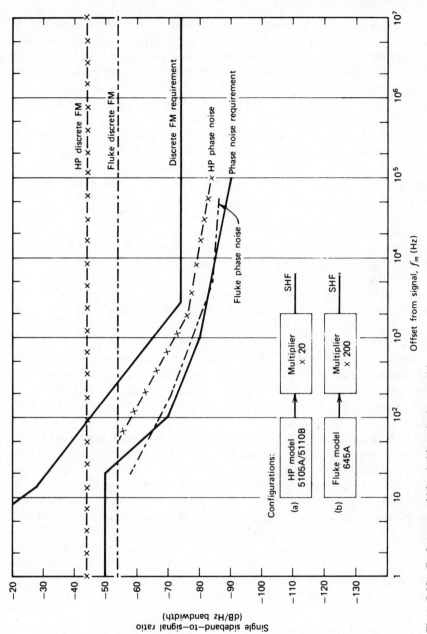

Figure 7-28. Performance of hf and uhf synthesizers at shf.

473

noise degradation by the amplifier to the model 645A synthesizer and rf power amplifiers driving two vhf multipliers (see Fig. 7-32 for a plot of the distribution amplifier residual phase noise). The hf synthesizer signal at 22.5 to 47.5 MHz is upconverted to uhf and then multiplied to shf. The double conversion is necessary to achieve an in-band spurious-output performance of better than -110 dB before the synthesized signal is applied to the times-2 uhf multiplier.

The three major design problems and their solutions are as follows: (a) suppression of spurious signals, which is dealt with by selecting an appropriate frequency plan and by applying extensive filtering to the processed signal after every synthesis operation, with the result that all out-of-band spurious outputs are attenuated by more than -110 dB at uhf with additional filtering provided in and at the output of the times-10 multiplier; (b) ultralow phase-noise operation at hf and uhf, which is achieved by selecting the reference source, distribution amplifier, and hf synthesizer that exhibit the lowest possible phase-noise performance, by designing for the lowest multiplication factor associated with the circuitry following the hf synthesizer, and by applying known phase-noise reduction techniques to all circuits such as amplifiers and multipliers used in frequency synthesis; and (c) suppression of 60 Hz discrete PM sidebands, which is considered in some detail next.

The shf spurious FM, or PM, sidebands requirement is -41 dB below the desired signal at 60 Hz offset frequency (see Fig. 7-27). This requirement is -104 dB at 5 MHz at the input to both vhf multipliers, or approximately 63 dB more stringent because of a multiplication factor of 1410.

The first aspect of the problem arises from a frequency coherence of the spurious signal considered; that is, circuits such as amplifiers, unless driven by a battery power supply in an environment free of 60 Hz magnetic and electric fields, act upon the processed signal as phase modulators rapidly enhancing the 60 Hz sidebands of the signal. The analysis is accomplished by drawing a detail block diagram of the 5 MHz portion of the system (see Fig. 7-33), assigning a phase modulation index associated with every circuit under consideration, and summing up the individual contributions. For example, assume that the atomic standard output signal is phase modulated at 60 Hz with a pair of sidebands being -110 dB below the signal. Assume also that each amplifier modulates the processed signal at a 60 Hz

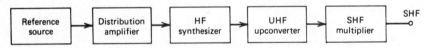

Figure 7-29.   SHF synthesizer. Courtesy of ITT Defense Communications Division.

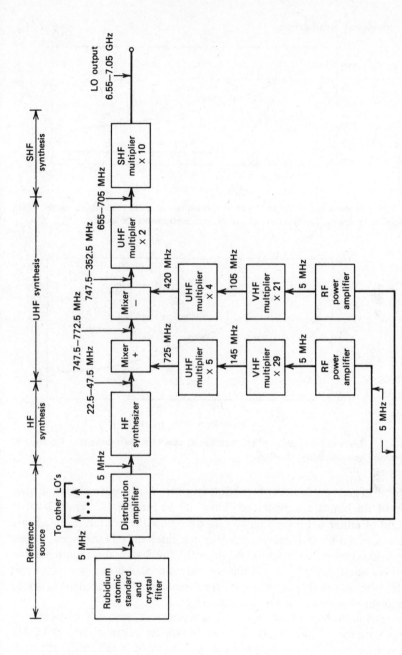

Figure 7-30. SHF synthesizer; functional block diagram. Courtesy of ITT Defense Communications Division.

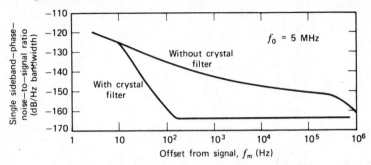

Figure 7-31. Phase-noise performance of the rubidium frequency standard, model 304D, manufactured by Tracor, Inc. Courtesy of Tracor, Inc., Instruments Division.

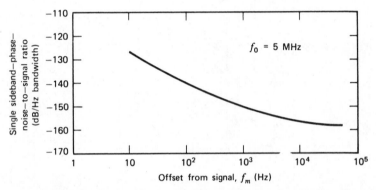

Figure 7-32. Residual phase noise of the distribution amplifier, shf synthesizer. Courtesy of ITT Defense Communications Division.

rate by a degree equivalent to modulating an ideally pure signal so that −110 dB sidebands are produced. Then the 60 Hz PM sidebands level at the input of either vhf multiplier is −98 dB below the desired signal. Table 7-8 shows a step-by-step enhancement of the sidebands as the signal passes the individual amplifier circuits. As this table indicates, the requirement is met with a small margin only if the equivalent contribution of each circuit is −120 dB. A phase coherence of the sidebands is assumed in order to arrive at the worst-case results.

The second aspect of the problem is associated with the difficulty in measuring levels of sidebands that are −120 dB below signal at 60 Hz offset frequency. Although test equipment exists with a capability for such measurements, it is questionable whether the data taken are valid in an environment that is not free from magnetic or electric fields produced by the 60 Hz power lines or test equipment or both. Some manufacturers of atomic standards and distribution amplifiers, therefore, do not guarantee

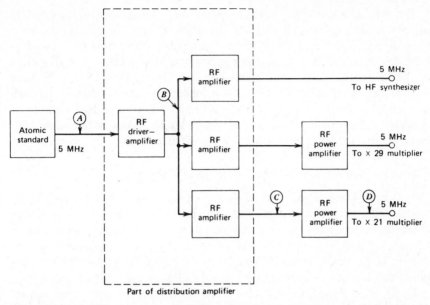

Figure 7-33. 5 MHz frequency distribution. Courtesy of ITT Defense Communications Division.

Table 7-8. Levels of 60 Hz Sidebands at Various Points in the 5 MHz Distribution System Shown in Fig. 7-33

The 60 Hz FM sidebands requirement assigned to each circuit in Fig. 7-33	Total level of 60 Hz sidebands at the points identified in Fig. 7-33 (dB)			
(dB)	A	B	C	D
− 110	− 110	− 104	− 100.5	− 98
− 115	− 115	− 109	− 105.5	− 103
− 120	− 120	− 114	− 110.5	− 108

to meet any 60 Hz PM requirement that calls for sidebands lower than − 110 dB. What is left to the designer is to make measurements at the output of either vhf multiplier, where the levels of the sidebands are approximately − 80 dB, and to determine, by elimination and substitution techniques, which circuit fails to perform properly.

In the case of the 5 MHz power amplifiers driving the vhf multipliers satisfactory performance is achieved by designing for a low 60 Hz ripple power supply and by providing further ripple attenuation with voltage regulators, type 723 manufactured by Fairchild, in conjunction with each power amplifier.

The mechanical design effort associated with this system consists mainly of the uhf upconverter packaging. The upconverter is located in a 19-in.-wide drawer with the rf amplifiers and multipole filters (not shown in Fig. 7-30, which follow the vhf and uhf multipliers and mixers occupying most of the space. Modular construction is used throughout. Double-braided coaxial cables, such as RG/U-142, provide rf interconnections between the modules. The − 110 dB isolation between the individual circuits or modules of the upconverter is obtained by taking basically the same steps as in the case of the synthesizers previously described.

UHF-SHF Synthesizer, Systron-Donner's 1600 Series

The Systron-Donner microwave system consists of a synthesizer driver, model 1600, and a set of rf units, models 1600-5 to 1617. The driver generates three signals that are utilized by the rf units: a programmable dc voltage, a spectrally pure 100 MHz reference, and a signal whose frequency is varied between 10 and 20 MHz in 0.1 Hz steps. The rf units synthesize frequencies between 0.5 and 18 GHz in 1 Hz increments.

Table 7-9 is a partial specification for the synthesizer. The requirements of special interest are an octave-band operation, a 1 Hz frequency-increment generation at shf, and a phase-noise performance that is virtually independent of synthesized frequency. These parameters are discussed in the system description.

The synthesizer circuitry is grouped into a fixed-frequency, a hf synthesis, and an uhf-shf synthesis section. The fixed-frequency section generates 100 kHz, 1 MHz, and 100 MHz signals from a spectrally pure 10 MHz VCXO, which is frequency locked by way of a frequency divider and phase comparator to either an internal 3 MHz or an external 1 or 5 MHz reference source as shown in Fig. 7-34. A rear panel INT/EXT REF switch determines the mode of operation. Independently of the reference source used, the reference signal utilized in frequency synthesis is always at 1 MHz. By making the loop bandwidth of the 10 MHz PLL very narrow, the phase-noise spectrum of the 1 MHz synthesized reference is almost identical to the spectrum of the 10 MHz VCXO signal, appropriately scaled down, at all times, which is an important advantage.

This approach is repeated in the generation of the 100 MHz signal. The loop bandwidth of the 100 MHz PLL is approximately 100 Hz. Hence the signal displays a phase-noise spectrum of the 100 MHz VCXO at offset frequencies above 100 Hz. A spectrally pure VCXO is employed.

The signal at 100 MHz is fed into the uhf-shf synthesis section for further processing. The signals at 100 kHz and 1 MHz are used in hf synthesis to generate a signal varying between 10 and 20 MHz in 0.1 Hz steps.

Table 7-9. Partial Performance Specification for Series 1600 Microwave Synthesizers, Manufactured by Systron-Donner[a]

Frequency range:	0.5–1.0 GHz	Model No. 1600-1600-5
	1–2	1600-1601
	2–4	1600-1603
	3.5–6.5	1600-1606
	4–8	1600-1607
	8–12	1600-1611
	12–18	1600-1617

Frequency increments:	1 kHz standard; 1 Hz optional

Nonharmonically related spurious outputs	− 60 dB below signal typical

Phase noise:

Single sideband phase noise-to-signal ratio (dB/Hz bandwidth)	Offset from signal (kHz)	Frequency range (GHz)
− 89	1	0.5–8.0
− 95	10	
− 115	100	
− 85	1	8.0–12.0
− 95	10	
− 115	100	
− 75	1	12.0–18.0
− 85	10	
− 105	100	

Switching time:	1 msec maximum to within 10 kHz of commanded frequency except when slewing to or through frequencies that are exact rationals of 100 MHz, in which case switching time is 50 msec

[a]By permission of Systron-Donner Corporation, Concord Instrument Division.

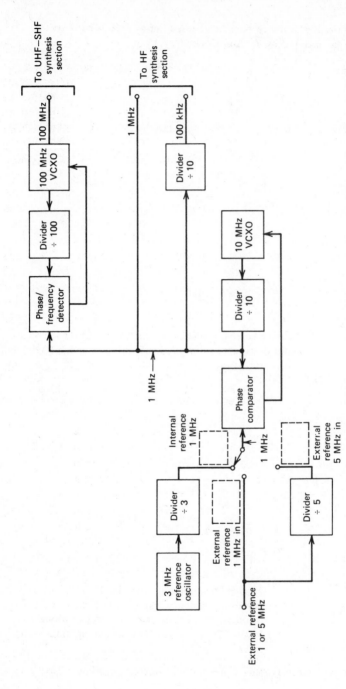

Figure 7-34. Fixed-frequency section of the series 1600 microwave synthesizer, manufactured by Systron-Donner Corporation, Concord Instrument Division. Reprinted by permission from *Instrumentation Manual for Model 1600 Synthesizer Driver*, Systron-Donner Corporation, November 1974.

A combination of analog and digital phase-locked loops, connected in the fashion shown in Fig. 7-35, comprises the hf synthesis section. A complete frequency-increment decade consists of a digital PLL operating between 9 and 18 MHz in 1 MHz steps, a divider that divides by 10 the frequency of the hf signal synthesized in the preceding decades, and an analog PLL that sums up the two signals. Two complete decades, 1 kHz and 1 MHz, are shown in Fig. 7-35. The decades generating 10 and 100 kHz are omitted from the illustration because they do not add any new information to the description of the hf synthesis.

To reduce the cost and complexity of the hf synthesis section, 0.1 to 100 Hz increments are generated by one analog and two digital PLLs, which basically are the same as the PLLs employed throughout the hf synthesis section.

The loop bandwidth of the hf phase-locked loops varies from 3 to 5 kHz. Within this bandwidth the phase noise associated with the synthesized hf signal is the 10 MHz VCXO noise appropriately scaled up with the maximum enhancement of 6 dB at 20 MHz. Above 5 kHz offset frequency the noise contribution of the output hf VCO predominates. A low-noise circuit design is used in implementing the hf synthesis section.

The synthesized hf signal is effectively multiplied by 10 in the uhf-shf synthesis section, and so is the smallest frequency increment. How this multiplication takes place is shown in Fig. 7-36.

The uhf-shf synthesis section consists of two PLLs (see Fig. 7-36). The output PLL accepts two signals: a signal at 10 to 20 MHz and a signal that is a multiple of 100 MHz. The auxiliary PLL selects an appropriate harmonic of 100 MHz, attenuating adjacent harmonics, and downconverts the signal generated by the output YIG-tuned oscillator to 100 to 200 MHz. The 100 to 200 MHz signal is divided by 10 and compared to the 10 to 20 MHz reference generated by the hf synthesis section.

Figure 7-36 demonstrates how two octave bands are produced by the system. As the output YIG-tuned oscillator is tuned from 1 to 2 GHz, the auxiliary PLL oscillator is tuned from 0.9 to 1.9 GHz in 100 MHz increments. Similarly, when the output oscillator is tuned over the 2 to 4 GHz band, the auxiliary PLL oscillator is tuned from 1.9 to 3.9 GHz. The dc tuning voltage for both oscillators is provided by the synthesizer driver, model 1600. With this type of frequency synthesis any frequency increments are generated as easily at 18 as at 0.5 GHz. For illustrative purposes the synthesizers described in Figs. 7-34 to 7-36 generate 1 Hz increments. Such operation is offered as an option. The frequency of the standard model is incremented in 1 kHz. This means a removal of three complete decades from Fig. 7-35.

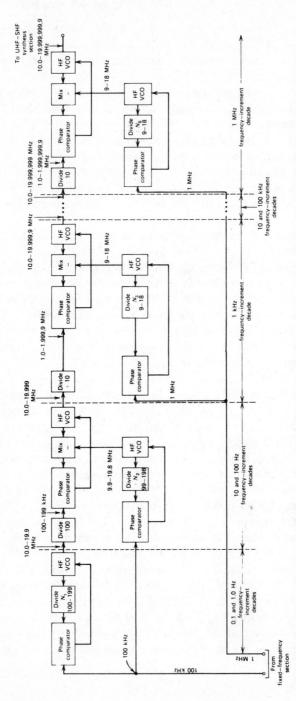

Figure 7-35. HF synthesis section of the series 1600 microwave synthesizer, manufactured by Systron-Donner Corporation, Concord Instrument Division. Reprinted by permission from *Instrumentation Manual for Model 1600 Synthesizer Driver*, Systron-Donner Corporation, November 1974.

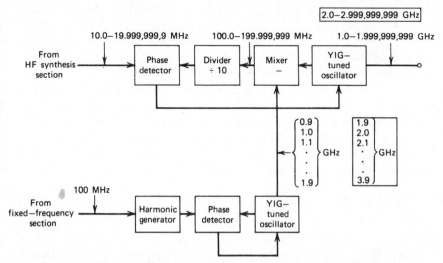

Figure 7-36. UHF-SHF synthesis section of the series 1600 microwave synthesizer, manufactured by Systron-Donner Corporation, Concord Instrument Division. By permission of Systron-Donner Corporation, Concord Instrument Division.

There are numerous other advantages of this approach. First, a YIG-tuned oscillator is a high-Q device. Quality factors of 2000 to 4000 are feasible at uhf and shf. The availability of high-Q circuits makes it possible to design an oscillator exhibiting virtually constant phase-noise performance up to and above 18 GHz, even though the oscillator is tuned in octave bands. This means that at offset frequencies above the output PLL bandwidth (which is 100 kHz) the synthesizers exhibit excellent phase-noise performance that is independent of the factor by which the 10 MHz reference oscillator noise is multiplied. At offset frequencies between 100 Hz and 100 kHz the 100 MHz VCXO noise multiplied to the final synthesizer frequency predominates, and at offsets below 100 Hz the 10 MHz reference noise is the dominant factor. The selection of various loop bandwidths in this system is governed by the requirement for optimum noise performance.

Second, extremely good tuning linearity (of the order of $\pm 0.2\%$) can be achieved with YIG-tuned oscillators over an octave band, simplifying the oscillator tuning circuitry. Such linearity accounts for the constant phase-noise level sustained over the whole octave band, which is not obtained with VCOs utilizing varicap (or varactor) tuning.

In addition, presently available YIG-tuned oscillators operate to 26 GHz, making an extension of the synthesizer operating range to this frequency feasible.

Finally, the YIG-tuned device is capable of delivering a high output rf power (+7 dBm to 18 GHz), which is relatively constant over an octave band. This not only provides the synthesizer with a constant output power but also results in a constant mixer drive power, which is independent of the power variations in harmonics of the 100 MHz signal.

Figure 7-37 is a set of photographs of the synthesizer driver and a rf unit. The system can be remotely controlled. These are general-type synthesizers, which find use in applications such as microwave communications, spectroscopy, alignment and calibration of microwave equipment, and measurements of group delay and Doppler phase, as well as in other areas requiring stable, accurate, and spectrally pure signal sources.

UHF-SHF Synthesizer, Model WJ-1250, Manufactured by Watkins-Johnson Company

The Watkins-Johnson microwave synthesizer, model WJ-1250 with WJ-1251-X rf sources, is an excellent example of equipment operating over a multioctave frequency range by way of a main frame and a set of plug-in units (for a partial synthesizer specification see Table 7-10).

The synthesizer consists of a vhf and an uhf-shf digital phase-locked loop locked to a 5 MHz reference signal. The internal reference is a high-purity temperature-compensated crystal oscillator (TCXO) exhibiting a drift of 3 parts in 10^9 per day. A TCXO is used because it requires a short warm-up period after turn-on.

The vhf phase-locked loop generates signals at 1 and 100 MHz by locking a voltage-controlled crystal oscillator (VCXO) to the 5 MHz reference (see Fig. 7-38). The bandwidth of the vhf loop is very narrow so that both signals retain the spectral purity of the 100 MHz VCXO. The frequency of the 1 MHz output is divided by 10 before the signal is applied to the frequency-phase comparator of the uhf-shf phase-locked loop. The 100 MHz output is fed to a YIG-tuned harmonic generator that provides a signal 50.0 to 149.9 MHz below the final synthesizer frequency to the mixer of the uhf-shf loop.

The generator consists of a 100 MHz power amplifier, broadband step-recovery-diode (SRD) multiplier, and broadband YIG-tuned filter with extremely good tuning linearity and high selectivity. The power amplifier delivers approximately 1 W to the SRD multiplier. For proper harmonic generation a dc bias voltage is applied to the SRD. The voltage is varied with the order of selected harmonic, resulting in optimum multiplier performance across the full frequency range. The YIG-tuned filter attenuates all unwanted harmonics of 100 MHz generated in the SRD circuit. Upon a manual or remote control command the harmonic generator produces a signal that is a multiple of 100 MHz. The outstanding

Table 7-10.   Partial Performance Specification for Watkins-Johnson Synthesizer, Model WJ-1250 and WJ-1251 Series rf Plug-in Units

Frequency range:	0.5–1.0 GHz	Model No. WJ-1250/WJ-1251-1
	1-2	WJ-1250/WJ-1251-2
	2-4	WJ-1250/WJ-1251-3
	4-8	WJ-1250/WJ-1251-4
	8.0-12.4	WJ-1250/WJ-1251-5
	12.4-18.0	WJ-1250/WJ-1251-6
	8-18	WJ-1250/WJ-1251-7
	1-4	WJ-1250/WJ-1251-8
	18.0-26.5^b	WJ-1250/WJ-1251-9
Frequency increments:	100 kHz	
Nonharmonically related spurious outputs:	60 dB below signal maximum	
Phase noise:	Single sideband-to-signal-phase-noise ratio measured in 1 Hz bandwidth 100 kHz removed from signal is -90 dB maximum in any specified frequency band	
Switching speed:	Less than 40 msec to within the accuracy of internal reference for 100 MHz frequency increments; typically 50 msec for 1 GHz frequency increments	

[a]Reprinted by permission from Watkins-Johnson *Technical Data Sheet for Frequency Synthesizer WJ-1250 with WJ-1251 Series RF Sources,* June 1974.
[b]Under development at the time when the manuscript of this book was prepared.

feature of the multiplier-filter configuration is a multioctave tuning range that spans from 0.4 to 17.9 GHz.

The constituent parts of the uhf-shf phase-locked loop are a YIG-tuned oscillator (YTO), a mixer that converts the oscillator signal to 50 to 150 MHz, a variable-ratio divider, and a frequency-phase comparator.

The YTO is a microwave oscillator operating in the fundamental mode. For frequencies between 0.5 and 8.0 GHz the oscillator active element is a transistor. Above 8 GHz the rf power is supplied by gallium arsenide (GaAs) Gunn-effect devices. The resonating element in the oscillator is a YIG sphere that is tuned by varying the strength of the magnetic field in which the sphere is placed. Two coils produce the desired field. A coarse-tuning coil carries most of the tuning current, which is an analog of the

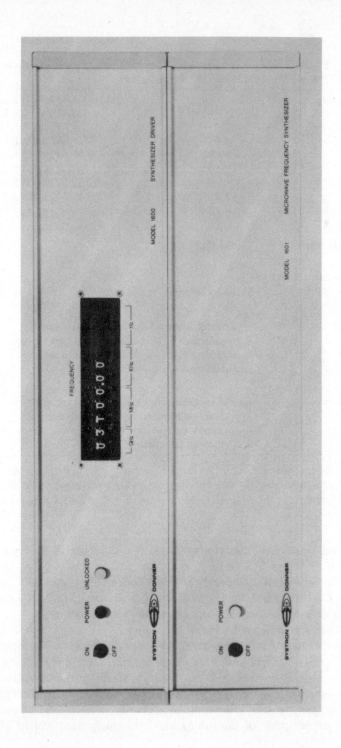

486

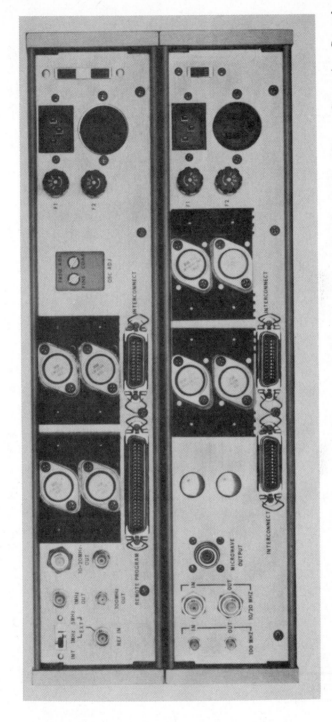

Figure 7-37. Front and rear panel views of the series 1600 microwave frequency synthesizer. By permission of Systron-Donner Corporation, Concord Instrument Division.

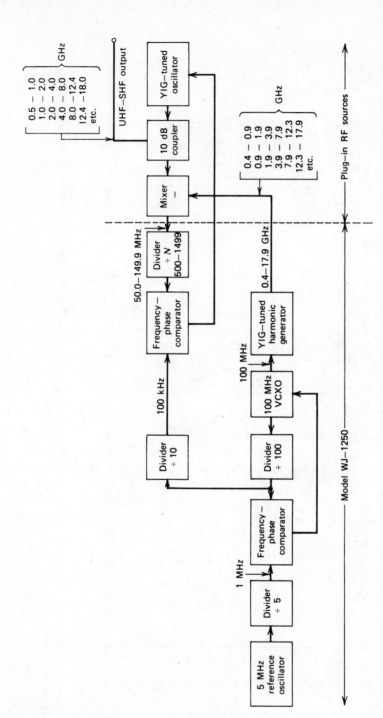

Figure 7-38. Microwave frequency synthesizer, model WJ-1250, manufactured by Watkins-Johnson Company. Reprinted by permission of Watkins-Johnson Company from *Operational and Maintenance Manual for WJ-1250 Microwave Frequency Synthesizer and WJ-1251-X RF Plug-in Units.* June 15, 1974.

programmed synthesizer frequency. The current is derived from the output voltage of a D/A converter. (The frequency error of a free-running YTO ranges from a few megahertz at uhf to 20 to 30 MHz at shf.) A fine-tuning coil carries a relatively small amount of current which is derived from the error voltage generated by the 100 kHz frequency-phase comparator. Both tuning currents are converted from the corresponding voltages by high-gain linear-transconductance amplifiers. Hence the YTO is frequency locked to two low-phase-noise signals, at 100 kHz and 0.4-17.9 GHz, by being externally coarse tuned close to the desired frequency for phase locking to take place.

Whenever previously considered, a VCO, a VCXO, and a YTO were shown as having two rf outputs. However, the presence of rf amplifiers or directional couplers following the oscillators was implied in all cases and examples were given to point out the important functions assigned to such circuits. To bring this subject once more to the attention of the reader, a 10 dB coupler is shown following the YTO in Fig. 7-38. The coupler provides an isolation between the mixer and the YTO and delivers part of the synthesizer output power to the local oscillator input of the mixer.

The signal input to the mixer is provided by the YIG-tuned harmonic generator previously described. The signal power varies from -20 to -40 dBm. The mixer output is at 50.0 to 149.9 MHz, when the uhf-shf loop is locked, at a level that primarily depends on the harmonic generator output power and mixer loss. The mixer output power varies typically from -25 to -50 dBm.

A low-pass filter following the mixer suppresses all undesirable signals occurring above 200 MHz. The device, not shown in Fig. 7-38, is a five-pole tubular filter with a 1 dB corner frequency at 200 MHz.

The signal at 50.0 to 149.9 MHz is amplified and applied to a variable-ratio frequency divider, $\div N$, which provides a signal at 100 kHz for phase comparison. Both the divider and frequency-phase comparator are digital circuits.

Within the bandwidth of the microwave loop the synthesizer spectral purity approaches that of the 100 MHz VCXO enhanced to the final synthesizer frequency. Outside the loop bandwidth the spectrum of the synthesized signal is essentially that of the free-running YTO. Figures 7-39 and 7-40 are typical double-sided phase-noise plots of the synthesizer output signal at 4.05 and 7.05 GHz, respectively.

The synthesizer frequency is either set from a 13-button keyboard located on the front panel (see Fig. 7-41) or programmed by way of remote BCD data input to a rear panel connector. A six-digit light-emitting-diode (LED) display indicates the synthesized frequency. The ENTER pushbutton, when pressed, transfers the new frequency word from the input register to

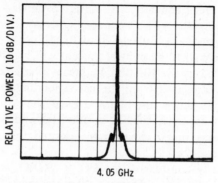

4. 05 GHz

FREQUENCY (50 kHz/DIV.)

Figure 7-39. Typical synthesizer rf signal spectrum (from WJ-1251-4, 4 to 8 GHz), 1 kHz measurement bandwidth. Courtesy of Watkins-Johnson Company.

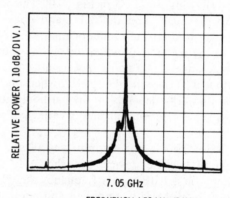

7. 05 GHz

FREQUENCY (50 kHz/DIV.)

Figure 7-40. Synthesizer rf signal spectrum (from WJ-1251-20, 7 to 1i GHz), 1 kHz measurement bandwidth. Courtesy of Watkins-Johnson Company.

the programming register. If the selected frequency is outside the range of the rf source plug-in unit installed in the synthesizer main frame, the ENTER strobe (a positive TTL pulse) sets an error latch, which causes the LED display to flash on and off. When this happens, the CLEAR pushbutton should be pressed to clear the display in order that the synthesizer can be set to a new frequency.

A single LED lamp at the left of the digital readout indicates, when it is lit, that a new input word has not been entered. If the lamp is not lit, the synthesizer is programmed to the displayed frequency.

The synthesizer can be locked to an external 5 MHz reference whenever

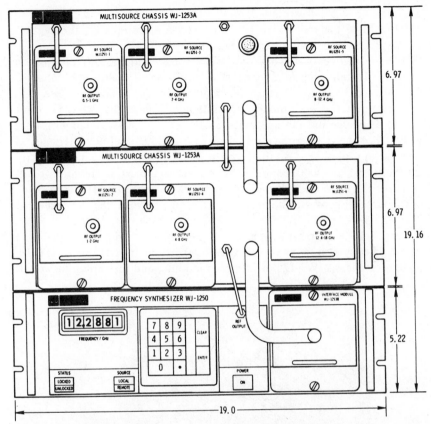

Figure 7-41. Multisource chassis WJ-1253A with interface modules WJ-1253B associated with model WJ-1250 microwave synthesizer, manufactured by Watkins-Johnson Company. Courtesy of Watkins-Johnson Company.

the proper signal is applied to an appropriate BNC connector located on the rear panel and the reference select switch is set to EXT. A status indicator lights green when the synthesizer is locked to the internal or an external reference. The indicator lights red when either a vhf or an uhf-shf phase-locked loop loses synchronism.

The use of plug-in rf sources is dictated by the frequency range over which a YTO can be tuned. Any frequency range between 0.5 and 18 GHz can be provided in a single plug-in as long as a YTO is available to cover that range. Presently, the synthesizer frequency range (0.5 to 18.0 GHz) is covered by six plug-in rf sources, models WJ-1251-1 to WJ-1251-6 (see Table 7-10), which are accessible from the front panel of the model WJ-1250 main frame. Two additional broadband rf sources, models WJ-

1251-7 and WJ-1251-8, are available for the user's convenience. When the development of the model WJ-1251-9 rf source is completed, the synthesizer frequency range will be extended to 26.5 GHz.

The convenience of having one main frame and various plug-ins is extended to multi-plug-in operation by utilizing a multisource chassis, model WJ-1253A, and an interface module, model WJ-1253B (see Fig. 7-41), in conjunction with model WJ-1250 main frame. The WJ-1253A unit provides dc power and tuning signals to the rf sources associated with it. It also switches the rf referrence signal to, and the 50.0 to 149.9 MHz signal from, the appropriate rf source. The uhf-shf output is obtained directly from the front panel connectors of the rf sources utilized in the multi-plug-in operation. The WJ-1253B module is installed in the main frame in place of a rf source. It provides the necessary interface between the WJ-1250 and the WJ-1253A units.

Figure 7-42 is a photograph of the synthesizer with a multisource chassis. The system consists of the WJ-1250 main frame, WJ-1253A multisource chassis, WJ-1253B interface module, and three plug-ins, mod-

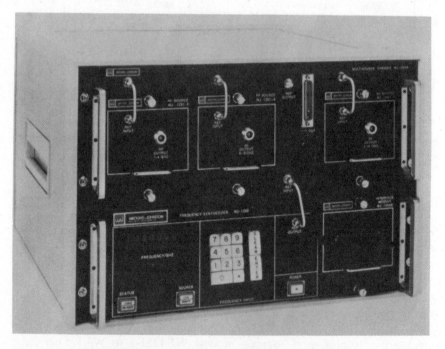

Figure 7-42. Model WJ-1250 synthesizer, manufactured by Watkins-Johnson Company, in a multi-plug-in chassis. Courtesy of Watkins-Johnson Company.

Figure 7-43. Front view of WJ-1250 microwave synthesizer and associated rf sources, manufactured by Watkins-Johnson Company. Courtesy of Watkins-Johnson Company.

els WJ-1251-8, WJ-1251-4, and WJ-1251-7. The system provides a coverage of the 1 to 18 GHz frequency range in 100 kHz frequency increments.

Figure 7-43 is a front view of the Watkins-Johnson microwave synthesizer, model WJ-1250, and associated rf sources.

Applications of the synthesizer include research and development laboratory instrumentation, calibration of microwave equipment, radar systems, wideband reconnaissance systems, and communication systems.

7-4   Direct Digital Synthesizers: Wavetek, Former Rockland Systems, Model 5100

The model 5100 frequency synthesizer, manufactured by Wavetek, former Rockland Systems, Inc., is a direct digital synthesizer operating between dc and 2 MHz in 1 mHz increments. Table 7-11 gives a partial list of the synthesizer requirements. The system is basically the same as that described in Section 1-4, Figs. 1-19 and 1-20, so that a complete description of the particular model is not necessary. Only the differences between the approach presented in Chapter 1 and the model 5100 synthesizer are considered below.

The phase accumulator of the model 5100 synthesizer consists of a modulo 8 and three modulo 1000 accumulators serially coupled to each other so that each accumulator feeds its overflow bits (carry) to the next subsequent accumulator (see Fig. 7-44).

The modulo 8 accumulator consists of a 3-bit adder and 3-bit register, Fig. 7-45. It accepts the most significant bit of the desired frequency setting corresponding to 0 or 1 MHz, the carry signal from the preceding modulo 1000 accumulator, and the clock, and it generates the sign bit, the quadrant bit, and the most significant bit (MSB) of the ROM address. The adder sums up the digital signals from the frequency register and 3-bit register and updates the 3-bit register with the most recent sum. The 3-bit register transfers the updated digital data from the output of the adder to its input at every clock pulse so as to make the modulo 8 accumulator overflow periodically with a period determined by the desired output frequency setting.

A typical modulo 1000 accumulator consists of a 7-bit adder, 10-bit adder, 10-bit register, and flip-flop, Fig. 7-46. It accepts the data corresponding to the assigned decade of the desired frequency setting (100, 10, 1 kHz or 100, 10, 1 Hz or 100, 10, 1 mHz), A_0 through A_9, the carry from the preceding modulo 1000 accumulator, and the clock, and it generates a carry fed into the adder of the following accumulator at each overflow. The modulo 1000 accumulator preceding the modulo 8 accumulator also generates the ROM address. The 7-bit adder effectively eliminates the first 24 counts from a base 1024 of the 10-bit binary digital codes, thus

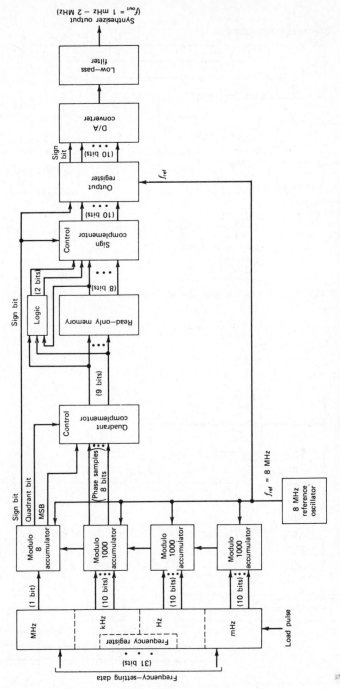

Figure 7-44. 1 mHz to 2 MHz direct digital synthesizer, model 5100, manufactured by Rockland Systems Corporation. By permission of Wavetek, former Rockland Systems Corporation.

495

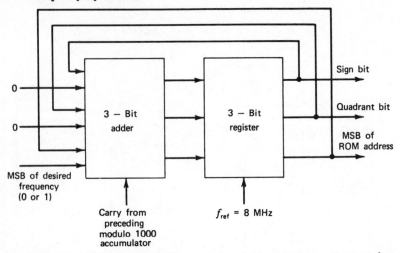

Figure 7-45. **Modulo 8 accumulator. By permission of Wavetek, former Rockland Systems Corporation.**

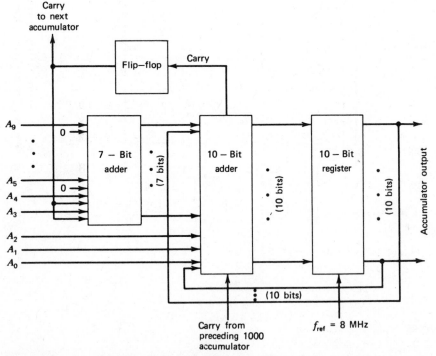

Figure 7-46. **Modulo 1000 accumulator. By permission of Wavetek, former Rockland Systems Corporation.**

Table 7-11. Partial Performance Specification for Wavetek Synthesizer, Model 5100[a]

Frequency range:	dc to 2 MHz (3 MHz optional)
Frequency increments:	1 mHz

Nonharmonically related
  spurious outputs:

Output frequency	Level of spurious signals (dBc)
1 mHz to 100 kHz	-70
100 kHz to 500 kHz	-60
500 kHz to 2 MHz	-50

Phase noise in 30 kHz bandwidth,
  excluding 1 Hz centered on the
  output signal (dB):
	-50

Switching time:	1.5 μsec for BINARY-WORD with phase and amplitude continuity maintained

[a] Material reproduced with permission of Wavetek, former Rockland Systems, Inc.

providing an accumulator with a base of 1000. The flip-flop provides the carry signal to the next accumulator and at the same time generates the fixed "1" signal applied to the first two input positions of the 7-bit adder to achieve the incrementing of the accumulator by 24. The outputs from only the 10-bit register, which corresponds to the 1, 10, and 100 kHz decades, are utilized in frequency synthesis, because these are the most significant phase bits.

The logic circuit (see Fig. 7-44) is used to save hardware. By generating the two most significant bits of the digital number representing the synthesized waveform from various ROM inputs and outputs, the logic circuit makes possible the synthesis of a signal with a 10-bit accuracy, using an 8-bit capacity storage for each ROM location.

The major design effort associated with the model 5100 synthesizer is concentrated on implementing the approach with as few integrated circuits as possible without significantly affecting the performance of the synthesizer. This consideration governs the selection of the ROM as well. To achieve the required D/A converter linearity for a low-spurious-output operation, a highly linear digital-to-analog converter is designed and successfully implemented.

The synthesizer is phase-locked to an internal 8 MHz clock, and it also can be locked to an external 1 MHz reference source.

Figure 7-47. Top view of model 5100 synthesizer. Courtesy of Wavetek, former Rockland Systems Corporation.

As is typical of this type of synthesizer, model 5100 exhibits excellent tuning time characteristics (see Table 7-11). Frequency, phase, and amplitude of the model 5100 output signal can be remotely programmed, that is, the phase and amplitude of the synthesized signal are maintained or set to zero by the selection of appropriate control logic commands as the synthesizer frequency is changed (negligible switching transients).

The mechanical configuration of the model 5100 synthesizer is shown in Figs. 7-47 and 7-48, which are the top and bottom views of the synthesizer

Figure 7-48. Bottom view of model 5100 synthesizer. Courtesy of Wavetek, former Rockland Systems Corporation.

with covers removed. As these illustrations show, most of the circuits are located on one printed circuit board with no shielding employed between the sections of the synthesizer operating at different frequencies. Figures 7-49 and 7-50 show the front and rear panels with appropriate controls, inputs, and outputs.

Model 5100 is a general type of synthesizer used as an accurate and stable rf signal source in such applications as modem testing, navigation, radar, tone generation for touch-tone-keyboard telephone, and analysis of chemical compounds (nuclear magnetic resonance). However, because of

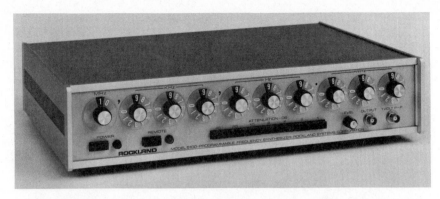

Figure 7-49. Front view of model 5100 synthesizer. Courtesy of Wavetek, former Rockland Systems Corporation.

Figure 7-50. Rear view of model 5100 synthesizer. Courtesy of Wavetek, former Rockland Systems Corporation.

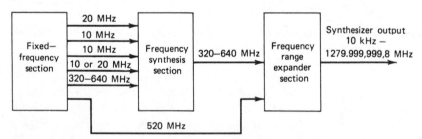

Figure 7-51. 10 kHz to 1280 MHz coherent synthesizer, model 8662A, manufactured by Hewlett-Packard Company. By permission of Hewlett-Packard Company.

the phase-continuous frequency switching and extremely short tuning time, the synthesizer is also used in such special applications as a local oscillator in frequency-hopping receivers.

7-5 Synthesized Signal Generators: Hewlett-Packard Model 8662A

Developments in communications, radar, automatic testing of electronic systems, and other areas requiring variable frequency sources prompted the need for a source that provides frequency and amplitude modulation of the output signal, wide-dynamic-range amplitude control, spurious-free operation, and low phase noise, and at the same time is frequency stable, accurate, and capable of being remotely controlled.

To meet the need of the industry a new line of test equipment called "synthesized signal generators" is being offered by several companies here and abroad. Model 8662A is a typical example of this type of equipment.

The synthesizer operates between 10 kHz and 1.28 GHz in steps of 0.1 or 0.2 Hz, depending on the output frequency. (See Table 7-12 for a partial list of model 8662A requirements.) This is accomplished by utilizing both direct and indirect synthesis. Figure 7-51 shows a breakdown of the synthesizer by sections. Several signals are generated in the fixed frequency section by frequency multiplication, division, and mixing. These signals are used in the frequency synthesis section in synthesizing the 320 to 640 MHz signal, variable in 0.1 Hz steps, by way of six PLLs, and in the frequency range expander section. Subtraction, multiplication, and division are employed in the frequency range expander to convert the 320 to 640 MHz band to the final synthesizer frequency range.

A discussion of the fixed frequency section of model 8662A is omitted in this text. A functional block diagram describing the rest of the synthesizer is shown in Figure 7-52. The diagram demonstrates the step-by-step synthesis used in generating the 320 to 640 MHz signal, as well as the technique of expanding this narrow band to the frequency range of the instrument.

Of special interest is the fractional-N PLL approach utilized in synthesizing several low-order decades of frequency increments and implemented in the 122 to 221 MHz and 100 to 200 MHz PLLs, which allows the generation of 1 MHz steps in a PLL locked to a 10 MHz reference and of 100 Hz steps in a PLL locked to a 100 kHz reference signal. (For a description of the design principles associated with fractional-N PLLs, see Chapter 1.) Notice also that only the 100 to 200 MHz PLL employs cancellation in reducing the level of the spurious beat note generated in the process of the fractional-N division, which requires a D/A converter and

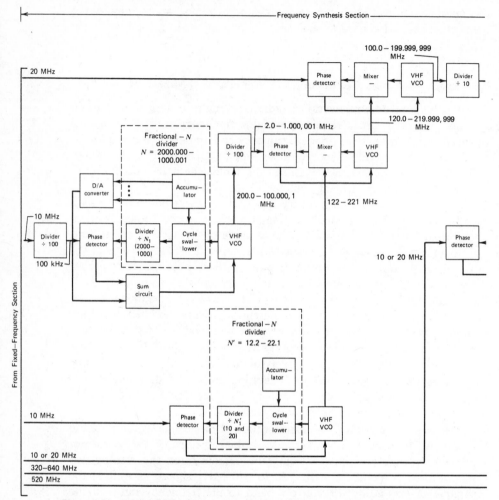

Figure 7-52.   Model 8662A frequency synthesis and range expander sections. By permission of Hewlett-Packard Company.

sum circuit, whereas the 122 to 221 MHz PLL does not. The cancellation of the beat note in the first PLL is necessary because the frequency of the beat note falls within the loop bandwidth of the PLL for several low-order decades of frequency increments, so that no attenuation of the beat note is provided by the PLL, and the attenuation of the FM sidebands resulting from the beat note provided by the divide-by-100 and divide-by-10 dividers following the PLL is not adequate for meeting the requirement of 90

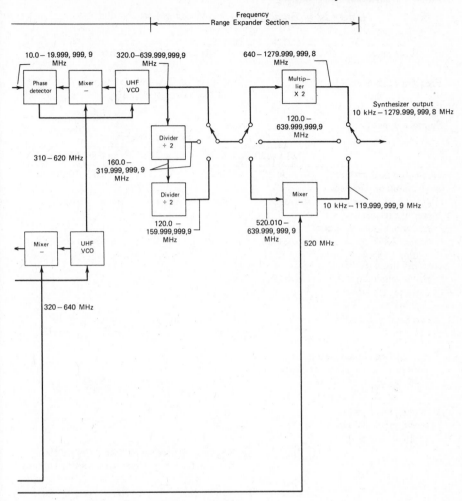

dB. The cancellation of the beat note in the second PLL is not necessary because the lowest frequency of the beat note is outside the loop bandwidth, and the effect of the PLL and divide-by-10 divider is to attenuate the FM sidebands by the required amount.

Great attention is paid by the designers of model 8662A to designing the system and circuits in such a manner as to arrive at the spurious outputs, phase noise, and tuning time performance displayed by the synthesizer. An appropriate frequency plan, in which many elements are assigned several functions, plays an important part in achieving this performance. For

Table 7-12. Partial List of Specifications for Model 8662A Synthesized Signal Generator[a]

Frequency range:	10 kHz to 1279.9999998 MHz
Frequency increments:	0.1 Hz for $f_{out} < 640$ MHz[b]
	0.2 Hz for $f_{out} > 640$ MHz

Spurious signals:

	Frequency range (MHz)				
	0.01 to 120	120 to 160	160 to 320	320 to 640	640 to 1280
Nonharmonically related	-90 dB	-100 dB	-96 dB	-90 dB	-84 dB
Subharmonically related ($f_{out}/2$, $3f_{out}/2$, etc.)	none	none	none	none	-75 dB
Power line related and microphonically generated (within 200 Hz)	-90 dB	-90 dB	-86 dB	-80 dB	-75 dB

Residual single sideband-to-signal-phase-noise ratio in 1 Hz bandwidth, CW and AM modes (320 MHz $< f_{out} < 640$ MHz)

Offset from signal (Hz):	10	10^2	10^3	10^4	10^5
Ratio (dB):	-100	-113	-122	-132	-133

Single-sideband broadband noise floor in 1 Hz bandwidth at 2 MHz (1 MHz $< f_{out} < 640$ MHz):   less than -150 dB

Switching time, typical (to be within 100 Hz of the final frequency):

Programming mode	Microprocessor time (msec)	Settling time (μsec)	Total switching time
String	12.1	400	12.5 msec
Character	8.3	400	8.7 msec
Remote sweep	In these modes,		700 μsec
Fast-learn mode	microprocessor time and rf time overlap.		420 μsec

[a]Material reproduced with permission of Hewlett-Packard Company.
[b]f_{out} = frequency of the synthesizer final output.

example, some PLLs are used not only to add or subtract signals but also to clean up the spectra of those signals. Similarly, frequency dividers are employed in the generation of low-order frequency increments, as well as in the attenuation of phase noise and spurious signals.

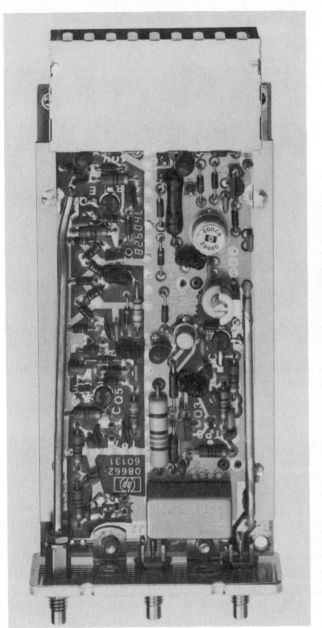

Figure 7-53. Typical printed circuit board with grounding brackets and shields used in model 8662A synthesized signal generator. Courtesy of Hewlett-Packard Company.

Figure 7-54. Typical modular extruded shielded enclosure with holes for air flow used in model 8662A synthesized signal generator. Courtesy of Hewlett-Packard Company.

Equal importance is assigned to an optimum system design. The two areas of special interest are the PLL design for optimum phase noise and tuning time performance and the utilization of the fractional-N approach mentioned above.

Similarly, attention is paid to the optimum circuit design. The circuit of special interest is the octave-band VCO used in the output PLL. The low-noise performance of this circuit is achieved by employing inductances to generate several subbands in order that the VCO gain constant will be low, as well as the noise associated with it. (There is a detailed discussion of some design aspects of the model 8662A synthesizer in Ref. 18.)

Special consideration is given to the packaging of the synthesizer circuits. Figures 7-53 and 7-54 are photographs of a typical printed circuit board and chassis displaying excellent shielding characteristics compatible with the stringent spurious outputs requirement. Figures 7-55 and 7-56 are top and bottom views of model 8662A with the covers removed, demonstrating the packaging techniques employed in the construction of the instrument. The digital control boards are located next to the front panel;

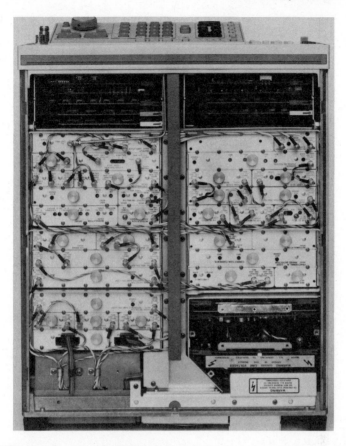

Figure 7-55. Top view of model 8662A synthesized signal generator. Courtesy of Hewlett-Packard Company.

the rf chassis, interconnected with coaxial cables for good shielding integrity, follows the digital printed circuit boards, and the power supply is mounted near the rear panel. Several "mother boards" provide all dc interconnections.

Omitted from Figure 7-52 are the circuits associated with the functions that make the synthesizer a bench-type signal generator. Model 8662A offers AM and FM capabilities, using either internal 400 Hz and 1 kHz tones or externally applied modulation signals, as well as a calibrated output level having a range between $+13$ and -139.9 dBm with a resolution of 0.1 dB.

Figure 7-56. Bottom view of model 8662A synthesized signal generator. Courtesy of Hewlett-Packard Company.

The instrument is both manually and remotely controlled. Manual control of all functions is provided by the front panel keyboard by way of a microprocessor-based controller. (See Figure 7-57 for a photograph of the model 8662A front panel.) All model 8662A functions except the line switch can be also remotely controlled; a rear panel connector is provided for that purpose. The interface offered for remote operation is HP-IB, which is the Hewlett-Packard implementation of IEEE Standard 488. The flow of messages is in both directions, that is, any number of manually entered front panel settings can be automatically transferred to an external controller and stored for use at a later time. In case of malfunction model 8662A can also request service and send the appropriate error message to the controller.

Figure 7-57. Front panel view of model 8662A synthesized signal generator. Courtesy of Hewlett-Packard Company.

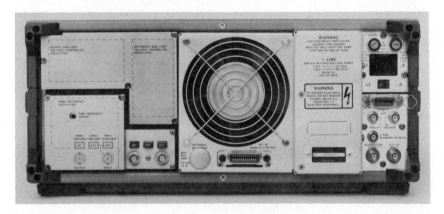

Figure 7-58. Rear panel view of model 8662A synthesized signal generator. Courtesy of Hewlett-Packard Company.

The use of a microprocessor-based controller makes available special functions not offered by other, comparable equipment. For example, the complete front panel setting can be stored and recalled for later use. Up to nine complete settings can be stored and recalled in a specified sequence. When the auto sequence mode is selected, the synthesizer automatically cycles through the sequence. The instrument is capable of being digitally swept either between two frequencies or symmetrically about a center frequency. A combination of these functions gives the designer a powerful means to analyze the performance of circuits or subsystems, such as displaying on an oscilloscope simultaneously both the narrowband and the broadband characteristics of a band-pass filter.

The reference source used in model 8662A is a low-noise crystal-controlled oscillator operating at 10 MHz. The synthesizer also can be locked to a 5 or 10 MHz frequency standard by way of the appropriate rear panel connectors. (See Figure 7-58 for a photograph of the model 8662A rear panel.)

The applications of the model 8662A synthesized signal generator are numerous: as a general purpose AM and FM signal generator; as a low-noise local oscillator in satellite communication, spectrum surveillance, and secure communication (slow hopping) receivers; as a variable frequency source in Doppler-shift radar; as a low-noise and accurate signal source in receiver testing of sensitivity, adjacent channel selectivity, and AGC characteristics; as a frequency source in nuclear magnetic resonance and in making peak shift measurement (bit jitter) on disks; as an accurate sweep generator for automatic testing of narrowband circuits such as crystal filters and discriminators; and in production automatic testing of electronic equipment. These are only a few of the many applications requiring the features offered by this versatile instrument.

7-6   Hybrid Synthesizers: Wavetek Model 5155A

The rapid rise in the cost of synthesizers has accentuated the need for high performance at low cost. Model 5155A is a typical product of the effort directed toward satisfying both of these requirements. The model uses a hybrid synthesis approach, modules of simple mechanical construction, and a printed circuit mother board for routing dc and some rf lines associated with the modules.

The synthesizer operates between 100 kHz and 1 GHz in steps of 0.1 Hz. (See Table 7-13 for a partial list of model 5155A requirements.) It consists of four sections: fixed frequency, hf-vhf, uhf, and frequency range expander (see Fig. 7-59).

The design of the fixed frequency section is based on brute-force synthesis. It provides signals at nine frequencies for use in hf-vhf and uhf sections (see Fig. 7-60). All signals are derived from an internal 10 MHz reference oscillator. Locking to an external 5 or 10 MHz reference is easily performed.

The hf-vhf section (see Fig. 7-61) generates 140 to 150 MHz in 0.1 Hz increments. To achieve fast tuning time, two decades of double-mix-divide

Table 7-13.   Partial List of Specifications for Model 5155A Synthesizer[a]

Frequency range	100,000,0–999,999,999,9 Hz
Frequency increments	0.1 Hz
Spurious signals	-70 dBc for $f_{out} < 500$ MHz[b] -64 dBc for $f_{out} \geq 500$ MHz

Residual single sideband phase noise at $f_{out} < 500$ MHz (at and above 500 MHz specification is reduced by 6 dB

Offset from signal (Hz)	100	1K	10K	100K
Ratio (dBc in 1 Hz bandwidth	-100	-110	-115	-120

Switching transient (to within 0.1 rad)	1 μsec for < 10 MHz frequency increments 100 μsec (typically 50 μsec) for 10 and 100 MHz frequency increments

[a] Material reproduced with permission of Wavetek San Diego, Inc., specifications are subject to change.
[b] f_{out} = frequency of the synthesized final output.

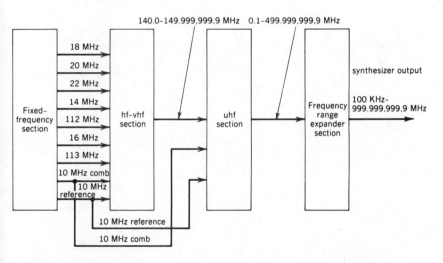

Figure 7-59.   0.1 to 1000 MHz coherent synthesizer, model 5155A, manufactured by Wavetek. By permission of Wavetek San Diego, Inc.

511

synthesis operate above 100 MHz. This insures wide bandwidth for the bandpass filters used in synthesis. Also, for fast tuning time, as well as low costs, a direct digital synthesizer replaces the six double-mix-divide decades which would be necessary for generation of 0.1 Hz increments if the double-mix-divide synthesis were used throughout.

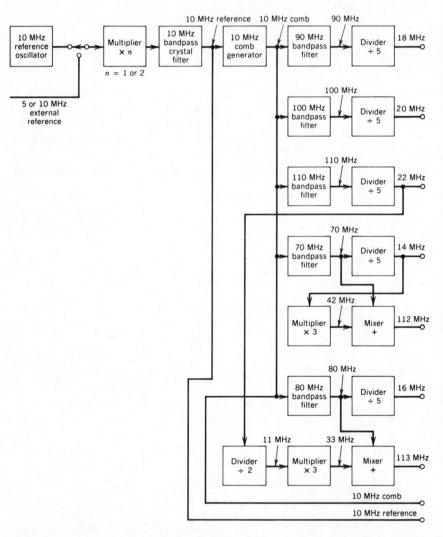

Figure 7-60. Model 5155A synthesizer; fixed-frequency section. By permission of Wavetek San Diego, Inc.

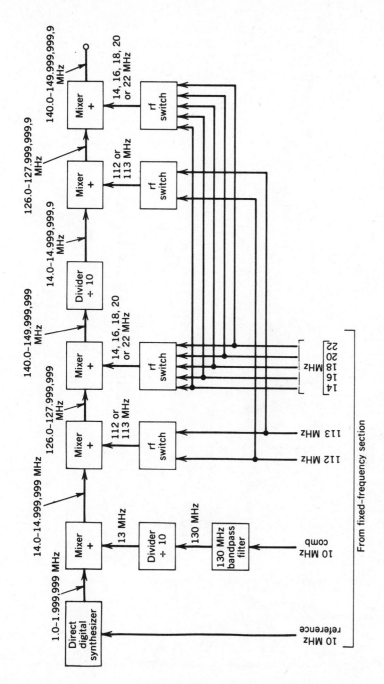

Figure 7-61. Model 5155A synthesizer, hf-vhf synthesis section. By permission of Wavetek San Diego, Inc.

513

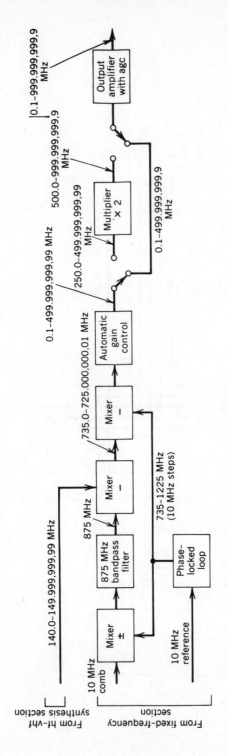

Figure 7-62. Model 5155A synthesizer, uhf synthesis and frequency range expander sections. By permission of Wavetek San Diego, Inc.

514

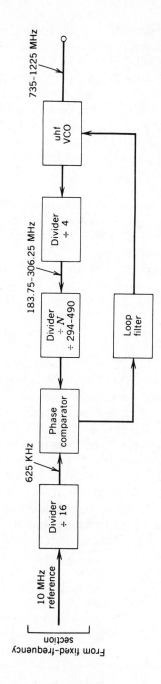

Figure 7-63. Model 5155A synthesizer, high-speed phase-locked loop. By permission of Wavetek San Diego, Inc.

The uhf section (Fig. 7-62) generates 10 and 100 MHz steps by way of triple-mix synthesis. The special feature of this part of the synthesizer is a broadband uhf phase-locked loop, PLL (Fig. 7-63), which acquires lock in less than 10 μsec, approaching the switching time of a free-running oscillator while providing a superior accuracy and long-term stability. The close-to-signal phase noise of the PLL is 40 dB lower than the noise of a typical broadband uhf oscillator, which helps account for the low synthesizer noise. (The theoretical phase noise of the PLL or tuned oscillator used in triple-mix synthesis is cancelled. However, the balancing provided by practical circuits is far from perfect, and care must be taken to keep the noise of the PLL and oscillator low.)

The frequency range expander consists of a uhf times-2 multiplier and a broadband amplifier with a high-speed automatic gain control. As the frequency of the basic synthesizer is multiplied by 2, so are the frequency increments. This problem is overcome in model 5155A by utilizing the capability of the direct digital synthesizer, located in the hf-vhf section, to generate frequency increments as small as 1 mHz. The selection of the appropriate frequency steps provided by the digital synthesizer which correspond to the desired 0.1 Hz steps at the final synthesizer output is accomplished with a microprocessor-based controller that controls all functions of the synthesizer.

The spurious signals requirement is satisfied by selecting an appropriate frequency plan which places all low-order spurious intermodulation products generated in mixers out of the band of the individual circuits of the synthesizer. High-rf-level mixers insure that inband spurious products are well below the desired signal. Radio frequency shielding is achieved by spacing the circuits operating at different frequencies or by separating them in shielded modules, such as those shown in Fig. 7-64, with module location playing an important role in reducing the leakage of the unwanted signals to susceptible-to-pickup circuits. Direct current power supply lines are rf-filtered at the input of each module, preventing leakages of unwanted signals along the path of dc lines. Whenever proven feasible, rf signals are routed by way of the mother board and dc connectors for low cost (see Fig. 7-65). In other cases rf semirigid cables and SMA connectors are used.

A unique performance feature displayed by this synthesizer is frequency hopping free of spurious signals over a limited frequency range at hf through uhf. This is made possible by utilizing direct digital synthesis, which permits changing frequency in a periodic manner without generating spurious amplitude or frequency modulation at the frequency of hopping. Another special feature, also attributed to direct digital synthesis, is phase continuous frequency hopping.

A significant characteristic of this synthesizer is very low phase noise, which is achieved in several ways. One is by proper selection of several frequency synthesis techniques, such as direct digital synthesis for generation of frequency increments smaller than 100 KHz and triple-mix synthesis for 10 and 100 MHz steps. Another is by a judicious placement of frequency sections, for example, by placing the digital synthesizer before a divide-by-10 frequency divider (Fig. 7-61), which improves the phase noise of the signal generated by direct digital synthesis by 20 dB. Still another

Figure 7-64. Typical modules used in model 5155A synthesizer, manufactured by Wavetek. Photograph courtesy of Wavetek San Diego, Inc.

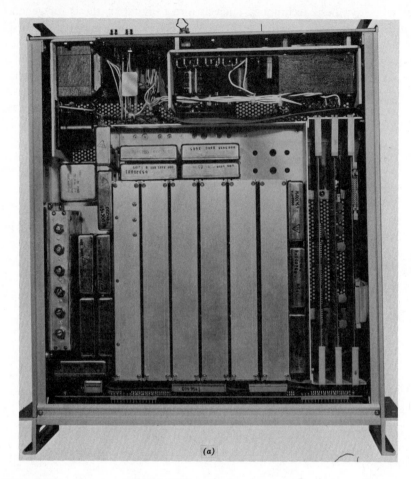

Figure 7-65a.   Top view of Model 5155A synthesizer. Photograph courtesy of Wavetek San Diego, Inc.

way of achieving low noise is by the selection of a frequency plan which allows high rf power levels to be used throughout the synthesizer. In conjunction with low-noise-figure rf amplifiers, this type of frequency plan results in low thermal-noise floor.

It is important to mention the contribution made by low-noise circuits to the excellent performance of model 5155A, such as the vhf comb generator in the fixed-frequency section and broadband uhf PLL, which were designed using the techniques described in this book.

A tuning time of less than 1 μsec to within 0.1 radian is achieved by using direct digital synthesis for generation of 0.1 Hz through 10 kHz frequency

Figure 7-65*b*.   Bottom view of model 5155A synthesizer. Photograph courtesy of Wavetek San Diego, Inc.

increments and by operating 100 kHz and 1 MHz decades at vhf. This provides 3 dB filter bandwidths of larger then 1 MHz throughout the hf and vhf synthesis. A larger, though still very short, tuning time associated with 10 and 100 MHz steps results from the settling time of the uhf phase-locked loop.

All functions of model 5155A synthesizer are controlled by way of the front panel pushbutton switches (Fig. 7-66) or rear panel remote control. Standard BCD programming for submicrosecond frequency switching and optional GPIB for simplicity of programming may be used. Separate multipin connectors are provided on the rear panel for that purpose (Fig.

7-67). A microprocessor-based controller permits fast frequency hopping, sweep, and frequency modulation in addition to fixed frequency and amplitude setting.

Applications of model 5155A synthesizer include laboratory instrumentation, fast automatic testing, frequency-hopping communications, frequency agile radar, and analysis of chemical compounds (nuclear magnetic resonance).

Figure 7-66.   Front panel view of model 5155A synthesizer. Photograph courtesy of Wavetek San Diego, Inc.

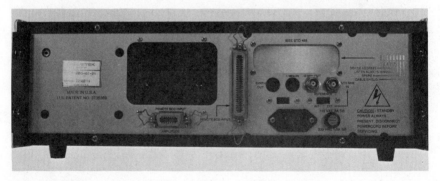

Figure 7-67.   Rear panel view of model 5155A synthesizer. Photograph courtesy of Wavetek San Diego, Inc.

References

1. Polarad Electronic Instruments, Division of Polarad Electronics Corporation. *Handbook of Operating and Maintenance Instructions for RF Synthesizers*, Models 1102 and 1102A.

2. John Fluke MFG Company, *Data Sheet for Frequency Synthesizer*, Model 645A. September 30, 1970.

3. D. G. Meyer. "An Ultra Low Noise Direct Frequency Synthesizer," *Application Note*, John Fluke MFG Company, July 1970.

4. Hewlett-Packard Company. *Data Sheet for Frequency Synthesizer*, Model 5105A/5110B. March 15, 1967.

5. Hewlett-Packard Company. "Frequency Synthesizers," *Application Note* 96. January 1969.

6. Systron-Donner Corporation. "Microwave Frequency Synthesizers with the Purest F_0. Systron-Donner's 1600 Series," *Application Note and Data Sheet for Series* 1600 *Frequency Synthesizers*. September 1974.

7. P. G. Tipon. "New Microwave Frequency Synthesizers that Exhibit Broader Bandwidths and Increased Spectral Purity," Systron-Donner Corporation, July 1974.

8. Systron-Donner Corporation. *Instruction Manual for Model* 1600 *Synthesizer Driver*. 1974.

9. Correspondence with Charles E. Foster, II. Watkins-Johnson Company.

10. Watkins-Johnson Company. *Operational and Maintenance Manual for WJ-1250 Microwave Frequency Synthesizer and WJ-1251-X RF Plug-in Units*. June 15, 1974.

11. Watkins-Johnson Company. *Technical Data Sheet for Frequency Synthesizer Model WJ-1250 with WJ-1251 Series RF Sources*. June 1974.

12. Watkins-Johnson Company. *Development Specification for Multi-Source Chassis WJ-1253A with Interface Module WJ-1253B*. June 6, 1974.

13. Watkins-Johnson Company. *Development Specification for WJ-1251-7, WJ-1251-8 and WJ-1251-9 RF Sources*. June 5, 1974.

14. Rockland Systems, Inc. *Data Sheet for Models* 5100 *and* 5110 *Programmable Frequency Synthesizers*. April 1979.

15. Personal contact with Joseph Flink and Roger H. Hosking. Rockland Systems, Inc.

16. Hewlett-Packard Company. *Data Sheet for Synthesized Signal Generator*, Model 8662A. October 1, 1978.

17. Correspondence with Brian D. Unter and Roland Hassun. Hewlett-Packard Company.

18. Dieter Scherer. "Design Principles and Test Methods for Low Phase Noise RF and Microwave Sources." Hewlett-Packard Company, Stanford Park Division.

19. Personal contact with J. Flink, D. Shamah, and R. MacKay. Wavetek Corporation.

8 Frequency Reference Sources

The most important element employed in coherent frequency synthesis is a frequency reference source. It determines the stability and accuracy of the synthesized signal and often contributes to phase noise and spurious outputs associated with the signal. Chapter 8 describes the basic design principles and operation of these devices.

8-1 Crystal-Controlled Oscillators

A reference source may consist of a crystal oscillator and a buffer amplifier. On the other hand, depending on the output frequency, stability, and phase-noise requirements, frequency multipliers with associated filters, dividers, and a crystal filter at the final output (see Fig. 8-1) may be employed as well. For example, if a reference source operating at 1 MHz and exhibiting a frequency drift of 2 parts in 10^9/day were required, a manufacturer of crystal oscillators would probably select a 5 MHz oscillator and use a divide-by-5 frequency divider to obtain a 1 MHz output because a typical high-grade 5 MHz crystal unit is more stable than its 1 MHz counterpart. Similarly, if a 10 MHz reference source with the same drift rate were required, a 5 MHz oscillator and a times-2 multiplier would probably be employed.

A typical set of specifications for such oscillators is given in Table 8-1. The oscillators are manufactured by Vectron Laboratories, Inc., and are among the finest of their kind made in this country. All oscillators of this

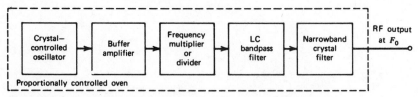

Figure 8-1. A crystal-controlled reference source; general block diagram.

522

Table 8-1. Specifications for Series CO-200, CO-211, and CO-206 Crystal-Controlled Oscillators, Manufactured by Vectron Laboratories, Inc.[a]

	CO-200 series	CO-211 series	CO-206 series
Stability			
Aging rate	CO-204 1×10^{-9}/day	CO-217 1×10^{-9}/day	CO-206 3×10^{-10}/day
	CO-203 3×10^{-9}/day	CO-216 3×10^{-9}/day	CO-207 1×10^{-10}/day
	CO-202 5×10^{-9}/day	CO-215 5×10^{-9}/day	
	CO-201 1×10^{-8}/day	CO-214 1×10^{-8}/day	
Over temperature			
0 to +50°C	$< \pm 3 \times 10^{-9}$	$< \pm 1 \times 10^{-8}$	$< \pm 1 \times 10^{-9}$
−20 to +70°C	$< \pm 5 \times 10^{-9}$	$\pm 3 \times 10^{-8}$	$< \pm 3 \times 10^{-9}$
−55 to +75°C	$\pm 3 \times 10^{-8}$	$\pm 5 \times 10^{-8}$	$\pm 1 \times 10^{-8}$
Output frequency	1, 5, or 10 MHz standard (other frequencies to 25 MHz optionally available)		
Output level	> 1 V rms into 50 Ω	> 1 V rms into 500 Ω	> 1 V rms into 50 Ω
	(TTL/DTL compatible output or other levels optional)		
Frequency offset versus supply ±5%	$< \pm 5 \times 10^{-10}$	$< \pm 1 \times 10^{-9}$	$< \pm 5 \times 10^{-10}$
Short-term stability	$< 3 \times 10^{-11}$/sec	$< 3 \times 10^{-11}$/sec	$< 5 \times 10^{-12}$/sec

Table 8-1 (*Continued*)

Phase noise (single sideband—dBc/Hz)			
100 Hz from carrier	-130	-130	-135
1 KHz from carrier	-140	-140	-145
	(15 dB improvement optionally available)		
Input voltage	24 V dc standard (any specified voltage between 5 V dc and 32 V dc optional)		
Input power			
Turn-on	6 W	6 W	6 W
Stabilized at 25°C	3 W	3 W	3 W
Frequency adjustment	Range sufficient to compensate for 5 to 10 years of crystal aging. Settable to $<1 \times 10^{-8}$; use VCXO option for improved settability.		
Oven control	Proportional	Proportional	Proportional
Humidity	100% (all units are fully sealed)		
Size/weight	$2 \times 2 \times 4$ in., 10 oz.	$2 \times 2 \times 3$ in., 7 oz.	$2\frac{1}{4} \times 2\frac{1}{4} \times 4\frac{1}{4}$ in., 10 oz.

"Courtesy of Vectron Laboratories, Inc.

class are placed in proportionally controlled ovens and constitute a compact package such as the one shown in Fig. 8-2. The reason for proportional control is to prevent any transients due to the oven control circuitry, which can contaminate the oscillator output spectrum, from being generated. These specifications completely define the oscillators from the manufacturer's point of view. However, a synthesizer designer should prepare his own specification on the basis of the system requirements. A typical reference oscillator specification may consist of the following.

1.0   Center frequency

Here the designer specifies the nominal frequency of the reference source if it is a fixed-frequency oscillator or a center frequency if it is a VCXO.

2.0   Long-term frequency stability

Usually, long-term frequency drift (aging) is specified per 24 hr. Some system requirements include also total drift per year or over a longer period. In any event, it is essential to specify the time period of continuous operation after which a desired drift rate is to be achieved.

3.0   Short-term frequency stability

Some users of synthesizers express short-term stability in terms of a frequency drift of less than x Hz in y sec, the time in question extending from 5 min to $\frac{1}{2}$ hr. Others express it in terms of a fractional frequency deviation, $\Delta f/f$, for one or a number of averaging periods, τ. Sometimes by "short-term stability" one means phase jitter, in which case the requirement is expressed in terms of a phase difference averaged over any two successive periods of x sec that does not exceed y degrees (or rad) when measurements are performed for at least z sample time periods.

4.0   Frequency stability with temperature

5.0   Frequency stability with power supply variations

6.0   Frequency accuracy

If frequency accuracy is specified with no restrictions, the requirement is interpreted as the accuracy at which the oscillator is set at the time of shipment at room temperature, nominal power supply voltage, and nominal rf load. In most practical cases, however, this is not the accuracy of the oscillator when it is received by the user, and the longer the period during which the oscillator is not energized the greater is the discrepancy. Formulation of this requirement is closely related to warm-up time and is further discussed in the following paragraph.

7.0   Warm-up time

The warm-up time of a crystal oscillator placed in an oven is greatly dependent on the response characteristic of the oven temperature-control circuit. As is the case with any feedback system, there is a set of control-circuit parameters that results in a compromise between speed of response and circuit stability. An overdamped system is stable but slow, and an under-

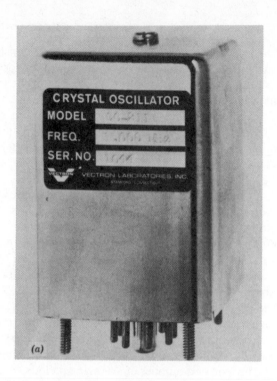

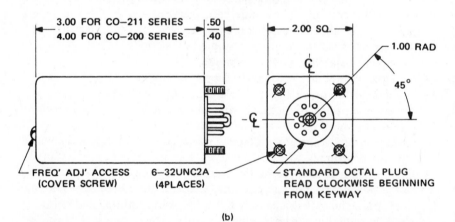

3.00 FOR CO–211 SERIES

4.00 FOR CO–200 SERIES

.50

.40

2.00 SQ.

1.00 RAD

45°

FREQ' ADJ' ACCESS
(COVER SCREW)

6–32UNC2A
(4PLACES)

STANDARD OCTAL PLUG
READ CLOCKWISE BEGINNING
FROM KEYWAY

(b)

Figure 8-2. Mechanical specification for series CO-200 and CO-211 crystal-controlled oscillators, manufactured by Vectron Laboratories, Inc. (*a*) Photograph of a typical oscillator; (*b*) mechanical drawing for series CO-200 and CO-211 oscillators. Courtesy of Vectron Laboratories, Inc.

damped system is fast but has a response characteristic displaying ringing, during which the oven temperature as well as the oscillator frequency is not stable.

If fast warm-up time and good accuracy are desired, the specification should include the time after turn-on at which a certain accuracy is to be achieved and the limits of short-term frequency drift (ringing) thereafter. In addition one should specify the turn-off period, after which the oscillator is to meet the accuracy requirement.

It may well happen that, having defined a warm-up time, frequency accuracy, and turn-off period, the synthesizer designer learns that these parameters cannot be satisfied simultaneously. What is usually done in such cases is to provide the oscillator oven with dc power at all times independently of the synthesizer operational status.

8.0   Frequency tuning

To maintain frequency accuracy over a long period of time, it may be necessary to tune the oscillator on frequency periodically. Hence the resolution, total range, and type (manual or electronic) of the oscillator tuning mechanism should be specified.

9.0   Spurious outputs

The synthesizer designer should be familiar with the approach used to derive the oscillator output frequency. On the basis of his own synthesis approach and the spurious-signal requirement for the synthesizer, he should compute and specify the allowable level of spurious signals at the reference oscillator output. If the requirement is stringent, calling for extensive filtering, the oscillator manufacturer may not satisfy it because of oven space limitations, in which case the synthesizer designer may have to provide additional filtering of his own external to the oscillator. One should also identify the type of spurious signal, such as AM, FM, or nonharmonically related, if such identification results in any cost reduction of the reference source.

10.0   Phase noise

It is important to define the phase-noise performance of the reference oscillator at offset frequencies of interest to the synthesizer designer. If the designer discovers that none of the standard models evaluated by him meets the requirement, he should investigate the capabilities of various oscillator manufacturers with respect to narrowband crystal filter design. Depending on the manufacturer, the variation in phase noise of reference oscillators otherwise equal in performance and price has been found by the writer to be more than 6 dB.

Figure 8-3 is a set of curves describing the measured phase noise of a 5 MHz oscillator, CO-200 series, manufactured by Vectron Laboratories, Inc. Curve A is the noise spectrum of the standard model. Curve B is the

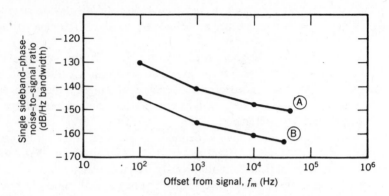

Figure 8-3.   Phase-noise performance of series CO-200 crystal-controlled oscillators, manufactured by Vectron Laboratories, Inc. Courtesy of Vectron Laboratories, Inc.

spectrum of this model optimized for low noise (the 15 dB improvement option). The technique for developing a low-noise signal is to use a very narrow bandwidth crystal filter followed by a low-noise amplifier.

11.0   Levels of harmonics
12.0   Output impedance
13.0   VSWR (voltage standing-wave ratio)
14.0   Output rf power
15.0   Direct-current power-supply voltage and maximum dc drain current
16.0   Power-supply voltage variations

This requirement, in conjunction with paragraph 4.0 of the specification, assists the oscillator manufacturer in designing the voltage regulator associated with the oscillator.

17.0   Temperature range
      17.1   Operating
      17.2   Nonoperating
18.0   Altitude
      18.1   Operating
      18.2   Nonoperating
19.0   Size

A detailed mechanical drawing of the oscillator should be included in the specification to prevent the oscillator manufacturer from making any mechanical modifications affecting the size of the oscillator and the type of mounting without approval by the synthesizer designer.

20.0   VCXO requirements

If the reference source is a voltage-controlled crystal oscillator, the following paragraphs are added to the specification.

20.1   Control voltage range
20.2   Frequency deviation
20.3   Linearity

The originator of a reference oscillator specification may add to, subtract from, or modify the formulation of paragraphs 1.0 to 20.3 to comply with the system requirements. A general rule to follow is to include in the specification only the requirements called for by the synthesizer specification or dictated by the synthesizer design considerations. However, any related paragraphs may be added to the specification if it is verified that their addition does not increase the reference oscillator cost.

The best long-term frequency stability that is attained in the single-oven oscillators specified in Table 8-1 is 1 part in 10^{-10}/day. When the oscillator is placed in a double oven, a model FS-321 type of crystal-controlled oscillator provides stablilty better by almost an order of magnitude. In the case of model FS-321, manufactured by Vectron Laboratories, Inc., the improved stability is achieved by using a high-stability 5 MHz crystal unit, driving the crystal at a low current, which is held constant by an oscillator AGC circuit, placing the oscillator in a proportionally controlled double oven (i.e., keeping the oscillator temperature at the crystal turnover point virtually constant with respect to both ambient temperature and time), and aging the oscillator for a long period of time.

Table 8-2 is a partial specification for model FS-321 frequency standard. The basic crystal oscillator operates at 5 MHz. The 1 MHz and 100 kHz outputs are derived from 5 MHz by frequency division.

The frequency of the 5 MHz oscillator is set to ± 3 parts in 10^{10} of the

Table 8-2.   Partial Performance Specification for Frequency Standard, Model FS-321, Manufactured by Vectron Laboratories, Inc.[a]

Output frequencies	5 MHz, 1 MHz, and 100 kHz
Frequency accuracy	Set to ± 3 parts in 10^{10} at the time of shipment
Long-term stability	(a) Better than 3 parts in 10^{10}/day at time of shipment (b) Better than 1 part in 10^{10}/day typical after 60 days of continuous operation
Nonharmonically related spurious outputs	Better than -80 dB referred to the desired output
Phase noise	See Fig. 8-4

[a] Courtesy of Vectron Laboratories, Inc.

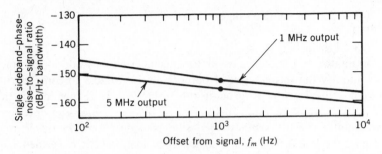

Figure 8-4. Phase-noise performance of frequency standard, model FS-321, manufactured by Vectron Laboratories, Inc. Courtesy of Vectron Laboratories, Inc.

nominal frequency at the time of shipment. As mentioned, this value is not the accuracy of the standard at the time when its user first switches it on. This difference, however, presents no problem because most users of model FS-321 type reference sources utilize appropriate equipment to calibrate periodically the frequency of their standards against an atomic reference maintained and made available for public use by the National Bureau of Standards.

The oscillator attains an aging rate of less than 3 parts in 10^{10}/day 4 hr after turn-on, following a 24-hr turn-off period, and 1 part in 10^{10} after 60 days of continuous operation. Units with better aging rates are available upon request.

Figure 8-4 is the phase-noise spectrum of the model FS-321 frequency

Figure 8-5. Front view of frequency standard, model FS-321, manufactured by Vectron Laboratories, Inc. Courtesy of Vectron Laboratories, Inc.

standard at two frequencies. The 5 MHz signal displays the best noise spectrum of the three output signals. It is this output that should be used as a reference for low-noise frequency synthesizers.

Figure 8-5 is a front view of the model FS-321 frequency standard. When used in conjunction with a rechargeable battery pack, the frequency standard identified as model FS-321E can operate for more than 35 hr without an external power source. In the event of prime power failure, automatic switchover to internal battery operation is designed in model FS-321E.

8-2   Atomic Frequency Standards

An atomic frequency standard can be described by the frequency-stabilizing feedback loop shown in Fig. 8-6. A spectrally pure 5 MHz output is generated by a voltage-controlled crystal oscillator whose frequency is stabilized by an atomic frequency discriminator. By making the loop bandwidth very narrow, one preserves the low noise spectrum of the VCXO and achieves exceptionally high degrees of accuracy and long-term stability, which closely approximate the accuracy and stability of an atomic energy transition.

Rubidium Frequency Standard, Model 304D, Manufactured by Tracor, Inc.

Atoms exist in discrete energy states, and each energy state is characterized by a specific value of internal energy. An atom can be raised from a low or stable to a high or an unstable energy state by exciting it with an external energy field at a frequency that is equal to the hyperfine resonant frequency of the atom, f_h, a precisely known quantity. Under these conditions some of the energy contained in the field is absorbed by the atom. The hyperfine atomic resonant frequency is related to the change in the

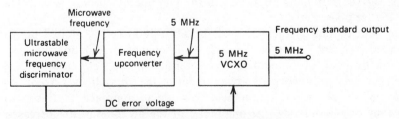

Figure 8-6.   An atomic frequency standard; basic block diagram.

internal energy of the atom by the equation

$$f_h = \frac{E_2 - E_1}{h},$$ (8-1)

where E_1 = the initial, stable energy state of the atom,
E_2 = the unstable energy state,
h = Planck's constant (6.24×10^{-27} erg-sec).

Since the higher energy state is unstable, the atom falls back to the initial state and the process repeats as long as the external field provides energy at the atomic resonant frequency.

The discriminator operation in a rubidium frequency standard is based on the energy absorption characteristic of rubidium-87. The reference element is an optically pumped rubidium-87 vapor cell located in a microwave cavity. Optical pumping is used because absorption of optical radiation is easier to detect than absorption of electromagnetic radiation. The energy absorption phenomenon is used to control the frequency of a 5 MHz VCXO in the following manner.

A light beam from a rubidium lamp is applied to a filter cell, which blocks the energy at the undesirable wavelength of 7947 Å and passes the energy at the 7800 Å wavelength to the rubidium absorption cell unattenuated (see Fig. 8-7). The rubidium vapor cell absorbs some of the light energy. The unabsorbed light leaving the cavity is monitored by a photodiode (photocell). At the same time an electromagnetic field is generated in the cavity. When the frequency of the electromagnetic field is made to approach the resonant frequency of the rubidium vapor (6834.685 MHz), the number of energy level transitions in the rubidium-87 gas is increased, more of the light emitted by the rubidium lamp is absorbed by the rubidium vapor cell, and the photodiode current decreases. The maximum absorption of light from the rubidium lamp occurs when the frequency of the excitation electromagnetic field exactly matches the rubidium resonant frequency. This absorption peak, shown in Fig. 8-8 as a dip in the light intensity at the output of the microwave cavity, which results in a corresponding decrease in the photodiode current, is detected by a 155 Hz phase detector.

The light output characteristic of the absorption cell is plotted on an expanded frequency scale in Fig. 8-9. By frequency modulating the electromagnetic field, one modulates the intensity of the light beam at the same rate. This produces an ac component of the photodiode current. As the frequency of the excitation field is made to approach the resonant frequency of rubidium-87, the amplitude of the ac component decreases.

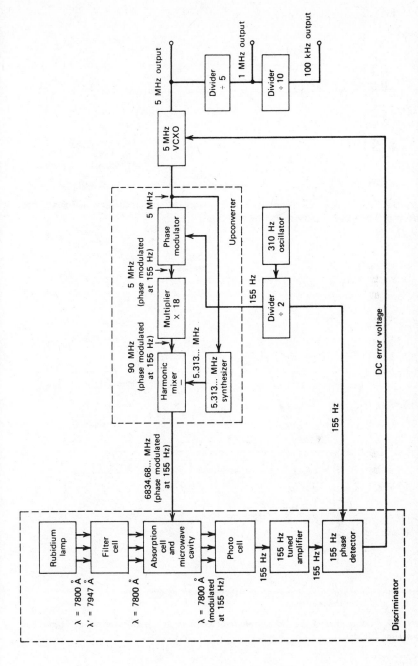

Figure 8-7. Rubidium frequency standard, model 304D, manufactured by Tracor, Inc. Courtesy of Tracor, Inc., Instruments Division.

533

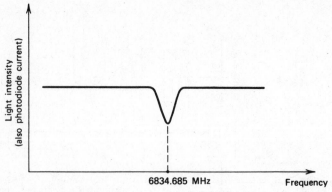

Figure 8-8. Absorption cell output versus frequency of excitation electromagnetic field. Courtesy of Tracor, Inc., Instruments Division.

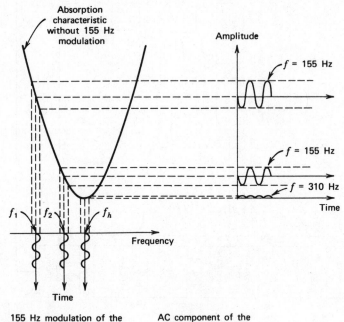

155 Hz modulation of the absorption characteristic

AC component of the photodiode current

Figure 8-9. Generation of the 155 Hz phase-detector input signal. Courtesy of Tracor, Inc., Instruments Division.

As the two frequencies are made to coincide, the ac component at the fundamental modulation rate vanishes and the photocell output consists mainly of the second harmonic at 310 Hz. When the frequency of the electromagnetic field is raised above the rubidium resonant frequency, the same process takes place except that the diode ac current is now 180 degrees out of phase with respect to the current generated when the frequency of the field is below the rubidium resonant frequency. A change in the amplitude of the photodiode ac current at 155 Hz is reflected as a change in the magnitude of the dc error voltage at the output of the 155 Hz phase detector. The phase of the ac current determines the sign of the error voltage. The error voltage causes the VCXO to change frequency in such a direction as to reduce the difference between the rubidium resonant frequency and the frequency of the electromagnetic field injected into the microwave cavity. When this frequency difference is zero, the VCXO is exactly locked to the atomic reference.

Figure 8-7 demonstrates how the electromagnetic field is produced. The VCXO signal at 5 MHz is phase modulated at the 155 Hz rate and multiplied by 18. The 90 MHz signal is applied to one input of a harmonic mixer. A signal at 5.313... MHz generated by a synthesizer, which is also locked to the VCXO, is applied to the other input of the mixer. The function of the harmonic mixer is to produce a signal at a frequency that is the difference between the 76th harmonic of 90 MHz and the 5.313... MHz synthesized frequency. The harmonic mixer output is coupled to the microwave cavity, which is tuned to the rubidium resonant frequency. A frequency multiplication enhances the level of the FM sidebands without affecting the modulation rate. Hence the electromagnetic field applied to the rubidium vapor cell is also frequency modulated at 155 Hz. An audio-oscillator is used to modulate the field. The audio-oscillator signal is also applied to the 155 Hz phase detector for comparison with the photocell ac signal.

Table 8-3 is a partial list of model 304D requirements, and Fig. 8-10 is a set of photographs of the rubidium standard. The frequency accuracy and long-term stability of this reference source are higher by orders of magnitude than the accuracy and stability of a crystal-controlled oscillator. The phase noise of the atomic standard at 5 MHz is mainly the noise of the VCXO; it is given in Fig. 7-31. Spurious signals at the output of the atomic standard, which are -80 dB below the desired output, are attributed to various frequencies utilized in the error voltage generation.

The frequency of an atomic transition is affected by external magnetic fields. A magnetic field of 1 G pulls the frequency of model 304D by less

Table 8-3 Partial Performance Specification for Rubidium Frequency Standard, Model 304D, Manufactured by Tracor, Inc.[a]

Output frequencies:	5 MHz, 1 MHz, and 100 kHz
Frequency accuracy:	(a) Set at factory to within 1 part in 10^{11} of specified time scale (b) 1 hr after turn on: less than 1 part in 10^{10} (c) 4 hrs after turn on: less than 5 parts in 10^{11}
Long-term stability:	Less than 2 parts in 10^{11}/month
Nonharmonically related spurious outputs:	Better than -80 dB referred to the desired output
Phase noise:	See Fig. 7-31

[a]Courtesy of Tracor, Inc., Instrument Division.

than 5 parts in 10^{12} even though double magnetic shields surround the optical-microwave unit. A dependence of atomic frequency on an external magnetic field is utilized in fine frequency tuning of the atomic standard. Model 304D is tuned over a frequency range of -0 to $+2$ parts in 10^9 with a resolution of ± 2 parts in 10^{12} by varying the dc current through a coil wound around the microwave cavity so that the coil axis coincides with the rubidium lamp beam.

The 5 MHz VCXO is located in a proportionally controlled oven and displays excellent accuracy and stability of its own. Manual switchover from locked to free-running VCXO operation is incorporated in model 304D. When the free-running operation is exercised, all three output signals, at 5 MHz, 1 MHz, and 100 kHz, exhibit the frequency accuracy and long-term stability of the VCXO.

In case of line power failure automatic switchover to a standby battery is designed in model 304D. A battery with approximately 10 min of standby power capacity is available as an option for this purpose.

A rubidium frequency standard is a secondary standard. At present the most accurate and stable reference source is a cesium beam frequency standard, which is described next.

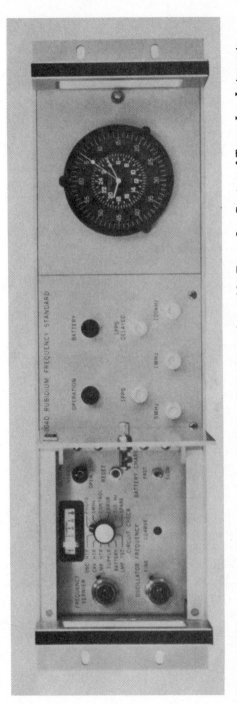

Figure 8-10-a. Front view of rubidium frequency standard, model 304D, manufactured by Tracor, Inc. Courtesy of Tracor, Inc., Instruments Division.

537

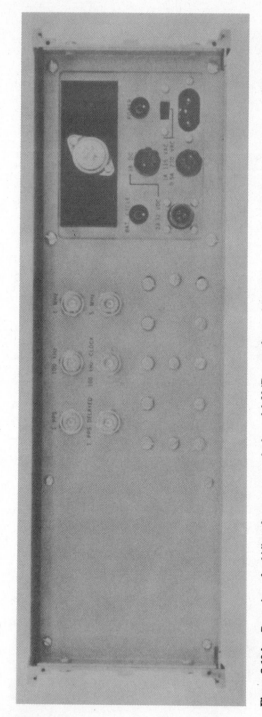

Figure 8-10-b. Rear view of rubidium frequency standard, model 304D, manufactured by Tracor, Inc. Courtesy of Tracor, Inc., Instruments Division.

538

Cesium Beam Frequency Standard, Model 5061A, Manufactured by Hewlett-Packard Company

The operation of the frequency discriminator in a cesium beam standard is based on the external electromagnetic excitation of cesium atoms, which results in energy transitions, and the detection of atoms that have undergone a specific transition.

A functional block diagram of the model 5061A cesium beam standard is shown in Fig. 8-11. The output signal of a 5 MHz VCXO is phase modulated at 137 Hz and applied to a times-18 frequency multiplier. The multiplier output signal at 90 MHz is used to drive one port of a harmonic mixer, whereas the synthesized signal at 12.631... MHz drives the other port. The harmonic mixer produces a 9192.631... MHz signal, phase modulated at 137 Hz, which is the sum of the 12.631... MHz synthesized frequency and the 102nd harmonic of 90 MHz. The harmonic mixer output is coupled to the microwave cavity of the cesium beam tube. Hence the electromagnetic field generated in the cavity is phase modulated at 137 Hz. This field in conjunction with a weak dc magnetic field, called the C-field, induces hyperfine energy level transitions in the cesium atoms, forming a beam that is directed into the microwave cavity.

When the 5 MHz VCXO is not locked to the atomic reference, the cesium-beam-tube output current contains an ac component at 137 Hz, the amplitude of which depends on the magnitude of the difference between the atomic frequency of cesium-133 (9192.63177139 MHz) and the frequency of the electromagnetic field, Δf_d. The phase of this component depends on the sign of Δf_d, that is, on whether the frequency of the electromagnetic field is below or above the atomic frequency. When the VCXO is locked to the atomic reference, the 137 Hz signal at the output of the tube vanishes.

The audio-oscillator signal used to phase-modulate the electromagnetic field is applied also to the 137 Hz phase detector, where it is compared to the 137 Hz component of the tube output. A change in amplitude of the 137 Hz component is reflected as a change in magnitude of the dc error voltage produced by the phase detector, whereas the phase of the 137 Hz component determines the sign of the error voltage. The error voltage tunes the 5 MHz VCXO so that Δf_d is minimized. Insight into the mechanism of the 137 Hz signal generation by the tube is gained by studying the operation of the tube in detail.

Cesium-133 has a single valence electron. The electron spin has two possible orientations and two energy levels associated with these orientations. The corresponding energy states are designated by the quantum numbers $F=3$ and $F=4$. Within an oven located in the cesium beam tube

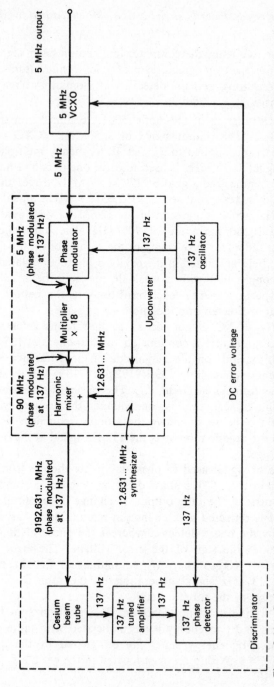

Figure 8-11. Cesium beam frequency standard, model 5061A, manufactured by Hewlett-Packard Company. Courtesy of Hewlett-Packard Company.

540

(see Fig. 8-12) liquid cesium is vaporized to a gaseous state. A collimator forms the cesium atoms into a narrow beam and directs the beam toward an input selector magnet which deflects $F=4$ atoms into the microwave cavity. When the atoms enter the C-field generated in the cavity, the $F=4$ energy level splits into a number of discrete levels, the Zeeman hyperfine levels designated by the quantum numbers m_F, which correspond to allowed orientations of the cesium atom with respect to the C-field. The transition used in the frequency standard is the one separating $F=4$, $m_F=0$ and $F=3$, $m_F=0$ states, designated as the $(4,0) \rightarrow (3,0)$ transition. This particular transition is selected because the atomic frequency associated with it is nearly independent of an external magnetic field, whereas other allowable transitions are linearly dependent on the field. With the exception of the magnetic field effects, the frequency associated with the $(4,0) \rightarrow (3,0)$ transition is completely independent of all environmental conditions.

Upon the application of a linearly polarized ac magnetic field at appropriate frequency applied parallel to the C-field, transitions $(4,0) \rightarrow (3,0)$ take place in the microwave cavity. The maximum number of transitions is

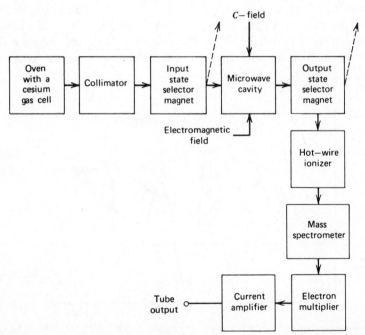

Figure 8-12. A cesium beam tube; functional block diagram. Courtesy of Hewlett-Packard Company.

induced when the frequency of the ac magnetic field exactly matches the atomic frequency of cesium-133.

An output state selector magnet deflects all atoms that underwent $(4,0) \rightarrow (3,0)$ transitions to a hot-wire ionizer, where the cesium atoms are ionized and deflected by an electric field into a mass spectrometer. The spectrometer acts as a selector also and directs only cesium ions that are positively charged to the first dynode of an electron multiplier. The electron multiplier converts the beam of ions into a usable electron beam by successive secondary emission stages. The multiplier drives a current preamplifier from which the cesium-beam-tube output is obtained.

Figure 8-13 shows a dc response characteristic of a cesium beam tube, known as the Ramsey curve. The main response of the curve is used in the frequency discriminator action. The secondary responses are useful in calibration of the frequency standard. The width of the main response and the distance between peaks are functions of the length of the microwave cavity and the average velocity of cesium atoms. The width of the main response in model 5061A is 550 Hz.

Figure 8-14 is an expanded view of the peak portion of the main response. By frequency modulating the ac magnetic field, one generates an ac component of the tube output current. The case considered in this illustration is for positive Δf_d. When the frequency of the ac magnetic field is above the cesium atomic frequency (negative Δf_d), the same process takes place except that the 137 Hz component of the tube output is 180 degrees out of phase with respect to the 137 Hz component generated when Δf_d is positive.

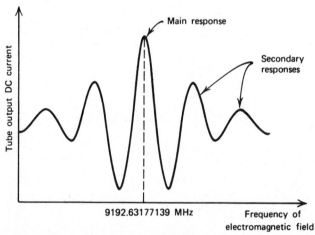

Figure 8-13. Direct-current response characteristic of a cesium beam tube (a Ramsey curve). Courtesy of Hewlett-Packard Company.

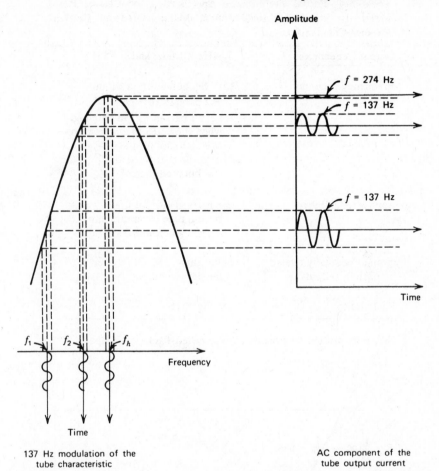

Figure 8-14. Generation of the 137 Hz phase-detector input signal. Courtesy of Hewlett-Packard Company.

Table 8-4 is a partial list of model 5061A requirements, and Fig. 8-16 shows the front and rear views of the frequency standard. The long-term frequency stability of this reference source is better by orders of magnitude than the stability of any other type of frequency reference presently in use. A cesium beam frequency standard is self-calibrating and at present is the primary frequency reference source.

In the event of a line power failure automatic switchover from the internal power supply to an external self-charging battery is designed in model 5061A. A power supply, HP model 5085A, with 8 hrs of standby power capacity is available for this purpose.

Table 8-4 Partial Performance Specification for Cesium Beam Frequency Standard, Model 5061A, Manufactured by Hewlett-Packard Company[a]

Output frequencies:	5 MHz, 1 MHz, and 100 kHz
Frequency accuracy:	(a) Set at factory to within ± 7 parts in 10^{13}
	(b) ± 1 part in 10^{11} maintained when subjected to temperatures from 0 to 50° C, magnetic fields up to 2 G, or any combination thereof
Long-term stability:	± 5 parts in 10^{12} for life of cesium tube (operating life: 3 years guaranteed)
Nonharmonically related spurious outputs:	Better than − 80 dB referred to the desired output
Phase noise:	See Fig. 8-15

[a]Material printed by permission of Hewlett-Packard Company.

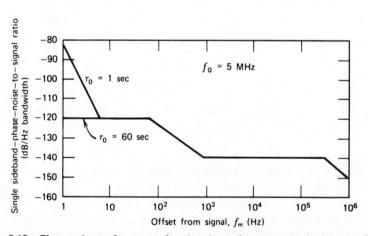

Figure 8-15. Phase noise performance of cesium beam frequency standard, model 5061A, manufactured by Hewlett-Packard Company. Courtesy of Hewlett-Packard Company.

Figure 8-16-a. Front view of cesium beam frequency standard, model 5061A, manufactured by Hewlett-Packard Company. Courtesy of Hewlett-Packard Company.

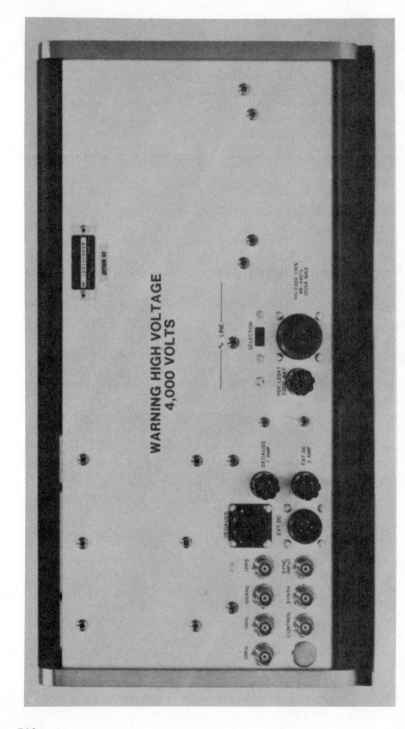

Figure 8-16-b. Rear views of cesium beam frequency standard, model 5061A, manufactured by Hewlett-Packard Company. Courtesy of Hewlett-Packard Company.

The model 5061A frequency standard can be made portable by combining it with a power supply, HP model K02-5060A. The power supply provides approximately 7 hr of standby power for the frequency standard from a set of sealed nickel-cadmium batteries.

References

1. Vectron Laboratories, Inc., *Data Sheet for Crystal Oscillators, CO-200, CO-211, and CO-206 Series.*
2. Vectron Laboratories, Inc., *Data Sheet for Frequency Standards, Models FS-321 and FS-323.*
3. Tracor, Inc. *Operational and Service Manual for Rubidium Frequency Standard, Model 304D.* April, 1970.
4. Hewlett-Packard Company. *Operational and Service Manual for Rubidium Vapor Frequency Standard, Model 5065A.* 1969.
5. Hewlett-Packard Company. *Data Sheet for Cesium Beam Frequency Standard, Model 5061A.* September 1, 1969.
6. Hewlett-Packard Company. *Training Manual for Cesium Beam Frequency Standard, Model 5060A.* February 1967.
7. Hewlett-Packard Company. *Data Sheet for Power Supply, Model 5085A.* August 15, 1967.
8. Atomichron, Inc. *Technical Manual for Cesium Beam Primary Frequency Standard, Model 3702.* May 1, 1970.

9
Troubleshooting of Synthesizers

A widely used technique for troubleshooting electronic equipment consists of replacing the modules, circuits, or components suspected of failure, one at a time, by proven modules, circuits, or components until satisfactory performance is obtained. This is the approach that works well in simple linear and digital systems. Feedback and hybrid systems, such as PLLs or a combination of analog circuits, digital circuits, and PLLs, however, in general do not lend themselves to such techniques, and more sophisticated test procedures have to be employed to arrive at an accurate diagnosis of a problem without unnecessary delay. The intent of this chapter is to make the reader familiar with these procedures.

The troubleshooting associated with the developmental stage of a synthesizer, intended to disclose the reasons why a circuit or complete system is not functioning according to the initial-design goals, is presented here jointly with the troubleshooting that is done at the manufacturing phase of the product and that deals with discovering human errors and locating defective components, because there is no clear-cut separation between the two procedures and because they are governed by the same basic principles. A thorough knowledge of these principles is as advantageous to a design engineer as it is to a manufacturing technician.

9-1 Basic Principles

A successful troubleshooting procedure consists of the following six steps:

1. Analysis.
2. Subdivision.
3. Selection of test equipment.
4. Elimination by substitution.
5. Taking data.
6. Comparing measured data to specifications.

The analysis consists of studying the synthesizer performance data that indicate a fault, such as a low rf output power or a high level of spurious signals or phase noise, and deciding upon the module or circuit most likely to have failed. This requires a thorough knowledge of the synthesizer block diagram and circuits.

The next step is to prepare a test plan intended to verify the assumptions made in the analysis. In most cases of system testing, the plan consists of subdividing the synthesizer into logical blocks of subsystems, modules, or circuits, considering each block a "black box" with several inputs and outputs, and performing a series of tests on individual blocks.

Selection of proper test equipment is extremely important because the identification of a failed black box is based on the test data, and incorrectly taken data will lead to wrong conclusions or no conclusions at all.

Measurements are limited to taking the data associated with the parameters that define a specific black box. Taking irrelevant data makes troubleshooting unnecessarily difficult.

Finally, the performance of the black box is compared with the performance recorded when the unit was operating properly, and a decision is made either to pass the black box as meeting all requirements or to reject it as failed. Each of the steps briefly described above is considered in detail in the following pages.

Analysis

A troubleshooting procedure is best illustrated with a practical example.

Consider the three-PLL synthesizer shown in Fig. 7-19. Assume that the symptom of the problem is a lack of frequency stability of the output signal, detected after a 24 hr burn-in of the synthesizer.

We base our analysis on the nature of the problem and prepare a mental summary of all possible places where a failure that would result in lack of frequency stability at the synthesizer output could have taken place. In the given example these are the three PLLs and the reference oscillator with crystal filter, which also contains a distribution amplifier driving the three fixed-ratio dividers, not shown in Fig. 7-19. We verify our assumptions by performing the following tests.

Subdivision

Most synthesizers are complicated systems. To find a failed circuit rapidly, we subdivide the synthesizer into subsections in some logical manner and test each subsection individually. Figure 9-1 is a functional block diagram of the synthesizer considered in the example, showing a module subdivision of the system—a good packaging technique successful from the point of view not only of rapid troubleshooting but also of easy rf shielding.

It is essential that the module breakdown of a synthesizer, done during the design phase, include considerations of easy alignment and testing of the subsections and individual circuits, such as 50 Ω input and output impedances compatible with the impedances of the test equipment required for that purpose. In the case of modules, the break points should carry rf connectors that can be easily disconnected. Circuit separation

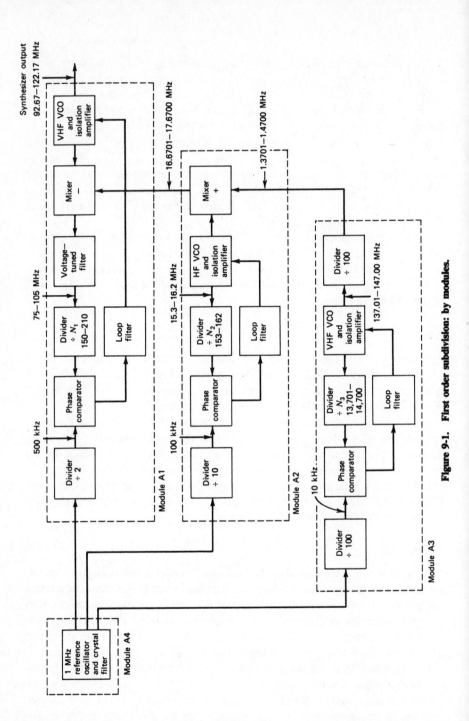

Figure 9-1. First order subdivision: by modules.

within a module is usually achieved by way of wire jumpers. It is also essential to provide easily accessible test points judiciously located throughout the system. Equally important is the ready availability of test data defining the satisfactory performance of each circuit. The data will be discussed more thoroughly below.

Returning to our example, assume that further testing on the module level, using a frequency counter, discloses that the signals at the outputs of both auxiliary PLLs are also frequency unstable. Having checked the dc power supply voltages and found them to be as specified, we conclude that module $A4$ has failed and turn our attention to that part of the synthesizer.

Were the outputs of modules $A2$ and $A3$ frequency stable, the problem would have been associated either with that output of the distribution amplifier $(A4)$ which drives the divide-by-2 frequency divider $(A1)$ or with module $A1$ itself, so that out testing might have proceeded as shown in Fig. 9-2a through e, which demonstrates subdivision by groups of circuits.

Finally, once a group of circuits is identified, subdivision by circuits is employed to limit the search to a particular circuit.

Selection of Test Equipment

The basic test equipment used in synthesizer troubleshooting is as follows:

1. Direct-current multimeter.
2. Radio-frequency voltmeter.
3. Frequency counter.
4. Oscilloscope.
5. Radio-frequency spectrum analyzer.
6. Two signal generators.
7. One or two dc power supplies.

The model numbers of this equipment and the manufacturers' names are not specified here because, with the exception of the dc multimeter and power supplies, selection of a specific instrument depends on the frequency of operation, and it is assumed that the reader has all the test equipment necessary to perform the tests. What will be discussed in this chapter is, given a symptom of a problem, what test equipment to use among the instruments mentioned above and how to use it to make the measurements correctly.

This leads us to the first consideration important in the troubleshooting of synthesizers. To make a proper choice of the instrument appropriate for the required measurement, one has to have a rough idea of what kind of frequency spectra the signals at the input and output of every circuit have, not only when the synthesizer is functioning properly but also when it does not. For example, one would expect to see all intermodulation products, defined by $f_{m,n} = mf_1 \pm nf_2$, at the output of the mixer, Fig. 9-2b, at all

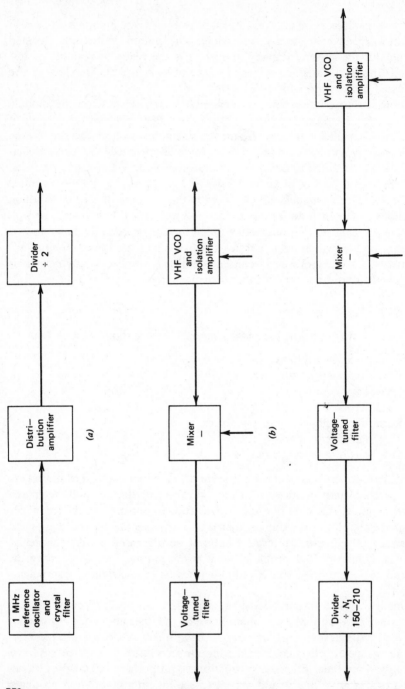

(a)

(b)

552

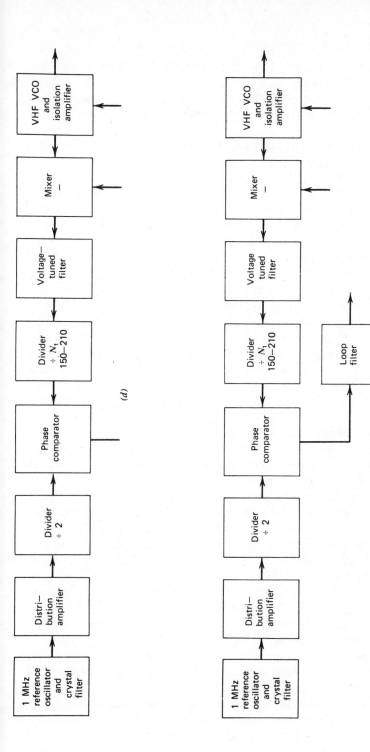

Figure 9-2. Second order subdivision: by groups of circuits. (*a*) 1 MHz reference and divide-by-2 divider; (*b*) vhf VCO, mixer and voltage-tuned filter; (*c*) same as in (*b*) and divide-by-N_1 divider; (*d*) sketches (*a*), (*b*), and (*c*) combined and phase comparator; (*e*) same as (*d*) and loop filter.

553

times, whereas the products would appear at a level higher than the specified one at the voltage-tuned filter output only if the mixer or filter were not functioning properly. Hence troubleshooting the rf mixers and filters with a spectrum analyzer rapidly gives a correct indication of what the problem is, whereas, say, an rf voltmeter or counter does not. If the frequency spectra cannot be estimated beforehand, the signals should be examined with a spectrum analyzer before any other type of test equipment is used.

The second consideration has to do with correct impedance matching between the circuitry and test equipment. Heavy loading or misalignment of the circuits by test equipment always indicates faulty operation, even in the absence of failure. Impedance matching requires two actions: (a) circuit parameters should be made compatible with the test equipment at the designated test points previously mentioned, and (b) when such compatibility is impracticable, impedance matching pads should be designed and employed. Usually, these pads have a high reactance (10 kΩ or higher) over the operating frequency range.

The use of long coaxial cables is not advisable, because the input/output impedances of the test instruments may have a 50 Ω resistive part shunted by an 11 Ω reactive part (30 pF at 500 MHz), thus providing a poor termination for the cable. If the use of long cables cannot be avoided, they should be terminated into a 6 dB resistive pad before being connected to the circuit.

Oscilloscope probes or any probes with long ground leads should not be used. A 3-in.-long wire with a diameter of 0.032 in., 20 AWG, has an inductance of approximately 0.079 μH, which at frequencies above 50 MHz may be high enough to result in an incorrect reading.

A mistake commonly made in troubleshooting is to use test equipment that does not have the required dynamic range or resolution. For example, it is a frequent error to use the same 70-dB-dynamic-range spectrum analyzer in measuring the levels of spurious signals at both the input and the output of the divide-by-N_1 frequency divider, Fig. 9-2c, without appreciating the fact that at the output of the circuit the spurious signals are at least 40 dB lower than at the input, so that a -50 dBc spurious signal referred to the input of the divider cannot be detected at the output of the circuit by using 70-dB-dynamic-range test equipment.

Similarly, hunting for spurious signals with a spectrum analyzer with, say, a 2-kHz-per-division dispersion without realizing that none can be detected up to at least 1 kHz away from the desired signal under such conditions is of common occurrence.

To repeat, it is important to estimate beforehand the rf levels and frequencies of the signals one is dealing with in selecting the test equipment. This, in turn, requires a thorough knowledge of the synthesizer circuitry.

Elimination by Substitution

In many instances the measurement of an rf voltage, frequency, or wave shape of a signal at the input and output of a subsystem is sufficient to determine whether this or a preceding group of circuits has failed. In cases involving spurious signals and phase noise, however, this is not always so. For example, what seems to be a harmless out-of-band spurious signal at the input of a mixer (caused by a poorly tightened cover or a failed power supply bypass capacitor) may result in an in-band spurious signal, violating the requirement, at its output. Substituting all input sources, one at a time, by external, "clean" signal generators permits one to evaluate the circuit performance under "ideal" conditions, thereby often saving time.

Substitution is particularly useful when several failures are taking place in the synthesizer simultaneously or when an interaction exists between two circuits, so that the problem can be identified only by eliminating this interaction.

In all cases substitution of the system signal sources by external generators gives one the freedom of setting the rf levels and frequencies of the signals to any desired value and testing the circuits under the conditions specified in test procedures—a useful approach in both the development of a new synthesizer and the troubleshooting of a manufactured product.

Taking Data

At the beginning of this chapter it is pointed out that the troubleshooting of synthesizers by replacement is an ineffective technique seldom leading to rapid identification and solution of a problem. It is shown that analysis and subdivision are the first two logical steps in the troubleshooting of any complicated electronic equipment. The same basic principles apply to troubleshooting on the circuit level; and, since taking data is closely related to circuit analysis and subdivision, these topics are considered simultaneously.

Every synthesizer circuit can be described by one or several blocks with every block performing a function defined by a set of design parameters. For example, a digital frequency divider consisting of one integrated circuit may be described by one functional block, $\div N$; see Fig. 9-3a. An rf amplifier may be described by two blocks; a frequency multiplier, by four blocks; a voltage-controlled oscillator, by six blocks; a phase comparator, by five blocks and two capacitors; and so on (Fig. 9-3b through e, respectively).

If the synthesizer design is performed in a meaningful manner, every function of a circuit is decided upon and described in the initial design stage, and adequate test data are taken during system optimization to verify the design and to record the information essential in the alignment and troubleshooting of the equipment.

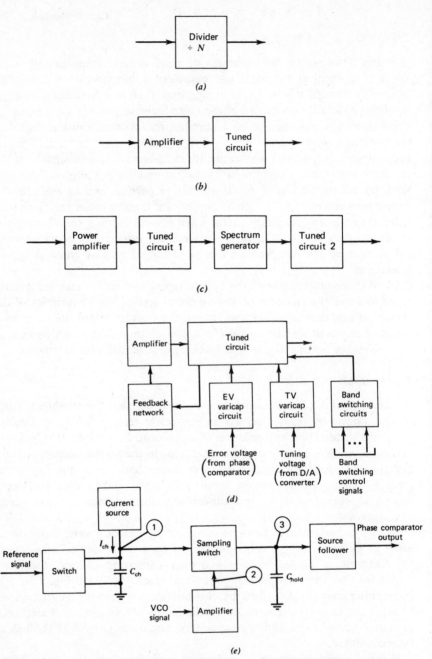

Figure 9-3. Third order subdivision: by functions of individual circuits. (*a*) Divide-by-*N* divider; (*b*) rf amplifier; (*c*) frequency multiplier; (*d*) VCO; (*e*) sample-and-hold phase comparator.

556

The parameters of importance for the basic circuits shown in Fig. 9-3 are given below.

DIGITAL FREQUENCY DIVIDER (SINE WAVE INPUT)

1. Minimum rf input power required for proper operation of the circuit.
2. Maximum rf input power (for those circuits that do not operate under an overdrive condition).
3. High and low voltage levels of the output pulse into a specified load.
4. Duty cycle of the output pulse.
5. Rise and fall times of the output pulse.
6. Maximum frequency of operation at nominal rf input power.

RF AMPLIFIER

1. Nominal rf input power.
2. RF output power into a specified load at the nominal input power.
3. 3 dB bandwidth of the tuned circuit.
4. X dB bandwidth of the tuned circuit (if the amplifier is designed to attenuate a spurious signal by X dB).

FREQUENCY MULTIPLIER

1. Nominal rf input power.
2. RF power delivered by tuned circuit No. 1 into a specified load at the nominal input power.
3. 3 dB bandwidth of tuned circuit No. 1.
4. High and low voltage levels of the spectrum generator pulse waveform at the nominal input power.
5. Duty cycle of the pulse (in cases where it is difficult to determine the duty cycle, the waveshape of the pulse is sketched).
6. Rise and fall times of the pulse.
7. 3 dB bandwidth of tuned circuit No. 2.
8. RF output power into a specified load at the nominal input power.
9. Level of the fundamental component and harmonics of the input signal at the multiplier output.

VOLTAGE-CONTROLLED OSCILLATOR

1. RF output power into a specified load (at least at the low and high ends of each frequency band).
2. VCO constant with respect to the error voltage input (at least at the low and high ends of the overall VCO frequency range).
3. VCO constant with respect to the tuning voltage input (at least at the low and high ends of each frequency band).
4. Nominal direct current required to drive the band-switching circuits.
5. Phase noise (at least at the low and high ends of each frequency band).

6. Starting voltage, that is, the minimum dc power supply voltage at which oscillations take place.
7. VCO frequency at the nominal error voltage and several tuning voltages in each frequency band.

PHASE COMPARATOR, SAMPLE-AND HOLD TYPE

1. Nominal rf reference signal power, if a sine wave (or voltage levels, duty cycle, rise and fall times of the reference pulse, if the signal is provided by a fixed-ratio divider).
2. Charging current, I_{ch}.
3. Minimum and maximum voltages of the sawtooth wave across C_{ch}, Fig. 9-3e, point 1.
4. Nominal rf VCO signal power, if a sine wave (or voltage level, duty cycle, rise and fall times of the pulse, if provided by a variable-ratio divider).
5. Voltage levels, duty cycle, rise and fall times of the sampling pulse, Fig. 9-3e, point 2.
6. DC voltage range at point 3, Fig. 9-3e, when the circuit is operating as a phase comparator.
7. Peak-to-peak ripple voltage at the sampling frequency across C_{hold} at point 3, Fig. 9-3e, when the circuit operates as a phase comparator.
8. DC voltage range at the phase comparator output.
9. AC beat note at the phase comparator output when the frequency of one input signal is offset by an amount that is less than the loop bandwidth.

To explain the principles involved, only a few circuits are described in detail. However, all circuits used in a synthesizer should be broken down into functional blocks during the synthesizer design, the parameters of every block specified, and the data taken and recorded.

It should be routine for every design engineer also to record (and not to rely on his memory) the dc voltage levels at every junction where such levels differ from the power supply voltage, with no rf power applied to the circuit. (In case of an rf oscillator, oscillations should be stopped by loading the tuned circuit of the oscillator when measurements are made.) Measurements under such conditions are important because there often is no rf voltage, or the voltage level is very low when a circuit fails.

Comparing Measured Data to Specifications

The last step in troubleshooting is self-explanatory. As measurements are made, the data are compared to the specified parameters; and, if the data associated with one or several parameters defining the "black box" undergoing testing are outside the specified limits, the black box is rejected as failed. It is important that the ambient temperature effects be included in the specified limits.

9-2 Spurious Products in Mixers

In synthesizers signals at different frequencies, and their addition or subtraction, whether designed for in frequency synthesis or accidental (i.e., unwanted), are common and usually numerous. The multiplicity of the origins of spurious products is often so extensive as to make the task of tracking down the circuits where the accidental mixing takes place time consuming. However, the techniques employed in identifying locations at which unwanted mixing processes are taking place and in deriving the constants of the intermodulation products equation describing these processes (the essential information used in discovering the origins of spurious signals) are easy to understand and apply in practical situations. Having this valuable information available, one can systematically check all possible leakage paths of signals (such as dc power supply lines) to the place of mixing until the actual leakage paths are identified.

In view of the importance of tracking down spurious signals to successful synthesizer development and production, special attention is given to this topic, and ample examples are presented to illustrate the techniques.

Single Mixer

Consider the single-mixer case, Fig. 9-4a. Assume that the mixer is part of a synthesizer such that the two input signals at f_1 and f_2 are generated in the process of synthesis. Under these conditions a spurious signal appearing at the output of the filter may be produced as a result of several processes:

1. Power supply leakage (if the mixer utilizes active devices, such as field-effect transistors).
2. Radiation from a circuit operating at the frequency of the spurious signal or at such frequencies as to generate an in-band spurious product when combined with f_1 or f_2.
3. Electrostatic coupling to another circuit.
4. Injection into one of the mixer input ports (the spectrum of one or both input signals may have its own in-band spurious signals which are translated into the mixer output band by the mixer itself).
5. Intermodulation expressed by $f_{spur} = mf_1 + nf_2$, where $m, n = \pm 1, \pm 2, \pm 3, \ldots$.

Power supply leakage is tested for by driving the circuit with an external power supply. (Bypassing the power lines with a large-value capacitor often has no effect on the leakage because the line impedance is already very low and cannot be further significantly reduced by putting another low-impedance component across it.)

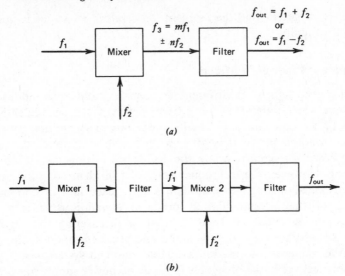

Figure 9-4.   Identification of spurious products. (*a*) **Single mixer;** (*b*) **two mixers in cascade.**

Radiation and coupling are identified by observing that some other circuit in the synthesizer operates at the frequency of the spurious signal and by physically moving the mixer away from that circuit.

Injection by way of one of the input ports is identified by replacing the two synthesized input signals (one at a time) by two external signal generators set to the same frequencies, f_1 and f_2, and rf power levels such as those provided by the system, and observing whether or not the spurious output disappears. If it does, further investigation should be directed toward the circuits preceding the mixer. In the event that the spurious output does not disappear after the substitution and the tests show that the cause of the problem is not a power supply leakage, radiation, or electrostatic coupling, the spurious output is probably generated in the mixer itself. In this case it is essential to use a powerful technique for identifying mixer intermodulation products, the most common cause of spurious signals in synthesizers. Such a technique is considered in some detail below.

Returning to the single-mixer case, assume that none of the tests described above led to the elimination or partial level reduction of the spurious signal. To derive the constants of the intermodulation products equation, we drive both mixer input ports with external generators, vary the frequency of one of the input signals, f_1, by a known increment, Δf_1, and measure the change in the frequency of the spurious signal, $\Delta f_{\text{spur},1}$; then

$$\frac{\Delta f_{spur,1}}{\Delta f_1} = m. \qquad (9\text{-}1)$$

Similarly, varying the frequency of the other input signal, f_2, by a known increment, Δf_2 (Δf_2 may be equal to Δf_1), we measure the change in the frequency of the spurious signal, $\Delta f_{spur,2}$; then

$$\frac{\Delta f_{spur,2}}{\Delta f_2} = n. \qquad (9\text{-}2)$$

The minimum frequency increment, Δf_1 or Δf_2, is determined by the resolution of the spectrum analyzer used in the test. The maximum value of the increment is limited by considerations of convenience; we would like the spurious signal to remain on the screen of the analyzer during the test.

Example 1.   Assume that

$$f_1 = 100 \text{ to } 110 \text{ MHz},$$
$$f_2 = 500 \text{ MHz, a constant,}$$

and the desired output is

$$f_{out} = f_1 + f_2,$$

that is,

$$f_{out} = 600 \text{ to } 610 \text{ MHz}.$$

It is disclosed during the synthesizer testing that, when

$$f_1 = 101.378 \text{ MHz} \qquad \text{and} \qquad f_2 = 500 \text{ MHz},$$

there exists an in-band spurious output at

$$f_{spur} = 608.268 \text{ MHz}.$$

Following the procedure outlined above, we set

$$\Delta f_1 = +1 \text{ kHz } (f_1 = 101.379 \text{ MHz}),$$

keeping f_2 constant. The recorded change in f_{spur} is $+6$ kHz, that is, $\Delta f_{spur,1} = +6$ kHz, and from Eq. 9-1

$$m = +6.$$

Next, setting $\Delta f_2 = +1$ kHz ($f_2 = 500.001$ MHz) and keeping f_1 constant, we notice no change in f_{spur}. Hence, $\Delta f_{\text{spur},2} = 0$, and (Eq. 9-2),

$$n = 0,$$

so that

$$f_{\text{spur}} = 6f_1.$$

Example 2.   Further testing of the mixer described in Example 1 leads to the disclosure of another spurious product. When

$$f_1 = 100.036 \text{ MHz},$$

the frequency of a close-to-band spurious output is

$$f_{\text{spur}} = 599.856 \text{ MHz}.$$

As in Example 1, we set $\Delta f_1 = +1$ kHz ($f_1 = 100.037$ MHz). The frequency of the spurious signal decreases by 4 kHz ($\Delta f_{\text{spur},1} = -4$ kHz). Hence

$$m = -4.$$

Setting $\Delta f_2 = +1$ kHz ($f_2 = 500.001$ MHz), we increase f_{spur} by 2 kHz, that is, $\Delta f_{\text{spur},2} = +2$ kHz, so that

$$n = +2$$

and

$$f_{\text{spur}} = 2f_2 - 4f_1.$$

Two Mixers in Cascade

Slightly more involved testing takes place when troubleshooting two mixers in cascade, located in a module, Fig. 9-4b. Assume that, without opening the cover of the module, we would like to identify a spurious output as to the location of its generation, either in mixer 1 or in mixer 2.

To determine whether the signal is generated in mixer 1 or in mixer 2, we vary f_1 and f_2 simultaneously in such a manner as to keep f_2' constant. If during this test f_{spur} remains constant, the spurious signal is generated in mixer 2. If it does not, the spurious signal is generated in mixer 1. When it is known in what mixer the spurious signal is generated, the derivation of the equations is reduced to a single-mixer case.

Unfortunately, this procedure does not work when $m=n$ and

(a) $f_1'=f_2+f_1$ and m,n are positive integers,

or

(b) $f_1'=f_2-f_1$ and m,n are negative integers.

The procedure also does not work when spurious products result from signal leakages taking place between mixers, such as the signal at f_1 leaking to mixer 2 and combining with the signal at f_2'. Nevertheless, in many instances this series of simple tests leads to easy identification of the processes and their locations and thereby saves time.

Example 3. Let

$$f_1 = 100 \text{ to } 110 \text{ MHz,}$$

$$f_2 = 500 \text{ MHz,}$$

$$f_1' = 600 \text{ to } 610 \text{ MHz } (f_1'=f_1+f_2),$$

$$f_2' = 140 \text{ to } 230 \text{ MHz, in 10 MHz steps,}$$

$$f_{\text{out}} = 740 \text{ to } 840 \text{ MHz } (f_{\text{out}}=f_1'+f_2').$$

The frequency settings at which a spurious output was first discovered are as follows:

$$f_1 = 100.036 \text{ MHz,}$$

$$f_2 = 500 \text{ MHz,}$$

$$f_1' = 600.036 \text{ MHz,}$$

$$f_2' = 150 \text{ MHz,}$$

$$f_{\text{out}} = 750.036 \text{ MHz,}$$

$$f_{\text{spur}} = 749.856 \text{ MHz.}$$

Assume that, when $\Delta f_1 = +1$ kHz ($f_1=100.037$ MHz), $\Delta f_2 = -1$ kHz ($f_2=499.999$ MHz), and $f_1'=100.037+499.999=600.036$ MHz, constant, whereas $f_{\text{spur}}=749.850$ MHz, not constant. Hence the spurious signal is generated in mixer 1. From this step on, we follow the procedure associated with the single-mixer case and find that, when $\Delta f_1 = +1$ kHz, f_{spur} is decreased by 4 kHz, so that $\Delta f_{\text{spur},1} = -4$ kHz and $m=-4$. Setting $\Delta f_2 = +1$ kHz, we increase f_{spur} by 2 kHz, that is, $\Delta f_{\text{spur},2} = +2$ kHz and $n=+2$, so that the equation describing this spurious product is

$$f_{\text{spur}} = 2f_2 - 4f_1,$$

our old friend from Example 2.

Example 4.   Assume that the same frequencies are used in this example for f_1, f_2, f_1', f_2', and f_{out} as in Example 3. Also assume that a spurious output is first discovered at the following frequency settings:

$$f_1 = 102.358 \text{ MHz,}$$
$$f_2 = 500 \text{ MHz,}$$
$$f_2' = 150 \text{ MHz,}$$
$$f_{out} = 752.358 \text{ MHz,}$$
$$f_{spur} = 754.716 \text{ MHz.}$$

When f_1 and f_2 are changed so that f_1' remains constant, the frequency of the spurious signal remains constant. Hence the signal is generated in mixer 2. To determine m' and n', where $f_{spur} = m'f_1' + n'f_2'$, we increase f_1' by 1 kHz, $\Delta f_1' = +1$ kHz, by changing either f_1 or f_2, and observe that f_{spur} increases by 2 kHz. hence $m' = +2$. Varying f_2' by $+1$ kHz, we decrease f_{spur} by 3 kHz, so that $n' = -3$, and

$$f_{spur} = 2f_1' - 3f_2'.$$

When the appropriate numbers are inserted in the equation, it is verified that $f_{spur} = 754.716$ MHz. Were this not the case, each mixer would have to be tested individually.

9-3   Troubleshooting of Direct Synthesizers

Application of the techniques described in Sections 9-1 and 9-2 to the troubleshooting of direct synthesizers makes the troubleshooting a straightforward routine as long as a complete package of requirements defining important functions of every circuit is readily available.

The task is further simplified by the behavior of most state-of-the-art circuits utilized in direct synthesis—they generate no output signal in the absence of an input signal. This makes troubleshooting a matter of tracking down the circuits generating low or no output signal in the presence of a nominal input signal. This procedure has already been considered in such great detail in Sections 9-1 and 9-2 as not to warrant further discussion.

9-4   Troubleshooting of Indirect Synthesizers—A Phase-Locked Loop

So far no mention has been made of special problems associated with the troubleshooting of feedback systems. These often are quite severe, mainly because a feedback system, such as a PLL, consists of many complicated circuits which are closely interrelated with each other, so that several functional failures may result in a problem characterized by the same symptom, or one failure may manifest itself by several symptoms.

For example, a loss of frequency locking in a single-loop PLL may be caused by a failure in either the VCO fine tuning control (the error voltage varicap circuit) or the coarse tuning control (the tuning voltage varicap) or the band switching circuit. Similarly, there are several functional blocks of the phase comparator that, if failed, will cause lack of locking. On the other hand, a failure in one of the VCO band-switching circuits can cause a lack-of-lock condition in that band, as well as low VCO rf output power and high FM noise in all bands. A successful technique of troubleshooting PLLs, therefore, is a very useful tool.

Some of the problems encountered in PLL troubleshooting can be easily identified by observing the overall behavior of the PLL. Others require more extensive analysis and testing. For that reason the troubleshooting procedure described on the following pages is presented in two parts: (a) closed-loop tests, which have a preliminary, investigative nature, and (b) open-loop tests, which consist of an evaluation of the groups of circuits or the individual circuits that comprise the PLL.

The troubleshooting of only digital PLLs is considered. However, the technique can be applied directly to analog PLLs (or to any feedback system) by excluding those parts of the text that are associated with the variable-ratio divider.

Closed-Loop Testing of Phase-Locked Loops

The success of closed-loop troubleshooting depends greatly on how familiar one is with (a) the function of every circuit used in the PLL for a given design, (b) the parameters defining this function, and (c) the effect any changes in these parameters will have on the overall PLL performance.

For example, one should know what individual filters the loop filter consists of (a phase-lag, Twin-T, and low-pass or any combination of these circuits); what the parameters of the filters are, such as the cutoff frequency of the low-pass filter; and what changes in the capture range, loop bandwidth, spurious signals, phase noise, and tuning time are observed when these parameters are modified. Table 9-1 is a summary of such information. The table is far from all inclusive. It is presented here to illustrate the point made above and to serve as a useful guide.

Although there is no one "best" method of closed-loop troubleshooting, a recommended step-by-step procedure can be formulated in general terms. It is assumed that, as the test data are taken, the procedure will be modified or expanded, according to the findings, by the troubleshooter himself.

Table 9–1. The Effect of Circuit Parameters on PLL Performance

				Effect on:			
						Total Phase Noise of	
Change in circuit parameter	Hold-in range, $\Delta\omega_{hold-in}$	Capture range, $\Delta\omega_{capture}$	3 dB bandwidth, $B_{3 dB}$	Spurious outputs at the phase-comparator frequency	Reference signal multiplied to output PLL frequency	VCO	Acquisition time, $t_{acq, total}$
Decrease in VCO gain constant, K_{VCO}	Decreases	Decreases	Decreases	Decreases	Decreases	Increases	Increases
Decrease in phase-detector gain constant, K_ϕ	Decreases	Decreases	Decreases	Decreases	Decreases	Increases	Increases
Decrease in ac beat note at phase-detector output, $V_{beat}(t)$	No effect[a]	Decreases	No effect[a]	No effect[a]	No effect[a]	No effect[a]	Increases
Decrease in cutoff frequency of low-pass filter, f_c (part of loop filter)	No effect	Decreases	Decreases	Decreases	Decreases	Increases	Increases
Decrease in band reject center frequency of twin-T filter, f_0 (part of loop filter)	No effect	No effect	No effect	Increases spurious output level of fundamental component	No effect	No effect	Decreases $t_{acq, total}$ in high-speed PLL

[a]Unless K_ϕ is also changed by the failure that produces the decrease in $V_{beat}(t)$.

Step 1. Direct-Current Power Distribution. The correct distribution of the dc power supply voltages is verified directly at the inputs of the PLL circuits. This test checks both the power supply operation and the wiring of the dc power lines.

Step 2. General Pattern of Behavior. The hold-in and capture ranges are measured in terms of the dc error voltage at several VCO frequencies, including the low and high ends, in each VCO frequency band. Simultaneously with these measurements record the peak-to-peak beat-note voltage at the PLL loop filter input and output (points 6 and 7, Fig. 9-5, respectively) at only one VCO frequency. The last measurement is made with the VCO tuned out of lock so that the beat frequency is equal to or smaller than the loop bandwidth and the beat note is not attenuated by the PLL itself.

These simple tests are designed to check the performance of several circuits. For example, satisfactory hold-in range and amplitude of the beat note at the output of the loop filter establish proper operation of the

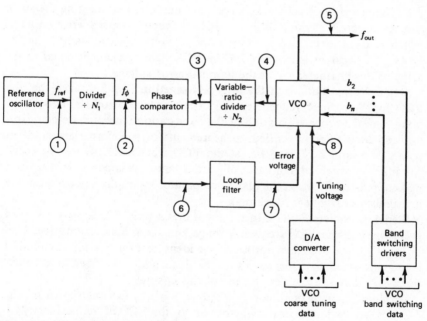

Figure 9-5.   Closed-loop tests. Digital PLL.

fixed-ratio divider, N_1, phase comparator, and loop filter. On the other hand, an unusually large voltage drop of the beat-note signal across the loop filter points to this circuit as a failed one; a low level of the beat note at the phase-comparator output indicates that a failure took place in the phase comparator; proper values of the beat-note voltage at the VCO error voltage input (point 7, Fig. 9-5) and of the hold-in range, together with a poor capture range, is likely to be the result of a low rf power applied to the variable-ratio divider, N_2, and so on.

The capture range, if expressed in terms of the error voltage, can be easily measured by first setting the VCO out of lock electronically (by means of the tuning voltage control) or mechanically (by varying the capacitance of the trimmer capacitor used in the VCO tuned circuit) above the VCO locked frequency and by slowly changing the VCO free-running frequency in the direction of the locked frequency. At the moment the ac beat note at the phase-comparator output (point 6, Fig. 9-5) disappears, the VCO tuning is terminated, and the error voltage corresponding to one end of the capture range is recorded. The other end of the capture range is measured in the same manner by tuning the VCO out of lock below the locked frequency.

The conversion of volts into hertz is accomplished by utilizing the static (i.e., open-loop) characteristic of VCO frequency versus error voltage, which is taken over a wide range of error voltage at the same time the VCO gain constant is measured. (The VCO gain constant is equal to the slope of this characteristic at the nominal error voltage.)

Step 3. Partial Failure. If the tests performed in Step 2 disclose that the VCO locks at some frequencies and does not lock at others, and at the same time the proper operation of the fixed-ratio divider, phase comparator, and loop filter is verified, measurements of waveforms are made at points 3, 4, and 8 of Fig. 9-5, at the VCO frequencies where "no-lock" condition is observed. (In cases of partial failure it is convenient to start making measurements at the phase-comparator input, the key circuit in the mechanism of frequency locking.)

Step 4. Complete Failure. In the event that a lack of locking is observed everywhere in the VCO frequency range and the measurements described in Steps 1 through 3 have not led to the identification of a failed circuit, it is recommended that further testing be done under the open-loop conditions considered in the next section of this chapter.

If one deals with multiloop synthesizer, such as the one shown in Fig. 2-2, it is worthwhile to check whether or not the VCO of the auxiliary PLL is locked at the very first frequency at which a lack of locking of the output PLL VCO is observed, before completing all the tests described above.

When phase noise or spurious signals are so high as to exceed system requirements and the tests have not provided a solution, reducing the PLL loop bandwidth by a factor of, say, 10 (20 dB) and observing the effect of the bandwidth reduction on the PLL performance allows one to deduce the approximate location of the circuit that does not function properly.

For example, if the symptom is a high phase noise which decreases with the loop bandwidth reduction, the likely sources to be examined closely are limited to four circuits: the reference oscillator, fixed-ratio and variable-ratio dividers, and phase comparator. The loop filter should be included in this group if it utilizes an active device, such as a dc amplifier. (This number of sources is further reduced by substituting a proven reference oscillator for the circuit initially used in the system, which usually is an easily replaceable module or a subassembly. The loop filter can also be removed from the loop without any difficulty by putting a short across it.) If the noise increases as the loop bandwidth is decreased, the most likely source of the trouble is the VCO with the associated circuitry, that is, band-switching drivers, D/A converter, and so on. (See the discussion of a PLL acting as a filter, pp. 24 and 25.)

Open-Loop Testing of Phase-Locked Loops

The method of open-loop troubleshooting presented here consists of performing tests on, and making function properly, the individual circuits that comprise a PLL in such an order as to permit their immediate use as driving sources in the evaluation of other circuits. This facilitates testing under the normal drive conditions, which often leads to rapid identification of the problem. If this approach fails, circuit isolation by substitution is necessary, as in cases of circuit interaction; see Example 3, p. 576.

For that reason the VCO is evaluated first with the error and tuning voltages provided by two variable dc power supplies, as shown in Fig. 9-6a.

Having established proper VCO operation, one proceeds with the testing of the variable-ratio divider (or the fixed-ratio divider or the rf mixer—whichever follows the VCO in a given system design), as shown in Fig. 9-6b.

The reference oscillator and fixed-ratio divider are tested next, Fig. 9-6c, to make both input signals of the phase comparator available for testing that critical circuit.

The phase comparator is evaluated in a mixer mode by measuring the dc output voltage and peak-to-peak beat-note voltage with the VCO frequency set so that the beat-note frequency is within the loop filter pass band, Fig. 9-6d. Evaluation of this circuit in a phase-detector mode takes considerably longer and is not necessary in this stage of testing.

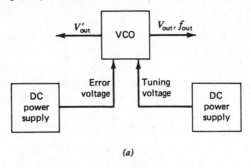

(a)

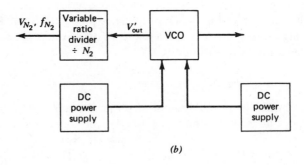

(b)

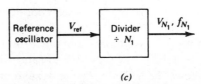

(c)

Figure 9-6. Open-loop tests. (a) VCO; (b) variable-ratio divider, N_2; (c) reference oscillator and divide-by-N_1 divider; (d) phase comparator and loop filter; (e) complete digital PLL, preliminary system measurements.

Finally, the magnitude of the loop filter transfer characteristic is measured by increasing the beat-note frequency and recording the changes in the beat-note voltage drop across the filter.

In such a simple manner every circuit in the PLL can be tested individually under open-loop conditions and the data compared with the

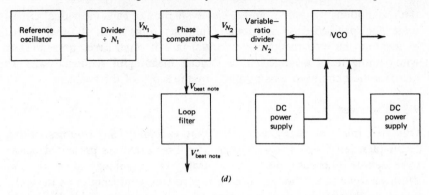

(d)

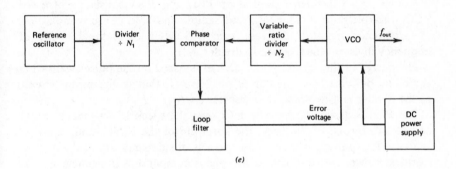

(e)

circuit specifications. After the proper operation of individual circuits is restored, the loop is closed again, and the capture range is measured at several VCO frequencies to verify satisfactory system operation, Fig. 9-6e.

If the closed- and open-loop tests are not sufficient to explain the problem, further troubleshooting consists in testing individual circuits by measuring the parameters defining each circuit under the specified drive and load conditions. (For a description of a basic breakdown by functions of the circuits frequently used in frequency synthesis, see the section of this chapter on taking data, pp. 555 through 558.)

A summary of the steps of the troubleshooting procedure described in this chapter may be of further assistance to the inexperienced engineer. These are (*a*) investigative measurements of governing parameters throughout the system that, if not leading to the disclosure of the causes of the problem(s), give enough information to allow assumptions of the approximate locations of failed circuits to be made and a detailed test plan for the verification of these assumptions to be prepared; (*b*) Verification of the assumptions by testing; and (c) an evaluation of each individual circuit under the specified drive and load conditions in the event all other means of locating the malfunctioning circuit(s) fail. It often saves time to start with simple tests, arranged in a logical, meaningful sequence, and to increase their complexity as required by the scope of the problem.

9-5 Examples

One can think of many examples that exemplify the troubleshooting techniques described in this chapter. The three examples presented below were selected because they demonstrate the techniques at work and illustrate some of the "tough" practical problems, confusing to an inexperienced troubleshooter, that too often take an unnecessarily long time to solve.

Example 1. Consider a single-loop PLL, Fig. 9-7, consisting of a vhf VCO, variable-ratio divider, N_2, and sample-and-hold phase comparator. The reference signal is derived from a 5 MHz oscillator by way of frequency division (the fixed-ratio divider, N_1).

The design of the variable-ratio divider is based on the pulse-swallowing technique described in Chapter 6, pp. 365-371, that is, the input circuit of the divider is a two-modulo prescalar ($\div N_U / \div N_L$).

The PLL is designed, made to lock over the specified frequency range, and tested. In one of the tests, the spectrum of the PLL output signal is analyzed for spurious outputs, and two sets of sidebands about the desired signal are recorded, one set at the phase comparator frequency, f_ϕ, the other at a frequency that varies with changes in the division ratio of the variable-ratio divider.

The presence of the sidebands at f_ϕ is anticipated, and steps are taken to reduce the level of the sidebands to a value below the requirement. The presence of the second set of sidebands comes as a surprise and requires an investigation.

It is observed that the sidebands occur at the rate at which the prescalar modulo is changed, that is, at the frequency of the modulo control signal. Making a measurement at the variable-ratio divider output discloses that the level of the sidebands follows the $20\log_{10}(1/N)$ characteristic, so that these are indeed FM sidebands. When the VCO is disconnected from the divider, the sidebands disappear, but the test is inconclusive because with no input signal the divider does not generate the modulo control signal.

Normal operating conditions are simulated by driving the divider with an external signal generator and letting the VCO run free. Under these test conditions no FM sidebands are observed at either the VCO or the divider output.

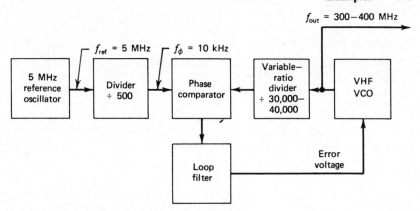

Figure 9-7.   Very high frequency digital PPL: Example 1.

The last test establishes that the problem is associated with the direct circuit connection between the VCO and the divider, not with a power supply leakage, magnetic radiation, or electrostatic coupling. The assumption is verified by reconnecting the circuit in the closed-loop configuration and inserting a variable attenuator between the VCO and the divider. The FM sideband level decreases with an increase in the isolation between the two circuits. It is deduced that, with the particular integrated chip used for the prescalar, the input impedance of the prescalar varies at the rate of the modulo control signal. This VCO load variation leads to the amplitude modulation of the rf voltage across the VCO tuned circuit—the low-frequency voltage variations that are followed by the VCO varicap capacitance.

The problem is solved by inserting an isolation amplifier with unity forward gain and the required reverse loss between the VCO and the variable-ratio divider.

Example 2.   Next consider a single-loop PLL operating between 200 and 220 MHz, Fig. 9-8. For low-noise operation the VCO frequency range is divided into three bands:

Band 1: 200.00 to 205.99 MHz.
Band 2: 206.00 to 212.99 MHz.
Band 3: 213.00 to 219.99 MHz.

An eight-bit D/A converter provides the tuning voltage that sets the VCO frequency within the capture range of the PLL. Twin-T and low-pass filters attenuate the reference signal at 10 kHz and its harmonics, leaking to the

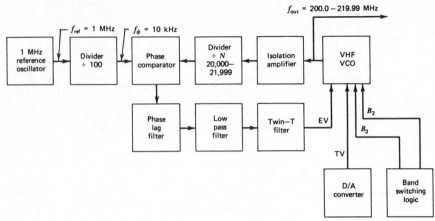

Figure 9-8. Very high frequency digital PPL: Examples 2 and 3.

VCO error voltage input, by the amount determined by the PLL spurious signal requirement, which is -80 dBc. Both phase-lag and low-pass filters reduce the loop bandwidth to approximately 10 Hz in the steady-state conditions. During locking the two filters are automatically removed from the loop to increase the capture range to its maximum value.

When the dc power is first applied to the PLL and the loop is checked for locking at the low and high ends of the frequency range, it is disclosed that no rf output is generated at 200 MHz, though the loop seems to work properly at 220 MHz.

The first step is to measure the PLL capture range and rf output voltage at the low and high ends of each band and at several other frequencies to establish a pattern of behavior throughout the frequency range, from which one can deduce the nature of the problem and the location of the failure, or at least be able to prepare a meaningful test plan. The following symptoms are observed:

1. In band 1 the VCO locks at some frequencies and does not lock at others. The maximum capture range is not as wide as the specified range, decreasing with the VCO frequency until the VCO stops oscillating at approximately 202 MHz.
2. In band 2 the VCO does not lock at all, and switching the VCO from band 1 to band 2 changes the free-running VCO frequency by a much greater amount than anticipated.
3. The VCO locks at some frequencies and does not lock at others in band 3. When it is locked, the measured capture range equals the

specified range but is not symmetrical with respect to the locked frequency except at the upper end of the band, where the symmetry is preserved.

When the test data are examined, three conclusions are reached immediately: (a) there is something wrong with the VCO circuit since it does not oscillate at the low end of band 1, (b) one of the failures is taking place in the switching circuit of band 2, and (c) this is a multifailure case with other, yet unidentified circuits contributing to the problem. At that phase of investigation it is not quite clear why the capture range behavior is erratic.

In the case of a multifailure, it is good practice to take care of the known failures before proceeding with the investigation; consequently, attention is directed toward finding out what is wrong with the band 2 switching circuit. The dc "on" voltage across the switching (PIN) diode in band 2 is measured and found to be equal to the power supply voltage instead of to 0.7 V. The diode is replaced, and the band-switching control of the VCO free-running frequency restored, but the frequency is not what it should be except at the high end of band 3, where all D/A converter switches are off and the converter output voltage nearly equals the power supply voltage. These data and considerations give one a clue as to where to look for the next failure.

The tuning voltage provided by the D/A converter is measured at several VCO frequencies and found not equal to the specified voltage. Further testing of the converter circuit discloses that, when bits 3 and 5 are used in generating the tuning voltage, the voltage is significantly different from that which is required to tune the VCO to that frequency. The dc voltages at the terminals of the switching field-effect transistors associated with bits 3 and 5 are measured and found incorrect. The FETs are replaced, and the D/A converter is checked again. The converter is now working properly. Remeasuring the capture range discloses that its erratic nature is no longer observed, but there still is no rf output at frequencies below 202 MHz.

So far, two problems have been identified and corrected by a few meaningful measurements and a deduction based on knowledge of the circuitry.

In concentrating on an obscure problem, such as a lack of oscillations, it often helps to ensure that no other failure is complicating the situation, so the next step is to perform the following checks at the high end of band 3 (the frequency at which the PLL seems to meet all the system requirements). The reference pulse wave at the phase-comparator input is examined with the help of an oscilloscope and found to be within the specified limits. The output of the variable-ratio divider and the rf voltage into the divider are also checked. The D/A converter is replaced with an external dc power supply; the VCO is tuned over the hold-in range of the PLL, and the corresponding phase-comparator dc output voltage is recorded. Finally, the VCO is tuned slightly outside the hold-in range, and the level of the beat note at the VCO error voltage input is recorded. Both the dc output voltage range and the level of the beat note are within the specified limits. (The loop filter response is not measured closed-loop because of the automatic switchover action associated with it, which was described at the beginning of this example.) These simple checks verify the performance of every PLL circuit utilizing active components and allow one to feel confident in assuming that the last major problem, at least, is caused by the VCO alone. Hence further troubleshooting is limited to that circuit, but before any open-loop measurements are performed, the oscillator circuit is analyzed as follows.

A vhf oscillator tends to stop oscillating over the low end of the operating frequency range either because the amplifier rf gain decreases with decreasing frequency and eventually becomes too low to compensate for the circuit losses and sustain oscillations, or the rf load associated with the tuned circuit increases with decreasing frequency, thereby decreasing the rf amplifier gain, or the feedback ratio changes with frequency, thus reducing the oscillator loop gain until the feedback voltage is insufficient to sustain oscillations. (For a functional block diagram of an oscillator see Fig. 9-3d.) Similarly, any permanent change in these parameters would produce the same effect: partial or complete failure of the oscillator.

It is time consuming to measure the rf amplifier gain or the load of the tuned circuit. It is advantageous, therefore, to check the oscillator electrical parts against the circuit schematic diagram, starting with the feedback network, which immediately discloses that the value of the collector-to-emitter feedback capacitor (this is a grounded-base oscillator) is, instead of 2.2 pF, only 1 pF, which is not enough to sustain the oscillations at the low end of the VCO band. Replacing the capacitor with the specified part restores the normal operation of the circuit.

Final system tests disclose no other problem afflicting the PLL. If the VCO low-end failure is not as simple as an incorrect value of the feedback capacitor, further tests will have to be performed under open-loop conditions with the isolation amplifier disconnected from the VCO and a rated load applied to the VCO output. These tests will consist of evaluating the performance of the oscillator functional sections shown in Fig. 9-3d.

Example 3. Let us assume that, when the dc power is first applied to the system shown in Fig. 9-8 and the loop is checked for locking at the low and high ends of the frequency range, no locking is observed at either frequency. Taking data at several other frequencies to establish a pattern of behavior throughout the frequency range meets with no success because the VCO does not lock at any frequency at which the measurements are made.

Under such conditions it is useful to start the closed-loop troubleshooting at the phase comparator, the key element in a PLL. A brief examination of the phase-comparator input signals with an oscilloscope discloses an absence of the pulse wave supplied by the variable-ratio divider. An rf voltmeter measurement of the rf voltage at the input of the divider provides no ready answer because, although the rf voltage level is lower than the nominal level, the voltage is within the specified limits.

Having isolated the problem to one circuit, that is, the variable-ratio divider, one continues the measurements with the loop open until the circuit is made to function properly. Hence the loop is opened at the input of the variable-ratio divider, and an external signal generator, instead of the isolation amplifier, is used as an rf driving source. Imagine the troubleshooter's surprise on discovering that, when the divider is driven by an external signal source, it functions properly even at an input rf voltage a couple of decibels lower than the level measured with the isolation amplifier connected to the divider.

At this time suspicion falls on the isolation amplifier, and the amplifier rf output signal is examined, using an rf spectrum analyzer. It is found that the second harmonic of the amplifier output signal is at the same power level as the fundamental component, instead of 20 dB lower, and that other harmonics are very high, as well. This means that the rf voltmeter measurement of the isolation amplifier output voltage is in error (because it is the sum of the powers of the fundamental and all harmonics which is measured) and that the actual power provided by the amplifier at the frequency of the fundamental component is at least 3 dB lower than the measured power, which is not enough to drive the divider.

To make sure that the isolation amplifier is not saturated by the preceding circuit, the VCO, the VCO rf output power is measured. It is found to be nearly equal to the nominal power.

The problem is easily explained when the values of the electrical components used in the isolation amplifier are compared with the values shown in the amplifier schematic diagram, at which time it is disclosed that the emitter resistance used in the output power stage is 10 times as high as the specified value, reducing the rf amplifier gain and introducing an excessive distortion.

10 Fast-Switching-Time Synthesizers

We have dealt with general considerations associated with synthesizer design and development in Chapters 1 through 9, at times pointing out the impact of one or another requirement on the selection of both a frequency synthesis approach and circuitry.

With the rising need for spectrally pure, precise signal sources capable of high-speed frequency hopping for such applications as spread-spectrum data links and radar, it becomes necessary to review the material in the light of a stringent switching-time requirement (20 μsec to submicroseconds) and to formulate the basic principles of fast-switching-time synthesis.

On a rough examination of the problem of fast frequency hopping, even a 20 μsec switching-time requirement seems to favor direct analog and digital synthesis. A closer look at the mechanisms that play an important role both in slowing down the time responses of the circuits used in frequency synthesis and in introducing unwanted side effects leads one to believe that there is no single approach free from limitations. The question that has to be answered is what the tradeoffs are when one approach is used instead of another in view of the overall system requirements.

10-1 System Approaches

There are four basic categories of coherent frequency synthesis which lend themselves to fast frequency hopping: (1) direct analog, (2) direct digital, (3) indirect, and (4) a combination of these approaches called hybrid. They are briefly reviewed next.

Direct Analog Synthesizers

Numerous approaches falling into the category of direct analog synthesizers are described in Chapter 1. Among the most popular are brute-force, harmonic, double-mix, triple-mix, and double-mix-divide frequency synthesis. Most frequently, several techniques of this type are used to generate

578

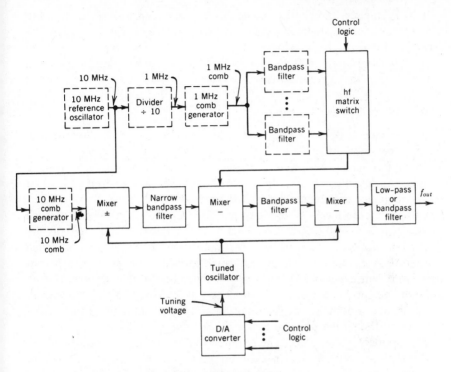

Figure 10-1.   Fast-switching-time direct analog synthesizer.

signals in the desired frequency band, as demonstrated in the example shown in Fig. 10-1.

The circuits drawn in solid lines in Fig. 10-1 are in the paths of the signals whose frequencies are switched, which affects the synthesizer switching time. The three mixers and hf matrix switch are broadband devices, and their effect on the switching time can be made negligible. An oscillator tuned by a low-output-impedance digital-to-analog converter (DAC) can change frequency in less than 1 μsec. If the selected frequency plan provides for wide-filter bandwidths, submicrosecond-tuning is possible with this approach. A low-phase noise performance can be achieved with such a synthesizer as well.

Problems experienced with direct analog synthesizers appear during frequency hopping. When the frequency hopping is periodic, spurious amplitude and frequency-modulation components at the hopping frequency are generated at relatively high levels. In the example discussed here, the hf-matrix switch and tuned oscillator are the two circuits where the spurious signals originate. Amplitude modulation in the rf switch is associated with the nonlinearity of the switching elements, such as diodes, used in the switch. Frequency modulation comes from the presence of an alternating current component at the hopping frequency in the tuning voltage waveform generated by the DAC.

Spurious amplitude or frequency modulation at the hopping frequency is not present in the synthesizer discussed next.

Direct Digital Synthesizers

This approach is described in Section 1-4 in some detail. A direct digital synthesizer is free of spurious amplitude and frequency components at the hopping frequency, and state-of-the-art synthesizers of this type can be switched in less than 1 μsec. It is also the least expensive approach when small frequency increments have to be generated.

At present the levels of inband spurious signals at the output of such synthesizers are -50 dBc at 1 to 2 MHz. This requires a frequency divider to provide necessary filtering to meet more stringent requirements, which further reduces both the frequency and frequency range of the output signal. Current experiments lead one to believe that in the very near future the same performance will be obtained at 5 to 25 MHz. This improvement, however, does not make one feel that vhf and uhf direct digital synthesizers are around the corner unless a major breakthrough in the design principles and components takes place. The frequency limitation is not important, and direct digital synthesis has played an important role in the design of high-frequency, fast-switching-time synthesizers.

Indirect Digital Synthesizers

In the past a digital phase-locked loop approach had been considered unsuitable for fast-switching-time frequency synthesis. It was felt that a digital PLL should be inherently slow because of its feedback control nature, that to make it lock faster, one had to sacrifice stability, spurious outputs, and phase noise, and that even then, frequency hopping faster than in milliseconds would not be practical. Recent work in fast-switching synthesizers using PLLs leads to different conclusions. Phase-locked loops frequency-tuned in tens of microseconds have been designed and tested. Ways of making a PLL hop fast are described below.

Phase-locked loops are prone to frequency modulation when the frequency hopping is periodic.

Hybrid Synthesizers—Best Choice for Fast Frequency Hopping

The lack of a frequency synthesis that satisfies all requirements of fast-switching-time synthesizers leads one to investigate the possibility of high-speed frequency hopping obtained with combinations of the frequency

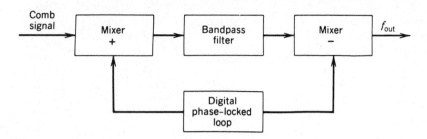

Figure 10-2. Digital phase-locked loop in double-mix synthesis.

synthesis approaches, that is, hybrid synthesis. If the design is directed by
certain principles a superior performance is achieved.

The first principle is to select a hybrid frequency synthesis which uses the
advantages of each known approach to reduce the effects of the shortcom-
ings of other approaches. A typical example is the combination of the
double-mix and digital PLL synthesis shown in Fig. 10-2. The stability and
accuracy of the PLL allows selection of the comb pickets of higher order
than that which is attained with a tuned oscillator, whereas the transient
frequency error of the PLL during the loop settling is reduced in the
double-mix synthesis.

The selection of a frequency plan which permits utilization of the circuits
with wide bandwidths in the path of frequency synthesis is the second
principle.

And finally the selection and design of circuits used in the paths of
frequency generation and control with fast response time, low phase noise,
and low spurious outputs characteristics is the third design principle.

A practical example of a fast-switching-time synthesizer designed according to these principles was given in Section 7-6.

10-2   System Optimization of Phase-Locked Loop Synthesizer

The subject of conflicting requirements controlling the design of PLL
synthesizers was dealt with in Chapters 2, 4, and 5, and examples were
given to demonstrate possible solutions. Here the problem is restated and
discussed in the light of PLL switching time.

Figure 10-3 is a block diagram of a basic phase-locked loop. The design
of the PLL is governed by the required output frequency range, frequency
increments, the level of spurious frequency modulation components at the
phase-detector frequency, phase noise, and switching time. The phase-
detector frequency, f_ϕ, is equal to the specified smallest frequency increment.

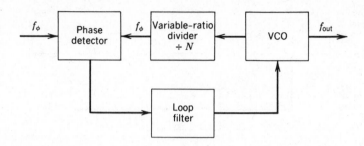

Figure 10-3. Digital phase-locked loop synthesizer.

The first restriction imposed on switching time stems from the frequency increments requirement. The loop bandwidth is always smaller than f_ϕ, so that in a simple single-loop PLL the smaller the frequency increments are, the slower is the PLL switching time. (Later in this chapter we discuss the ways of raising f_ϕ without increasing frequency increments.)

The second restriction originates from the spurious FM components requirement; the smaller f_ϕ is, the more difficult it is to suppress the leakage of the signal at f_ϕ from the phase detector to the VCO without reducing loop bandwidth. Again the solution comes from raising f_ϕ.

The third may or may not exist, depending on the phase-noise requirement, which may dictate the use of a narrow-loop bandwidth for optimum phase noise performance.

And finally, the loop stability considerations may not permit the use of a wide bandwidth. The loop bandwidth in the steady-state conditions should be smaller than the frequency at which the open-loop phase shift is -180 degrees. Since the phase detector is a major contributor of the phase shift, the restriction is relaxed by raising f_ϕ.

In three out of the four cases considered above, the restrictions imposed on the PLL switching time can be partially lifted or completely removed by raising f_ϕ. In many cases as we raise f_ϕ we relax the restriction associated with phase noise requirement as well.

Maximum Loop Bandwidth

A typical PLL switching-time requirement sets a limit to either phase error (i.e., the difference between the instantaneous phase at a given time and the steady-state phase of the VCO signal) or frequency error (i.e., the difference between the instantaneous and nominal VCO frequencies). The two errors are interrelated, since frequency is the rate of change of phase. This requirement includes both acquisition and settling times.

The first step in the design of a fast-switching-time PLL is to select the VCO frequency, if possible, to make the specified frequency range a small percentage of the PLL capture range. This eliminates the PLL acquisition time from further consideration, because the PLL always stays locked, and reduces settling time, because the change in the dc voltage generated by the phase detector that tunes the VCO on frequency in this case is small and is traversed in a shorter time than in the case when a complete capture range is used in VCO tuning. If this is not possible directly, for example, when the specified output frequency is 100 to 200 MHz, the VCO frequency higher than the specified output frequency is selected and converted down.

In all cases, after f_ϕ and VCO frequencies are decided upon, one selects a phase detector with large phase detector constant and then computes the VCO constant which gives the required maximum loop bandwidth and capture range.

Automatic Bandwidth Switchover

Considering switching time only does not usually produce the desired loop bandwidth required for loop stability, spurious frequency modulation at f_ϕ, and phase noise. To satisfy other requirements, a loop filter is designed. This filter is automatically removed from the loop during locking and is switched back into the loop after phase settling is completed.

A typical loop filter is shown in Fig. 10-4. A phase lag filter provides gain reduction without introducing phase shift at frequencies where the open-loop

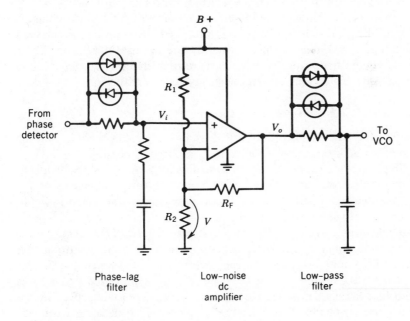

Figure 10-4. Typical loop filter using automatic switchover diodes.

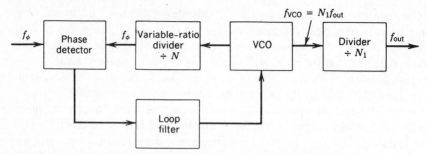

Figure 10-5. Digital phase-locked loop synthesizer followed by a frequency divider.

phase is excessive. This stabilizes the loop and reduces the leakage of the signal at f_ϕ to the VCO input. The low-noise dc amplifier expands the voltage range of the phase detector, and thereby effectively increases the phase-detector constant. The low-pass filter further attenuates the signal at f_ϕ, which leaks to the VCO. If more attenuation at f_ϕ is required, a multi-pole, sharp-rolloff, low-pass filter is used.

The diodes placed across the two filters are used to switch out both filters during VCO locking. The diodes are automatically switched on by the beatnote generated by the phase detector when the VCO goes out of lock. If the VCO stays locked during frequency tuning, the diodes are switched on by the step change in the voltage which tunes the VCO on frequency.

The brief, type 1 phase-locked loop is recommended for fast switching. A first-order PLL with no loop filter is preferred. If, however, the system requirements other than switching time dictate the use of a second-order PLL, automatic bandwidth switch-over is recommended for optimum loop bandwidth during locking and in the steady-state conditions.

Phase-Locked Loop with Frequency Divider

As stated, the loop bandwidth of a PLL is always smaller than f_ϕ, and the design effort is directed toward making f_ϕ as large as possible.

A simple way of raising f_ϕ is shown in Fig. 10-5. A frequency divider is used at the PLL output. Hence, for the same frequency increments at f_{out} the phase detector operates at N_1 times the frequency at which it would operate if the divider were not used.

This approach has many advantages. It results in faster switching time, both because higher f_ϕ allows wider loop bandwidth without introduction of instabilities and because frequency error is reduced when the divider follows the PLL. It improves the phase noise of the synthesized signal if it is verified that the divider noise is lower than the VCO noise improved by the divider. (See Appendix B for the typical phase noise of some frequency

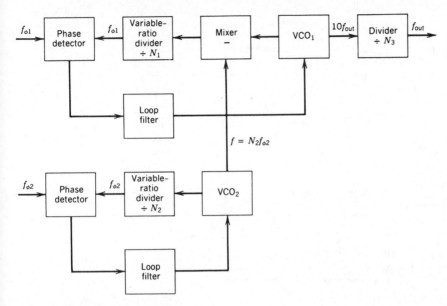

Figure 10-6.   Multi-loop synthesizer.

dividers.) The PLL followed by a divider has a relaxed requirement with respect to spurious FM components at f_ϕ because of the attenuation provided by the divider.

All of these improvements are achieved at a relatively low cost increase.

Multi-Loop Synthesizer

Further improvement in switching time is achieved by adding an auxiliary phase-locked loop as shown in Fig. 10-6. This approach is particularly advantageous for broadband PLLs generating small frequency increments. In such cases, if only one PLL is used, it often is not possible even by using a divider at the final output to obtain required switching time because f_ϕ is small and the range of the dc voltage generated by the phase detector which keeps VCO locked is large. By using an auxiliary PLL in such a way as to tune the auxiliary VCO over a small frequency range and to have the auxiliary PLL generate two to three least significant decades of frequency increments keeping the auxiliary VCO locked at all times, the phase detector frequency of the output PLL, $f_{\phi 1}$, is made high. This permits the output PLL to switch faster than without the auxiliary PLL, and the auxiliary PLL switches fast because it never breaks lock.

An example of the two-loop synthesizer with a frequency divider at the output is given in Fig. 10-7.

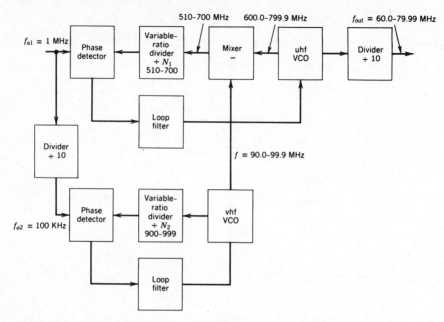

Figure 10-7.   Multi-loop synthesizer: an example.

Phase-Locked Loop and Direct Digital Synthesizer

A significant increase in f_ϕ can be arrived at by adding by way of an upconverter a direct digital synthesizer, as shown in Fig. 10-8. This approach is particularly useful when very small frequency increments are required. If the direct digital synthesizer generates small frequency increments, f_ϕ can be raised by several orders of magnitude.

The upconverter is necessary to keep the spurious intermodulation products generated in the PLL mixer out of the mixer frequency range. The divide-by-N_1 frequency divider reduces the level of the spurious signals and phase noise associated with the output signal of the direct digital synthesizer.

Phase-Locked Loop in Double-Mix Synthesis

Frequency error generated during PLL settling can be reduced, if the PLL is used in double-mix synthesis, as shown in Fig. 10-2. A steady-state frequency error associated with the PLL is cancelled in the right-hand-side mixer as long as it is not larger than half of the 3 dB bandwidth of the band–pass filter. (Thus a voltage-tuned oscillator can be used in this configuration instead of a PLL, without introducing frequency incoherence.) An instantaneous frequency error associated with a change in phase of the

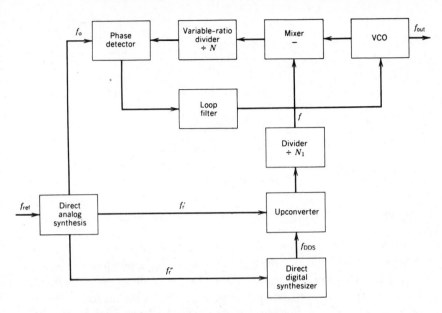

Figure 10-8.   Phase-locked loop and direct digital synthesizer.

PLL signal is cancelled only partially, because phase cancellation provided by the double-mix synthesis is partial.

The double-mix synthesis is advantageous also because the PLL phase noise is partially cancelled in the right-hand-side mixer. This provides for a convenient means of generation of large frequency steps without an increase in the phase noise of the synthesized signal.

10-3   Circuit Optimization of Phase-Locked Loop Synthesizer

The problem of conflicting requirements is dealt with by selecting an appropriate frequency plan and by designing PLL circuits with fast time response, low spurious signals and low phase noise.

Fast-Tuning-Time Voltage-Controlled Oscillator

The block diagram of a typical VCO is shown in Fig. 10-9. A phase detector provides error voltage by way of a loop filter. The tuning voltage is generated by a high-speed DAC. The band-switching logic is applied to that part of the VCO that decides which VCO band is used. All VCO inputs consist of broad-band circuits with fast transient response. Because of the wide-band requirement imposed on the VCO inputs, the error voltage,

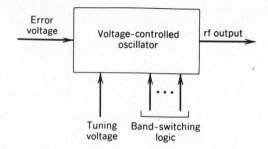

Figure 10-9.   Block diagram of a typical voltage-controlled oscillator.

tuning voltage, and band switching logic are low noise signals clean of spurious signals.

A 2R, R resistor-ladder-network DAC is described in Section 6-10. For fast response, the values of the ladder resistors are low. Hence to achieve high speed, the output impedance of the DAC is made as low as is feasible without introducing excessive shot noise. For a low noise, all resistors are of metal film type. The dc voltage driving the DAC is provided by a voltage regulator, such as LM723, designed for low noise operation.

The schematic diagram of a typical uhf VCO is shown in Fig. 10-10. Because of the wideband requirement imposed on the input circuits, the values of all bypass capacitors (Cs) are made as small as is possible without introducing temperature and noise problems. (Note another advantage of using a frequency divider at the PLL output: higher VCO frequency requires smaller values of bypass capacitors, which makes VCO tuning easier.)

The values of the feedback network components, C_1, C_2, and L_s, are optimized for maximum output rf power to achieve low noise. The source resistor, R_S, is optimized also for maximum output rf power until further increase in the drain current results in a rise in shot noise above the VCO noise floor.

The reverse-biased drain-to-gate junction behaves as if it were a voltage-tuned capacitance, which converts the amplitude noise associated with the drain dc power supply into phase noise. The conversion factor is minimized by making the drain dc voltage as high as possible without violating the rated drain-to-source power dissipation. Further noise improvement is achieved by using a low-noise voltage regulator, such as LM723, to provide dc drain voltage.

If the VCO phase noise is predominantly a noise voltage generated internally in the VCO and converted to phase noise in the error and tuning voltage varicap circuits, two varicaps connected as shown in Fig. 10-11

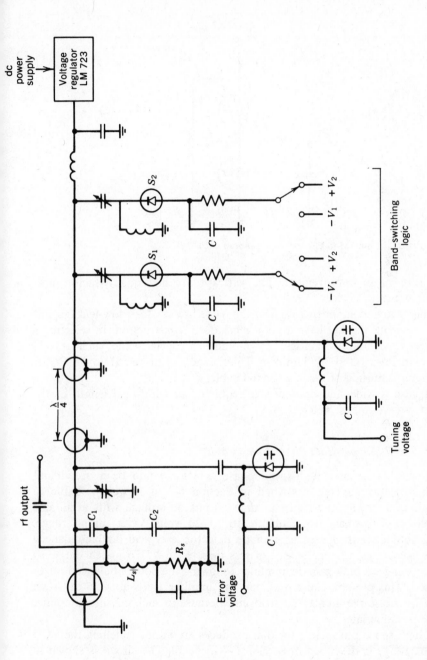

Figure 10-10. Fast-tuning uhf voltage-controlled oscillator.

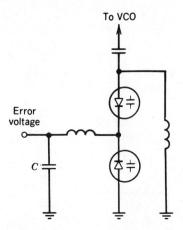

Figure 10-11. Low AM-to-PM conversion varicap circuit.

result in noise improvement. The varicaps are high-quality-factor, low-leakage-current devices.

All reactive components used in the VCO are low-loss, low-leakage-current components that have a good mechanical construction. In selecting a quarter-wave line, a semirigid, large-diameter coaxial cable, such as UT-0.141 manufactured by Uniform Tubes, Inc., is preferred to a smaller diameter semirigid or flexible coaxial cable.

In case a wideband operation is required, an additional length of the quarter-wave line is switched in.

High-Speed Sample-Hold Phase Comparator

The best choice of a high-speed phase detector would be a broadband double-balanced mixer, described in Section 6-2, if it were possible to design such a circuit to give a large output dc voltage with a suitable power-supply voltage to match a 10 to 20 volts VCO tuning range. Unfortunately, when used as a phase detector, a typical double-balanced mixer generates a linear output of less than ± 0.5 V. To match the VCO tuning range, a high-gain dc amplifier following the double-balanced mixer is used. This presents noise problems for amplifier gains above 6 dB when the PLL noise requirement is stringent, because in such case the amplifier noise predominates.

A high-speed phase detector that produces an output matching the VCO tuning range is described in Section 6-7 in detail. The circuit is shown in Fig. 6-60b with the associated waveforms given in Figs. 6-58 and 6-61. For convenience a portion of the schematic diagram and related waveforms are

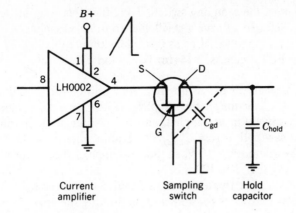

Figure 10-12. Sample and hold circuits.

reproduced in Fig. 10-12 to aid the discussion about making this circuit display low noise, respond fast, and exhibit a low level of ripple at f_ϕ.

Low phase-noise performance is achieved by using metal-film resistors and low-loss capacitors throughout and providing dc voltages by way of low-noise voltage regulators.

Fast response is obtained by supplying the current necessary to charge the hold capacitor, C_{hold}, to the required voltage in the time interval equal to the width of the sampling pulse, that is, by satisfying the following equation.

$$I_{ch} = \frac{C_{hold}(V_2 - V_1)}{T}, \tag{10-1}$$

where I_{ch} is the current which is provided by the current amplifier to charge, or discharge, C_{hold} from the original error voltage V_1 to the final error voltage V_2 in time T. For example, for the width of the sampling pulse, $T = 100$ nsec, $C_{hold} = 500$ pF, $V_1 = 4$ V, and $V_2 = 14$ V

$$I_{ch} = \frac{510 \times 10^{-12}(14 - 4)}{100 \times 10^{-9}} = 510 \text{ mA}.$$

This would be an excessively high current. A more practical number would be 100 mA. This means that under the conditions specified above it would take at least six samples to charge C_{hold}, so that for $f_\phi = 1$ MHz the PLL switching time could not be faster than 6 μsec.

The design of the phase detector becomes more difficult, but not unmanageable, when the third requirement, that of low ripple at f_ϕ, is added.

There are three major mechanisms aiding in the generation of the ripple. The first is a capacitive coupling of the high-level sampling pulse from the gate to the drain of the sampling switch. To reduce this coupling, the FET of the switch is selected to have a small gate-to-drain capacitance, C_{gd}, and the hold capacitor is made as large as possible without violating Eq. 10-1. A high-frequency FET, such as 2N4416, is a good choice for the sampling switch.

The second mechanism has to do with the width of the sampling pulse. The wider the pulse, the longer the sampling switch stays on, the longer the rising sawtooth waveform is sampled, and the greater is the amplitude of the reproduced sawtooth across C_{hold}. For example, let $f_\phi = 100$ kHz, the amplitude of the sawtooth = 10 V, so that the slope of the sawtooth, $S = 1$ V/μsec. Assuming that $T = 100$ nsec (0.1 μsec), the ripple sawtooth across $C_{hold} = ST = 0.1$ V. The ripple due to this source is reduced by making the sampling pulse as narrow as possible and by filtering the error voltage line. However, smaller T means higher charge current. A compromise is reached by making I_{ch} as large as feasible, so that T can be made as short as possible.

The third important mechanism of ripple generation, that of high-order harmonics of f_ϕ overlapping the VCO frequency range, is caused by the selection of a high frequency FET to satisfy the condition $C_{gd} \ll C_{hold}$. This is a high frequency problem and is solved in several ways.

First, the VCO circuit is shielded from the phase detector in an rf inclosure and all inputs to the VCO are filtered out. This reduces leakages due to rf radiation. Second, the phase detector is built on a metal chassis (not a printed circuit board) using point-by-point wiring and teflon standoffs. The metal chassis provides good rf grounds, and teflon standoffs provide approximately 0.7 pF to ground at every tie point, thus reducing the level of the high-order harmonics of f_ϕ generated at the sampling switch. And finally, if more attenuation is required, a low-pass filter with a sharp rolloff characteristic is placed at the error-voltage input of the VCO inclosure. The filter stop-band extends beyond the VCO frequency range. The filter is designed to introduce small phase shift at frequencies that fall within the PLL loop bandwidth.

Loop Filter

A typical loop filter is shown in Figure 10-4. The phase-lag filter is used to provide loop bandwidth optimum with respect to phase noise performance of the PLL, to stabilize the loop, and to reduce the leakage of the signal at f_ϕ. The equations associated with the filter design are given in Section 4-2. The examples of phase-noise optimization and stability analysis are presented in Sections 2-5 and 4-2 respectively and are not repeated here. In general, a phase-lag filter designed for more than 26 dB of attenuation presents switching time problems because of large RC time constant and stability problems because of large phase shift.

The diodes used to switch out the filter are high-speed, low-capacitance, low-forward-voltage, and high-off/on-resistance ratio devices. These characteristics, keep the amplitude of the error voltage transient generated during the time the diodes are turning off low. Schottky diodes are recommended for this application. The resistors are metal-film type, and capacitance is a low-loss, low-leakage component for low noise.

The direct-current amplifier is a low-noise device, such as NE5534 manufactured by Signetics. it is used primarily to expand the phase-detector voltage range to match the VCO tuning range. The useful amplifier design equations are as follows:

$$V_o = V_i \left(\frac{R_2 + R_F}{R_2} + \frac{R_F}{R_1} \right) - (B^+) \left(\frac{R_F}{R_1} \right), \qquad (10\text{-}2)$$

$$\text{gain} = 1 + \frac{R_F}{\left(\dfrac{R_1 R_2}{R_1 + R_2} \right)}, \qquad (10\text{-}3)$$

and

$$V = (B^+) \left(\frac{R_2}{R_1 + R_2} \right) \qquad (10\text{-}4)$$

where V is the dc bias voltage at the inverting input at which $V_o = V_i = V$. It is made equal to the center of the amplifier output range. The amplifier gain is made as high as possible without increasing VCO noise. This reduces the voltage range over which C_{hold} in the phase detector has to be charged, that is, reduces PLL switching time.

When a sharp rolloff, low-pass filter is used at the phase detector output, the phase-lag filter is moved to the output of the dc amplifier, so that the load presented to the low-pass filter is constant during locking and settling of the loop, as well as in steady-state conditions. It should be verified that enough current to charge the capacitance of the phase-lag filter fast is provided by the driving source, whether phase detector or dc amplifier.

All resistors used in the amplifier circuit are of metal film type.

Automatic switchover diodes may or may not be used with the RC low-pass filter. In cases when f_ϕ is significantly higher than loop bandwidth, the 3 dB cutoff frequency of the filter is selected such that the filter does not affect the PLL switching time. This selection eliminates the need of the diodes. It is a cost-effective way of getting an additional 6 to 10 dB of attenuation of the signal at f_ϕ, because the capacitance of the low-pass filter already exists (it is the bypass capacitance of the error voltage varactor circuit; see Fig. 10-10) and only a resistor is added to make up the filter.

Appendix A

This appendix describes the capabilities and use of a computer program that calculates the frequencies of mixer intermodulation products.

The program computes the frequency range of each intermodulation product, IM, generated by the mixing of a single input signal, IF, and the LO, and determines whether the product falls in a specified output band. The products computed are the $(M \times LO) + (N \times \text{input signal})$ and the $(M \times LO - N \times \text{input signal})$. For each product the following information is printed:

N
M
Lowest frequency of product
Highest frequency of product
"IN BAND" or "OUT OF BAND"

The program will handle mixer designs with either fixed or variable input bands and LO frequencies; however, it will only check if an IM product is in a single output band. It will compute all IM products up to any desired order (sum of M and N). In addition, the input, LO, and output frequency bands can be changed by fixed increments, and the IM products calculated for each case. This feature is useful when it is desired to select the optimum combination of frequencies for a mixer.

It should be noted that when the mixing process has a variable LO and the input carrier has a finite bandwidth the program must be run twice. This is illustrated in example B (Section A-4).

A-1 Program Description

The input data for the computer program are as follows:

FA   Lower input frequency I
FB   Upper input frequency I

ᵃCourtesy of ITT Defense Communications Division

FC	Lower input frequency II
FD	Upper input frequency II
FE	Lower LO frequency
FF	Upper LO frequency
FG	Lower output frequency
FH	Upper output frequency
FI	Input frequency increment for FA, FB, FC, FD
FJ	LO frequency increment for FE, FF
FK	Output frequency increment for FG, FH
L	Number of frequency increments
MAX	Highest intermodulation product order considered

The program first considers the $M \times LO + N \times IF$ intermodulation products. All combinations of M and N with $M + N \leqslant MAX$ are considered. The lower end of the output band is $M \times FE + N \times FA$, and the upper end is $M \times FF + N \times FD$. These frequencies are computed and listed with the corresponding M and N. Then they are compared to FG and FH to see whether any part of the two bands coincides. If $M \times FE + N \times FA \leqslant FH$ and $M \times FF + N \times FD \geqslant FG$, the IM product is in-band. The last column of output will state whether the IM product is in- or out-of-band and flag the in-band products with asterisks.

Next the program considers the $M \times LO - N \times IF$ products. It finds the maximum output frequency, which is either $M \times FE - N \times FA$ or $M \times FF - N \times FC$, and the minimum, which is either $M \times FE - N \times FB$ or $M \times FF - N \times FD$. These frequencies are printed with the corresponding M and N.

To determine whether the products are in-band, the program performs several logical checks. If the minimum frequency is negative and the maximum frequency is positive, then, if the larger of the two in absolute value is greater than or equal to FG, the IM is in-band. If both maximum and minimum are positive or negative, the same procedure as for the $M + N$ case is used to check for in-band products. A copy of the program is given in section A5.

A-2 Form of Data Cards

Four data cards are required to input values of FA to FK, L, and MAX. The form is as follows:

Data Card I

    FA FB FC FD (format 4F20.5)

Data Card II

    FE FF FG FH (format 4F20.5)

Data Card III

FI   FJ   FK (format 3F20.5)

Data Card IV

L   MAX (format 2I5)

A sketch of the cards is shown below.

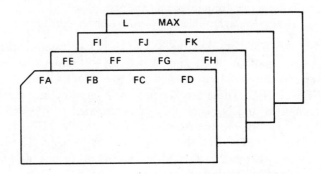

NOTE: In the present form up to 8 sets of 4 data cards each may be stacked together to yield 8 independent runs. Changing the first statement in the program allows batching more test runs.

A-3   Example A

It is desired to compare the IM products generated when a 50 to 90 MHz first IF is converted to a second IF centered between 500 and 2000 MHz, where IF stands for "intermediate frequency." The LO is always 70 MHz below the second IF center frequency, and the IM products of interest are those in the 40 MHz band centered at the second IF frequency. The following are the computer program inputs for the case in which the second IF and the LO are varied in 100 MHz increments and the highest IM product order is 15:

FA	50.0	FI	0.0
FB	90.0	FJ	100.0
FC	50.0	FK	100.0
FD	90.0	L	15
FE	430.0	MAX	15
FF	430.0		

A-4   Example B

It is desired to compare the IM products generated when a 40-MHz-wide carrier anywhere in the 7250 to 7750 MHz band is down converted to a first IF between 500 and 2000 MHz. The LO is tuned to a frequency that equals the carrier center frequency minus the first IF frequency. The IM products of interest are those in the 40 MHz band centered at the first IF frequency. The first IF and LO frequencies are varied in 100 MHz increments, and the highest IM product order checked is 15.

Two computer runs are required. The first run uses 40-MHz-wide carriers at the lower and upper ends of the input band as the limiting cases. The second run uses single-frequency carriers at the lower and upper ends of the input band as the limiting cases.

The computer inputs for the two runs are as follows:

	Run 1	Run 2
FA	7250.0	7250.0
FB	7290.0	7250.0
FC	7710.0	7750.0
FD	7750.0	7750.0
FE	6770.0	6750.0
FF	7230.0	7250.0
FG	480.0	480.0
FH	520.0	520.0
FI	0.0	0.0
FJ	− 100.0	− 100.0
FK	100.0	100.0
L	15.0	15.0
MAX	15.0	15.0

```
PAGE    1

// JOB  T

LOG DRIVE     CART SPEC     CART AVAIL    PHY DRIVE
  0000          0207          0207          0000
                              2507          0001

V2 M07   ACTUAL 16K   CONFIG 16K

// FOR
*LIST ALL
*IOCS(1403PRINTER,CARD)
C    INTERMODULATION PRODUCT
 1000 READ(2,666) IMP
  666 FORMAT(I5)
      READ(2,1)FA,FB,FC,FD
      READ(2,1)FE,FF,FG,FH
    1 FORMAT(4F20.5)
      READ(2,2)FI,FJ,FK
    2 FORMAT(3F20.5)
      READ(2,3)L,MAX
    3 FORMAT(2I5)
      WRITE(5,11)
   11 FORMAT(1H1,T15,'INTERMODULATION PRODUCT TEST')
      WRITE(5,12)FA,FB,FC,FD
   12 FORMAT(1H0,T3,'LOWER INPUT FREQUENCY 1 =',F7.1,3X,'UPPER INPUT FRE
     1QUENCY 1 =',F7.1//T3,'LOWER INPUT FREQUENCY 2 =',F7.1,3X,
     2'UPPER INPUT FREQUENCY 2 =',F7.1)
      WRITE(5,13)FE,FF
   13 FORMAT(1H0,T3,'LOWER LO FREQUENCY =',F7.1//T3,'UPPER LO FREQUENCY
     1= ',F7.1)
      WRITE(5,14)FI,FJ,FK
   14 FORMAT(1H0,T3,'INPUT FREQUENCY INCREMENT = ',F7.1//T3,'LO FREQUENC
     1Y INCREMENT = ',F7.1//T3,'OUTPUT FREQUENCY INCREMENT = 'F7.1)
      WRITE(5,413)L
  413 FORMAT(1H0,T3,'NUMBER OF FREQUENCY INCREMENTS =',I4)
      WRITE(5,412)MAX
  412 FORMAT(1H0,T3,'MAXIMUM INTERMODULATION PRODUCT TESTED =',I4)
      WRITE(5,817)
  817 FORMAT(1H0,T12,'M*LO FREQUENCY + OR - N*IMPUT FREQUENCY')
      K=L+1
      DO 99999 I=1,K
      A=FA+FI*(I-1.)
      B=FB+FI*(I-1.)
      C=FC+FI*(I-1.)
      D=FD+FI*(I-1.)
      E=FE+FJ*(I-1.)
      F=FF+FJ*(I-1.)
      G=FG+FK*(I-1.)
      H=FH+FK*(I-1.)
      WRITE(5,15)A,B,C,D
   15 FORMAT(1H1,T3,'LOWER INPUT FREQUENCY 1 =',F7.1,'MHZ'//T3,
     1'UPPER INPUT FREQUENCY 1 =',F7.1,'MHZ'//T3,'LOWER INPUT FREQUENCY
     22 =',F7.1,'MHZ'//T3,'UPPER INPUT FREQUENCY 2 =',F7.1,'MHZ')
      WRITE(5,16)E,F
   16 FORMAT(1H0,T3,'LOWER LO FREQUENCY =',F7.1,'MHZ'//T3,'UPPER LO FREQ
     1UENCY =',F7.1,'MHZ')
      WRITE(5,17)G,H
   17 FORMAT(1H0,T3,'LOWER OUTPUT FREQUENCY =',F7.1,'MHZ'//T3,'UPPER OUT
```

```
    1PUT FREQUENCY =',F7.1,'MHZ')
     WRITE(5,18)
 18 FORMAT(1H0,T15,'N',T25,'M',T28,'FREQUENCY BAND OF N+M INTERMOD PRO
    1DUCT')
     MAXIM=MAX+1
     DO 4 IM=1,MAXIM
     IK=MAXIM+1-IM
     DO 4 IN=1,IK
     IMM=IM-1
     INN=IN-1
     F2=IMM*E+INN*A
     F3=IMM*F+INN*D
     IF(F2-H)5,5,6
  5 IF(F3-G)6,7,7
  6 WRITE(5,9)INN,IMM,F2,F3
     GO TO 4
  7 WRITE(5,10)INN,IMM,F2,F3
  4 CONTINUE
  9 FORMAT(1H ,T5,2I10,2F15.3,5X,'OUT OF BAND')
 10 FORMAT(1H ,T5,2I10,2F15.3,5X,'****** IN BAND ******')
     WRITE(5,19)
 19 FORMAT(1H0,T15,'N',T25,'M',T28,'FREQUENCY BAND OF M-N INTERMOD PRO
    1DUCT')
     DO 20 IMM=1,MAX
     IK=MAX-IMM
     DO 20 INN=1,IK
     F1=IMM*E-INN*A
     F2=IMM*E-INN*B
     F3=IMM*F-INN*C
     F4=IMM*F-INN*D
     IF(F1-F3)91,91,92
 91 FMAX=F3
     GO TO 93
 92 FMAX=F1
 93 IF(F2-F4)94,94,95
 94 FMIN=F2
     GO TO 99
 95 FMIN=F4
 99 IF(ABS(FMAX)-ABS(FMIN))100,110,120
120 IF(FMIN)124,124,123
123 IF(FMIN-H)124,160,150
124 IF(FMAX-G)150,160,160
110 IF(FMAX-FMIN)111,112,111
111 IF(FMAX-G)150,160,160
112 IF(ABS(FMAX)-H)113,160,150
113 IF(ABS(FMAX)-G)150,160,160
100 IF(FMAX)101,102,102
101 IF(ABS(FMAX)-H)102,160,150
102 IF(ABS(FMIN)-G)150,160,160
150 WRITE(5,9)INN,IMM,FMIN,FMAX
     GO TC 20
160 WRITE(5,10)INN,IMM,FMIN,FMAX
 20 CONTINUE
99999 CONTINUE
     IF(IMP) 969,969,1000
969 CALL EXIT
     END
VARIABLE ALLOCATIONS
  FA(R )=0000        FB(R )=0002        FC(R )=0004        FD(R )=0006
  FG(R )=C00C        FH(R )=000E        FI(R )=0010        FJ(R )=0012
```

Appendix B

Figure B-1 presents data on the measured phase noise of some integrated-circuit frequency dividers.

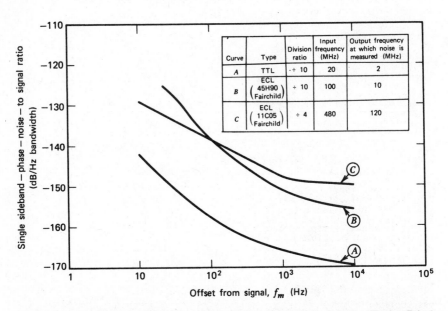

Figure B-1. Phase noise of digital frequency dividers. From Dieter Scherer, "Design Principles and Measurement of Low Phase Noise RF and Microwave Sources," presented at Hewlett-Packard RF and Microwave Measurement Symposium and Exhibition, Sheraton Heights Hotel, Hasbrouck Heights, N.J., April 3 and 4, 1979. Printed by permission of Hewlett-Packard Company.

Index